Salivary Gland Development and Regeneration

Seunghee Cha
Editor

Salivary Gland Development and Regeneration

Advances in Research and Clinical Approaches to Functional Restoration

Springer

Editor
Seunghee Cha
Oral and Maxillofacial Diagnostic Sciences
University of Florida
Gainesville, Florida
USA

ISBN 978-3-319-43511-4 ISBN 978-3-319-43513-8 (eBook)
DOI 10.1007/978-3-319-43513-8

Library of Congress Control Number: 2017933465

Printed on acid-free paper

This Springer imprint is published by Springer Nature
The registered company is Springer International Publishing AG
The registered company address is: Gewerbestrasse 11, 6330 Cham, Switzerland

Preface

Rome was not built in a day, as the English playwright John Heywood famously wrote. Innovation and advancement in the field of salivary gland regeneration is one of the great examples that reflects this sentiment. The first research article available on this topic through the US National Library of Medicine dates back to 1950. The article, entitled "Regeneration in the Submaxillary Gland of the Rat," by B.B. Milstein in 1950, cites Van Podwyssozki as the first to describe regeneration of organs of small animals as long ago as 1886 and regeneration of the salivary glands (Die Regeneration an den Speicheldrusen) in 1887.

Since the 1950s, an ever-expanding literature and diversified approaches aimed at functional restorations have mirrored strong interest and attention to this particular subject of research. Journal articles dealing with autoimmune Sjögren's syndrome, effects of radiation, and ductal ligation models in rats and mice appeared in the early 1980s, followed by research on neural regulation of secretion and effectiveness of epidermal growth factor in wound healing models and glandular regeneration in the late 1980s through the mid-1990s.

In 1995, the late Dr. Michael Humphreys published his well-known review article entitled "Saliva and growth factors: the fountain of youth resides in us all," which emphasized the vital importance of growth factors in oral/systemic health and glandular repair/regeneration. Histological analyses of glandular architecture and development were established by Dr. Robert Redman, the author of Chap. 4 of this volume. With the turn of a new century, molecular and cellular mechanisms of branching morphogenesis and glandular development were further investigated and pioneered by Drs. Kenneth M. Yamada and Matthew P. Hoffman, whose work provided foundations for the application of tissue engineering concepts and methodologies to salivary regeneration. Outstanding contributions by Dr. Bruce J. Baum to the field of tissue engineering and gene therapy have ultimately been solidified in applications such as clinical trials involving AAV2-mediated human aquaporin-1 delivery in recent years. Investigation of ductal ligation models, irradiation models, and Sjögren's syndrome NOD models dominated interest in the field until around 2010, when stem cell research in vitro and in vivo reignited research interest and passion in salivary gland regeneration.

In the current era, the authors and coauthors in this book, who are renowned researchers, dentists, and surgeons in the field, have spearheaded efforts to discover the underlying pathogenesis of xerostomia and innovative approaches to restore secretory function. I am proud to present their collective efforts and years of their research outcomes revealed in their book chapters, which will establish another significant milestone in the history and tradition of studies on glandular regeneration.

This book begins with the description of fundamental and molecular processes occurring during salivary gland organogenesis/branching morphogenesis and molecular communications among epithelial, mesenchymal, endothelial, and neuronal cells for cellular differentiation and organ development (Chap. 1, Dr. Lombaert). The importance of understanding the communications and simulating optimal environments in glandular repair and regeneration is further discussed under Future Prospects.

With rapidly advancing biotechnology, the application of systems biology has become an indispensable tool in this field. Chapter 2 discusses the definition and applications of systems biology for glandular tissues and saliva samples (Chap. 2, Dr. Larson et al.). Systems biology approaches in conjunction with traditional approaches unveil the complex molecular, cellular, and physical processes in development, disease processes, and regenerative medicine involving the salivary glands.

One of the underappreciated subjects in the field is the important role of a large family of mucins in oral health. In Chap. 3, the authors summarize the main structural and functional characteristics of salivary mucins, their expression patterns during salivary gland development and regeneration, and qualitative and quantitative changes in pathological processes in the salivary glands due to irradiation, autoreactive immune cells, neoplasm, or inflammation (Chap. 3, Dr. Castro et al.).

Changes due to radiation are not limited to mucin expression profiles but are also manifested in the parenchymal and stromal structures in the salivary glands. These changes are detailed in Chap. 4 with photomicrographs and transmission electron micrographs of rat and human salivary glands (Chap. 4, Dr. Redman). Understanding the damage occurring in the glands before and after radiation therapy will expedite the development of intervention strategies to protect the salivary glands from the harmful radiation.

In Chap. 5, Dr. Tran's group reviews recent advances from the years 2010 to 2015 in the treatment of salivary gland hypofunction with a special emphasis on mesenchymal stem cells (Chap. 5, Dr. Tran et al.). This chapter covers in detail adipose tissue-derived stromal cells, mesenchymal stromal cells derived from various sources, and finally the authors' experience with the soluble contents/factors in bone marrow soup extracted from a whole bone marrow cell lysate.

As differentiation-inducing factors are crucial for initiating stem cell differentiation from the state of quiescence, these extrinsic and intrinsic factors (transcription factors) involved in pancreas, liver, and salivary gland regeneration are further detailed in Chap. 6 with a focus on directed-cell differentiation and transdifferentiation (Chap. 6, Drs. Park and Cha).

Current cell models for bioengineering of the salivary glands are presented in Chap. 7, along with the pros and cons of utilizing various salivary cell lines. Practical tips on cell isolation and culture techniques in conjunction with the use of scaffolds complement the use of stem, progenitor, and acinar cells for salivary gland regeneration. Current trends in salivary gland bioengineering deliver great promise in functional restoration of the salivary glands (Chap. 7, Dr. Baker).

To explore further the subject of bioengineering, factors and elements needed for successful development of a functional salivary gland are discussed in detail in Chap. 8, emphasizing the dynamic nature of the basement membrane and the significance of the extracellular matrix and cell polarity in salivary gland development and reconstruction. In addition, studies utilizing the salivary-derived stem cells/gland progenitor and three-dimensional (3D) biomimetic scaffolds encompassing decellularization methods, various matrices, and polymers are summarized for 3D culture technique, which underpins current knowledge on bioengineering of the salivary glands (Chap. 8, Martinez et al.).

3D printing technology creates life-size body parts and tissues using living cells as the ink. This technology has revolutionized the field of regenerative and reconstructive medicine, enabling customized and personalized therapeutic approaches. In Chap. 9, Dr. Choi et al. describe basic principles and different types of 3D technologies, patient-specific modeling, bioprinting, and salivary gland regeneration (Chap. 9, Dr. Choi et al.).

A novel bioengineering method involves epithelial and mesenchymal stem cell manipulation to generate a bioengineered organ germ. In Chap. 10, Dr. Ogawa explains that the bioengineered glandular germs demonstrated reciprocal interactions between epithelial and mesenchymal cells in one day and invagination of epithelial tissue in three days in vitro. Once the germ was engrafted into the parotid gland duct of salivary gland-defective mice, the connection between the germ and the duct was established in a month, and the mice exhibited restored salivary secretion after transplantation. This innovative approach emphasizes that current advancement in the field promises a therapeutic intervention for patients suffering from xerostomia (Chap. 10, Drs. Ogawa and Tsuji).

Currently, functional restoration of the salivary glands is still challenging to accomplish even with successful reconstruction of salivary cellular components. Therefore, understanding the mechanisms of saliva secretion becomes critical for positive clinical outcomes that we desire. Chapter 11 covers considerations for establishing functional secretion by providing information on stimuli for secretion, neural connection along with neurotransmitters and receptors, protein secretion, and studies of neural agonists and antagonists. The chapter also clarifies myths surrounding this topic with recent research data (Chap. 11, Drs. Carpenter and Carvalho).

Thought-provoking renderings of the past, current, and future of gene therapy in salivary gland diseases are provided by Dr. Passineau in Chap. 12. In this chapter, current challenges in the field of salivary gland gene therapy, along with the author's proposals to circumvent or overcome the hurdles, are forthrightly discussed (Chap. 12, Dr. Passineau).

Last, but not least, the chapter on surgical management of salivary gland disease reveals the critical considerations for glandular regeneration from the perspectives of otolaryngologists and surgeons (Chap. 13, Drs. Varadarajan and Dziegielewski). The extensive description in this chapter includes, but is not limited to, glandular anatomy, pathology, surgical advances for neoplastic and nonneoplastic diseases of salivary glands, and recent discoveries in the field such as salivary gland transfer and salivary duct repositioning. The importance of understanding the expected sequelae in human patients following radiation or surgery cannot be overemphasized as none of the existing laboratory approaches would come to fruition for patients without such knowledge.

Based on the cutting-edge information offered in this book, it is undoubtable that many more innovative strategies for salivary gland regeneration will emerge in upcoming years. Research that unlocks the complex processes of organ development would be fundamental to develop such approaches. With the current enthusiasm and growing interest in the field, it will just be a matter of time before we build another Rome.

Gainesville
FL, USA

Seunghee Cha, DDS, PhD

Contents

Contributors

Sergio Aguilera, MD INDISA Clinic, Santiago, Chile

Olga J. Baker, DDS, PhD School of Dentistry, University of Utah, Salt Lake City, UT, USA

María-José Barrera, PhD Institute of Biomedical Sciences, Faculty of Medicine, University of Chile, Santiago, Chile

Petko Bogdanov, PhD Department of Computer Science, University at Albany, SUNY, Albany, NY, USA

Guy Carpenter, PhD Mucosal and Salivary Biology Division, King's College London Dental Institute, London, UK

Polliane Carvalho, MSc, PhD Mucosal and Salivary Biology Division, King's College London Dental Institute, London, UK

Isabel Castro, MSc, PhD Institute of Biomedical Sciences, Faculty of Medicine, University of Chile, Santiago, Chile

Seunghee Cha, DDS, PhD Department of Oral and Maxillofacial Diagnostic Sciences, University of Florida College of Dentistry, Gainesville, USA

Sridar V. Chittur, PhD Center for Functional Genomics, University at Albany, SUNY, Albany, NY, USA

Jong-Woo Choi, MD, PhD, MMM Department of Plastic and Reconstructive Surgery, Ulsan University College of Medicine, Asan Medical Center, Seoul, South Korea

Juan Cortés, PhD Institute of Biomedical Sciences, Faculty of Medicine, University of Chile, Santiago, Chile

Connor Duffy Biological Sciences, University at Albany, SUNY, Albany, NY, USA

Peter T. Dziegielewski, MD, FRCS(C) University of Florida Department of Otolaryngology, Gainesville, FL, USA

Dongdong Fang, BDS, MSc McGill Craniofacial Tissue Engineering and Stem Cells Laboratory, Department of Biomedical Sciences, Faculty of Dentistry, McGill University, Montreal, Quebec, Canada

Mary C. Farach-Carson, PhD BioScience Research Collaborative, Department of Biochemistry and Cell Biology, Houston, TX, USA

Departments of BioSciences and Bioengineering, Rice University, Houston, TX, USA

María-Julieta González, MSc Institute of Biomedical Sciences, Faculty of Medicine, University of Chile, Santiago, Chile

Sergio González, MSc Mayor University, Santiago, Chile

Daniel A. Harrington, PhD Departments of BioSciences and Bioengineering, Rice University, Houston, TX, USA

Shen Hu, PhD, MBA Division of Oral Biology and Medicine, School of Dentistry, University of California, Los Angeles, CA, USA

Chang Mo Hwang, PhD Biomedical Engineering Research Center, University of Ulsan College of Medicine, Asan Medical Center, Seoul, South Korea

Namkug Kim, PhD Department of Convergence Medicine, Biomedical Engineering Research Center, University of Ulsan College of Medicine, Asan Medical Center, Seoul, South Korea

Hae Ryong Kwon, PhD Biological Sciences, University at Albany, SUNY, Albany, NY, USA

Oklahoma Medical Research Foundation, Oklahoma City, OK, USA

Melinda Larsen, PhD Biological Sciences, University at Albany, SUNY, Albany, NY, USA

Isabelle M.A. Lombaert, PhD University of Michigan, School of Dentistry, Ann Arbor, MI, USA

Biointerfaces Institute, Ann Arbor, MI, USA

Mariane Martinez Department of BioSciences, Rice University, Houston, TX, USA

Deirdre A. Nelson, PhD Biological Sciences, University at Albany, SUNY, Albany, NY, USA

Miho Ogawa, PhD Organ Technologies Inc., Tokyo, Japan

Laboratory for Organ Regeneration, RIKEN Center for Developmental Biology, Kobe, Hyogo, Japan

Yun-Jong Park Department of Pharmacology, The University of North Carolina at Chapel Hill, Chapel Hill, NC, USA

Michael Passineau, PhD Gene Therapy Program, Allegheny Health Network, Pittsburgh, PA, USA

Sarah B. Peters, PhD Biological Sciences, University at Albany, SUNY, Albany, NY, USA

Department of Cell, Developmental, and Integrative Biology, University of Alabama at Birmingham, Birmingham, AL, USA

Robert S. Redman, DDS Oral and Maxillofacial Diagnostic Science, University of Florida, Gainesville, FL, USA

Washington DC VA Medical Center, Washington, DC, USA

Ravi Sood Department of Computer Science, University at Albany, SUNY, Albany, NY, USA

Yoshinori Sumita, DDS, PhD Department of Regenerative Oral Surgery, Nagasaki University, Nagasaki, Japan

Simon D. Tran, DMD, PhD McGill Craniofacial Tissue Engineering and Stem Cells Laboratory, Department of Biomedical Sciences, Faculty of Dentistry, McGill University, Montreal, Quebec, Canada

Takashi Tsuji, PhD Organ Technologies Inc., Tokyo, Japan

Laboratory for Organ Regeneration, RIKEN Center for Developmental Biology, Kobe, Hyogo, Japan

Ulises Urzúa, PhD Institute of Biomedical Sciences, Faculty of Medicine, University of Chile, Santiago, Chile

Varun V. Varadarajan, MD University of Florida Department of Otolaryngology, Gainesville, FL, USA

Danielle Wu, PhD Department of BioSciences, Rice University, Houston, TX, USA

Part I

Updates on Salivary Gland Development

1 Implications of Salivary Gland Developmental Mechanisms for the Regeneration of Adult Damaged Tissues

Isabelle M.A. Lombaert

Abstract

The convergence of the fields of tissue engineering and regenerative medicine provides a potential blueprint to repair damaged tissues. Accordingly, a range of therapeutic applications have emerged that hold great potential to regenerate branching organs, such as salivary glands. This unique saliva-secreting organ is required for proper oral health, lubrication, immunity, and food digestion but is susceptible to damage either by co-irradiation as a side effect of radiotherapy cancer treatment, autoimmune-related Sjögren syndrome, disease-related medications, or surgical resection. This chapter focuses on fundamental cellular and molecular processes occurring during organ ontogenesis and in developing branching glands. We cover the growth of the epithelial compartment, which is the major functional component of the gland, but also how surrounding niches such as mesenchymal, endothelial, and neuronal cells communicate, intertwine, and influence the formation of glands and other branching organs. Finally, we highlight how this key information has created new regenerative-related approaches and how these impact future clinical translation.

1.1 Introduction

Increasing our knowledge of how organs develop has profound implications for the design of therapies to regrow and/or repair injured tissues. Understanding the mechanisms regulating cell survival, expansion, specification, movement, communication with neighboring cells, as well as how they respond to damage is critical to navigating the landscape of future therapy designs.

I.M.A. Lombaert, PhD
University of Michigan, School of Dentistry,
2800 Plymouth Road, Ann Arbor, MI 48109, USA

Biointerfaces Institute, 2800 Plymouth Road,
Ann Arbor, MI 48109, USA
e-mail: lombaert@umich.edu

S. Cha (ed.), *Salivary Gland Development and Regeneration*, DOI 10.1007/978-3-319-43513-8_1

In order to appropriately translate information gathered from studies on organ development, we need to compare molecular and cellular processes during embryonic development with adult homeostasis and when repair initiates and/or fails after each damaging event. Each of these stages correlates with specific cellular responses, activation of specific signaling pathways, and accumulation of environmental cues. Thus, developmental-related information is instrumental to stimulate regrowth within an existing damaged in vivo organ or to initiate de novo growth.

The majority of our current knowledge on salivary gland organogenesis derives from experimental animal models, primarily mice and rats. While rodent biology is not identical to that of humans, many processes and pathways are very similar. As such, developmental biologists have been and continue to be a valuable resource to other disciplines such as engineering, oral surgery, and oncology to translate conceptual ideas into therapeutic designs.

The advantages of specific biomaterials, gene therapy, and surgical in vivo approaches are outlined in depth in the following chapters. In this review, we focus on different salivary gland cell types and their supportive environment that is needed to form the fully functional secretory branching organ. Subsequently, we outline how this knowledge can render future therapeutic implications and/or what potential complications might arise.

1.2 Epithelial Growth Driven by Stem Cells

Branching organs such as salivary, lacrimal, and mammary glands are comprised of different cell types, including epithelial and the surrounding mesenchymal, endothelial, and neuronal cells (Fig. 1.1). Intertwined within these tissues are circulating hematopoietic-related blood and immune cells. The major component of developing and adult salivary glands (SGs) is the epithelia, which is responsible for saliva secretion and transportation to the oral cavity. Here, we describe how the epithelial compartment of three

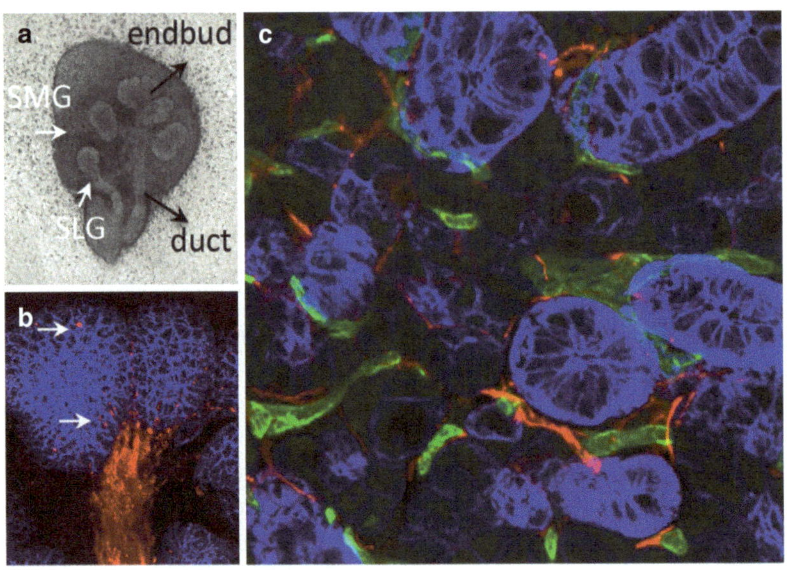

Fig. 1.1 Developing salivary glands in mice. (**a**) Bright field picture represents E13 submandibular (SMG) and sublingual gland (SLG). The epithelial compartment is comprised of a distal endbud and proximal duct area. (**b**) Epithelia (*blue*) innervated by the parasympathetic nerves (PSG, *red*) during embryonic SMG development. The PSG releases neurotransmitters via varicosities (*arrow*). Confocal image of E-cadherin stained epithelia and Tubbulin-3 stained PSG. (**c**) Different niches surrounding the epithelium in adult mouse submandibular gland. Confocal 30 μm projected image of stained SMG with epithelial marker E-cadherin (*blue*), neuronal marker Tubbulin-3 (*red*), and endothelial protein CD31 (*green*)

major glands, which provide 90 % of total saliva, becomes established by tightly controlled mechanisms of cellular interactions.

1.2.1 Morphological Development of Salivary Glands

Salivary glands originate as an invagination of the oral epithelium from a placode at embryonic day (E) 11.5 in mice or Carnegie stage 18 (~44 days, weeks 6–7) in humans. This thickening epithelium arises on the side of the tongue outside of the lamina dentalis at the anlage of the dental arch. Each major gland initiates at slightly different locations: the serous parotid gland (PAR) in the labiogingival sulcus, the mucous sublingual (SLG) in the paralingual sulcus, and seromucous submandibular gland (SMG) in the linguogingival sulcus. Even though glands arise in the tongue area, they grow out during development toward the back of the mouth below the ears, floor of the mouth near the mandibular bone, and the anterior floor of the mouth. While in mice, SMG, SLG, and PAR initiate around E11.5, E12, and E13, respectively, human SLGs initiate later than SMG and PAR at the ninth embryonic month. More detailed descriptions on anatomic locations of human glands have been recently reviewed [41]. The developmental origin of each gland has not been clear, with some classifying the PAR as ectodermal derived and SMG and SLG as endodermal, while genetic experiments in mice suggest they are all ectodermal [87].

Once the epithelial thickening arises, a cell population, termed endbud or tip, forms distally from an elongating stalk (Fig. 1.1a), which developmentally progresses to form major ducts termed Wharton's (SMG), Bharton's (SLG), and Stensen's (PAR) ducts. The unique SG branching pattern is created by repetitive clefting of the initial and subsequently formed endbuds. Ductal structures gradually mature by elongation, lumen formation, and expansion. The clefting endbuds mature by E16 in mice and 19–24 weeks (7th month) in humans to form polarized pro-acinar and pro-myoepithelial cells. While pro-acinar cells do express some secretory-related proteins, they are not yet fully functional and must still undergo specific acinar-lineage maturation so that by birth, in both humans and mice, the organ is comprised of functional secretory compartments.

1.2.2 Contribution of Stem Cells and Their Differentiating Progeny

The stem/progenitor cell theory asserts that all salivary gland cells are initiated from and maintained by stem and progenitor cells. By definition, these cells are characterized by their ability to expand themselves, i.e., self-renew, as well as to propagate multiple more defined cell types, such as acinar cells. Stem cells are further classified as being more potent than progenitor cells in their self-renewal and differentiation potential. When both these cellular processes are tightly controlled, stem/progenitor cells not only give rise to tissues but also maintain and repair organ structures during adulthood. Any deregulation in this regulatory network during development can lead to malformation/absence of the organ and in the adult may cause cancer formation. Over the past years, remarkable progress has been made wherein multiple stem/progenitors have been classified based on their ability to (1) form multiple cell types (mouse genetic lineage tracing, ex vivo culturing), (2) alternate quiescence with proliferation (BrdU incorporation or genetically labeled DNA tracing), and (3) restore radiation-induced damaged SGs (in vivo transplantation assay). One new consensus gathered from this data is that different stem/progenitor cells contribute to the growing SG and that these cells may originate at different time points during development. Importantly, this permits the organ to compensate for any losses in specific stem/progenitor cells and still allows proper development [88]. Known stem/progenitors contributing to SG organogenesis include cells marked by their expression of intracellular cytokeratin 5 (CK5, K5) and CK14 [50, 57], receptors KIT (c-Kit, CD117) and FGFR2b [57], and transcription factors SOX2 [3] and ASCL3 [10]. Remarkably, stem/progenitors contributing to development

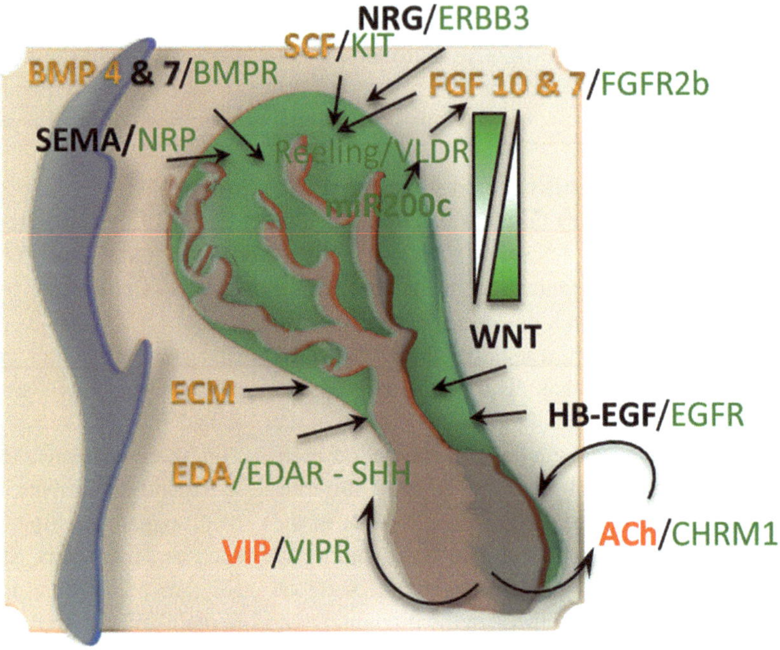

Fig. 1.2 Signaling pathways influencing epithelial growth. Cartoon represents known signaling pathways that influence SG epithelial cell survival, proliferation, expansion, and differentiation. *Green*: expressed by epithelial cells; *orange*: expressed by mesenchyme; *red*: expressed by neuronal cells; *black*: expression by multiple compartments

might not serve a similar role during adult homeostasis. Recent studies observed active proliferation of cells within specific compartments, such as acini and intercalated, striated, and excretory ducts, wherein these cells self-duplicate to replenish their own entity, as reviewed in [4]. To what extent these adult compartmental "reservoir" cells contribute to recovery after injury is a focus of ongoing research. At least after severe radiation-induced damage, which leads to irreversible hyposalivation, there is no active repair initiated by remaining SG cells. This is often a combinatorial result of (a) drastic loss of acinar and duct "reservoir" cells or stem/progenitor cells, (b) decrease in signaling pathways required to activate surviving "reservoir" or stem/progenitor cells, and/or (c) severely damaged cells that can no longer contribute to self-duplication or differentiation. In such cases, multiple strategies ranging from constructing a new gland to gene therapy and stem/progenitor cell transplantations may aid in restoring the functional and morphological components of the gland. Thus far, transplantations of cells selected for their expression of receptor KIT, EPCAM, CD24, and/or CD29 (Integrin β1, ITGβ1) were shown to restore acinar and ductal compartments, leading to significantly increased saliva levels [58, 62, 72, 103]. This does not, however, exclude the potential of other SG-specific epithelial cells, non-SG specific cells, and/or their bioactive cell lysate to contribute to the repair of damaged SGs. These options will be surveyed in following chapters, and their impact on when to use them in different damaging situations has been recently reviewed [61]. In this chapter, we will further outline our current understanding of how SGs are structurally built by various cell types (Fig. 1.1b, c) and how their continuous interactions are informing the design of current and future therapies.

1.2.3 Lessons from Developmental Regulatory Mechanisms Guiding the Epithelium

Often disorders in humans and genetic rodent model systems can provide critical information on what signaling pathways are essential for epithelial cell survival, proliferation, differentiation, and movement (Fig. 1.2). Major examples are *Fgf10*$^{-/-}$ and *Fgfr2b*$^{-/-}$ mice, which are related to human loss-of-function mutations in FGF10 and FGFR2 that result in hereditary diseases including lacrimal and SG-related aplasia of lacrimal and salivary glands (ALSG), lacrimo-auriculo-dento-digital (LADD) syndrome, and lung-related chronic obstructive pulmonary disease (COPD). In these conditions, SG development is stalled, as FGFR2b+ epithelium no longer receives survival and proliferative cues from the surrounding FGF10-producing mesenchyme [80]. Thus, when invading oral epithelial cells at gland ontogenesis receive FGF10, they initiate an endbud and duct formation. From then on, FGFR2b signaling expands KIT+ progenitors in the continuously clefting endbuds in combination with stem cell factor (SCF)/KIT signaling [57]. As FGF10 has a heparan-binding (HB) core, it evokes and expands more rapid responses once it is bound to specific 3-*O*-sulfated heparin sulfate (3-*O*-HS). This HS belongs to a group of heparan sulfate proteoglycans (HSPGs) located in the basement membrane or at cell surfaces. Interestingly, KIT+ endbud progenitors highly express HS3ST3, the 3-*O*-HS-specific modifying enzyme 3-*O*-sulfotransferase, to rapidly increase their expansion during development [80]. A similar function remains during adult homeostasis. Regulating this FGFR2b signaling pathway is of crucial importance so that every epithelial cell does not undergo extensive proliferation. Ductal cells therefore express FGF antagonists, Sprouty 1 and 2, to lower FGFR2b signaling and upregulate WNT [48]. Both canonical WNT/β-catenin and noncanonical WNT5b pathways drive ductal formation via upregulation of *Tfcp2l1* while inhibiting endbud development. In turn, endbuds repress duct development by FGF-mediated Wnt5b repression and secretion of WNT ligand-sequestering protein SFRP1 [78]. Evidently, a tight FGF-WNT gradient allows for KIT+ progenitor expansion in endbuds, while ductal cells prepare for upcoming lumenization and maturation. In this process, ERBB1 (EGFR)+ ductal K5+ progenitors proliferate in response to HB-EGF to give rise to maturing K19+ cells [50]. One mechanism of action is via induction of membrane-type-2 matrix metalloproteinase (MT2-MMP) and FGFR expression in epithelial cells. MT2-MMP is crucial to release bioactive NC1 domains from extracellular matrix (ECM) protein collagen IV, which in turn promotes branching via epithelial ITGβ1 [84]. MMPs also cleave pro-HB-EGF into an N-terminal and C-terminal fragment at the membrane so that the latter fragment can move to the nucleus to activate cell proliferation via cyclin A [95]. Conversely, another member of the EGF family, neuregulin (NRG), binds ERBB3 on endbuds to aid in their local expansion [70]. NRG1 is further essential for innervation as $Nrg1^{-/-}$ mice are devoid of nerves and show aberrant duct formation and lumenization [73].

Similar to FGF10, *Fgf8* hypomorphic and *Fgfr2c* heterozygous mice exhibit hypoplastic glands due to reduced communication between FGF8-producing epithelia and FGFR2c-receiving mesenchyme. In both FGF-deficient mice, initial epithelial invagination occurs, but subsequent SG growth does not occur. To date, FGF8 has been described as a potential target of the EDA pathway. Human mutations in ectodysplasin-A (EDA) or its receptor EDAR result in hypohidrotic ectodermal dysplasia (HED). Defects in teeth, hair, sweat, and salivary glands are noticeable due to reduced cell proliferation and differentiation [45, 74]. In SGs, EDA and downstream target NF-kB aid in ductal lumenization and endbud branching, presumably by inducing ductal maturation within the center of endbuds. Early on, mesenchyme-produced EDA is downstream of mesenchymal WNT and upstream of epithelial SHH (sonic hedgehog) signaling. As such, SHH treatment can rescue SGs deprived of EDA [100]. After E13, EDA does not seem to correlate with WNT anymore, based on their different expression pattern located in the epithelial or mesenchymal compartment [31, 78]. SHH's important role in SG development has been confirmed, as SHH-deficient SGs are hypoplastic with unpolarized epithelial cells and underdeveloped lumen formation [36, 43]. SHH is also linked to FGF8 as both can upregulate each other [43]. Therefore, FGF8 is able to rescue Hedgehog inhibition but surprisingly not EDA deficiency [43]. As such, EDA-FGF8's precise signaling interaction still needs to be determined.

FGF signaling, in particular via FGFR1 (variant b in the epithelium and c in the mesenchyme), can also upregulate bone morphogenetic protein

(BMP) ligands. BMPs are part of the TGFβ signaling family and signal via BMP receptors. FGFR1 signaling regulates BMP7 directly and BMP4 indirectly to regulate epithelial growth. BMP4, which is mesenchyme specific, inhibits epithelial branching, while BMP7, released by both epithelium and mesenchyme, increases it [90]. The role of another member of the TGF family, TGFβ1, is still inconclusive. While TGFβ1-deficient mice have normal SGs, overstimulation of TGFβ1 results in acinar loss, elongated ducts, and/or fibrosis [35, 42].

Additionally, ECM and epithelial integrin cell interactions are just as essential for branching morphogenesis. These ECM molecules line up the basement membrane (BM) separating the epithelium from mesenchyme. Interestingly, isolated epithelial cells can easily grow without the physical presence of mesenchymal cells but not without ECM component(s), such as laminin, fibronectin, perlecan, collagen, or mouse sarcoma-derived reconstituted BM "Matrigel." Deposition of unique ECM components along the clefting endbuds and elongating ducts plays a role in correct branching. Impairing these connections will lead to reduced clefting, endbud number, cell movement, and/or growth. Detailed descriptions of disruptive ECM cell outcomes were recently reviewed in [79].

Finally, an underexplored area contributing to SG formation are microRNAs (miRNAs), which are small, noncoding RNAs that specifically target mRNAs to globally regulate gene expression. Epithelial endbud progenitors highly express miR200c to reduce FGFR-dependent proliferation. miR200c downregulates the autocrine reelin/very low-density lipoprotein receptor (VLDLR) pathway, which positively regulates FGFR signaling [83]. Additionally, it was found that EGF can specifically induce mesenchymal production of miR-21, which decreases multiple target mRNA candidates. One of these, RECK, inhibits MMPs, which subsequently influences ECM degradation to enhance SG branching [37].

In conclusion, various signaling pathways instruct different cell types within the epithelial compartment and not surprisingly interact with and regulate each other to safeguard temporal-spatial proliferation, differentiation, and clefting. Initiation of some of those embryonic signaling pathways has been observed in active repair situations, such as ductal ligation settings where acinar atrophy and hyposalivation is temporarily induced by restricting salivary flow from the major duct [17]. We can thereby try to manipulate the activation and/or repression of specific developmental pathways to stimulate in vivo repair of damaged SGs, as is outlined further in this chapter.

1.3 Environmental Cues Patterning Epithelial Branching and Maturation

SGs are highly vascularized and innervated, all of which integrate within a condensed mesenchyme. Developmentally, SG epithelia invade a condensed mesenchymal placode already containing a complex endothelial network and parasympathetic neuronal bodies awaiting cues for innervation [48]. Both signals for epithelial invasion into the mesenchyme and subsequent branching are transmitted via direct cell-cell contact and/or indirect paracrine signaling pathways, which are discussed below.

1.3.1 Guiding Neurons

Different cranial nerves innervate the pre- and postnatal SG where they exert different functions. While the autonomic nervous system regulates the SG at an unconscious level and in stress conditions, sensory neurons respond to mechanical, thermic, and light signals.

For decades, both the parasympathetic and sympathetic nervous system have been acknowledged as the driving stimulant to release saliva from acinar cells into ducts. While parasympathetic stimulation results in serous secretion and ion release, sympathetic activation stimulates mucous or protein-containing saliva and can also play role in local inflammation and blood flow [23, 67]. Both parasympathetic and sympathetic nerves are part of the autonomic nervous system,

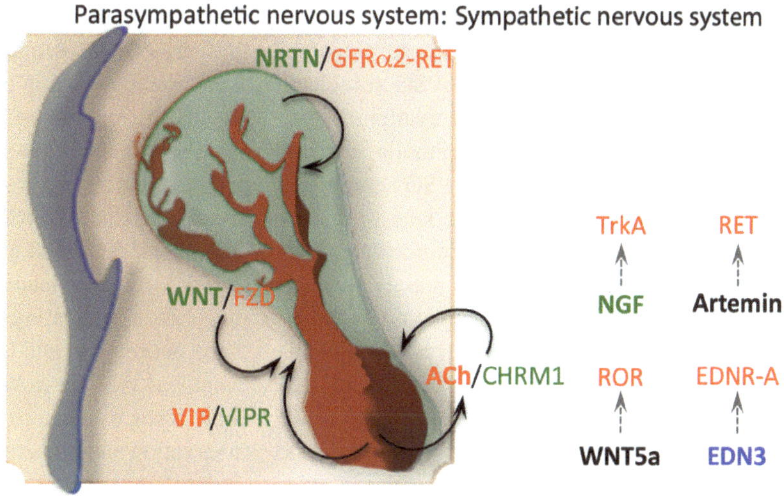

Fig. 1.3 Signaling pathways driving neuronal survival and innervations. Illustrated signaling pathways for parasympathetic nervous system were demonstrated in prenatal glands, sympathetic nervous system in postnatal SGs. *Green*: expressed by epithelial cells; *red*: expressed by neuronal cells; *blue*: expressed by endothelial cells; *black*: undefined which compartment takes part in it

and their neuronal guidance is ensured by axons that sprout along unique paths within the tissue. While the former innervates along the epithelia, the latter follows the vasculature. This directional guidance is driven by neurotrophic factors, secreted by cells in the periphery, as well as the presence of specific receptors on the axons that allow or block their adhesion to the adjacent ECM. The most notable trophic molecules include neurotrophins (e.g., NGF, BDNF, NT-3), netrins, semaphorins, ephrins, and myelin inhibitors. Similarly, axons secrete neurotransmitters in their proximity, exerting a variety of effects through specific receptors on their target cells (Fig. 1.3). Parasympathetic nerves signal via the cholinergic acetylcholine (ACh) pathway, targeting muscarinic receptors on neighboring cells, as well as water channels such as aquaporin 5 (AQP5). In contrast, sympathetic nerves release epinephrine and norepinephrine (i.e., noradrenaline, NA) that bind to β-adrenergic receptors (adrenoceptors) on acini. Other non-ACh, non-NA neurotransmitters can be produced by both parasympathetic and sympathetic nerves and may include vasoactive intestinal peptide (VIP), substance P (SP), neuropeptide Y (NPY), neurokinin A, pituitary adenylate cyclase-activating peptide (PACAP), and calcitonin gene-related peptide (CGRP).

Developmentally, parasympathetic ganglia (PSG) neuron cell bodies migrate along the branches of mandibular arteries [91] to cues from initiating SG epithelia to localize into ganglia around the primary duct and send out axons toward the endbuds [48]. Sympathetic nerves innervate SGs along the blood vessels during later stages of development when final epithelial maturation is needed. As such, developmental experiments can clearly dissect the role of neurotransmitters and neurotrophic factors affecting the PSG. It is now well appreciated that the PSG establishes a communication loop with specific epithelial cells to allow outgrowth of both compartments. When the PSG is absent, the pool of K5-expressing epithelial progenitors is significantly reduced [50], which influences downstream K19+ ductal luminal differentiation and subsequent epithelial outgrowth. This is mediated via a loss in ACh-CHRM1 (muscarinic receptor 1) signaling from the PSG to K5+ cells and resulting in a subsequent reduction of HB-EGF/EGFR pathway signaling that initiates maintenance and differentiation of K5+ progenitors. Lumenization, which marks further ductal maturation, is also coordinated by the PSG but not via the ACh pathway. The neurotransmitter VIP activates a cAMP/protein kinase A (PKA) pathway to induce epithelial duct cell proliferation and formation of a single lumen by the fusion of multiple microlumens. After initial lumen formation, VIP remains essential to expand the lumen size via the cystic fibrosis transmembrane (CFTR) pathway [73].

Organ development also requires proper bidirectional communication. Feedback signaling from epithelial cells toward the PSG stimulates cell survival, migration, and innervation. At SG ontogenesis, WNT-producing epithelia, particularly K5+ progenitors, maintain PSG neuron survival and proliferation [48]. At later stages of branching morphogenesis, the neurotrophic factor neurturin (NRTN), which is mainly secreted by endbud progenitors, not only promotes neuronal survival via GFRα2/RET but also maintains axon outgrowth along ducts toward the endbuds [49]. In the developing lung, there also appears to be a link between nerves and blood vessels. Denervation, in this case via physical cell ablation, resulted in reduced endothelial proliferation, leading to hypo-vascularized lungs [9]. It is unclear whether this is a direct or indirect neuronal-endothelial effect and whether similarities exist within the developing SG.

Detailed anatomical descriptions of nerves in adult SGs are outlined in a recent review [40]. It is assumed that similar communication between nerves and epithelium persists into adulthood as denervation of SGs, via ductal ligation or neurectomy, reduces epithelial content that regenerates after ligation removal if the nerve is intact or reconnected [46, 55, 65]. A morphological difference of early development with later stages and adulthood is that smaller ganglia are found dispersed within adult tissue [40], presumably to reach their target cells more easily as distances are much larger compared to embryonic development.

Even though tyrosine hydroxylase (TH)-expressing sympathetic ganglia are presumed not to be present at SG ontogenesis, some mRNA expression levels of its unique receptor neuropeptide Y receptor 2 (NPY2R) were detected early during development at low levels that increase before birth [23]. Since NPY2R is also present on endothelial cells, some, if not all, of the mRNA expression could be related to blood vessel formation within the SG. However, TH-expressing neuronal cells were detectable by E16.5, which might indicate there is a prenatal presence of sympathetic ganglia [89]. Nevertheless, postnatal sympathetic denervation does lead to hypoplasia of the gland [82] and thus must involve a direct or indirect role for sympathetic nerves in either epithelial cell maintenance or maturation. In the adult gland, RET signaling is also known to be essential for sympathetic neuron survival, but likely via the ligand artemin instead of NRTN. SGs also produce high amounts of NGF and genetic ablation of NGF or its TrkA receptor leads to defective sympathetic innervation, indicating its crucial role in sympathetic neuron survival [22, 27]. Depletion of non-canonical WNT5a in WNT1-derived neural crest cells further leads to incomplete sympathetic innervation and branching in prenatal SGs. While the authors suggest this is due to an autocrine WNT5a/retinoid-related orphan receptor (ROR) pathway in sympathetic neurons, it doesn't rule out that epithelial WNT5a-producing cells might be stimulating sympathetic neurons as well [89]. Similarly, endothelial-released endothelin 3 (EDN3) is also suggested to be a cue for a subset of EDN receptor A+ sympathetic neurons to innervate the prenatal SG along the nascent external carotid arteries [63]. The specific role of other neurotransmitters from the GDNF and NPY family as well as other neurotrophic factors are still being explored. While semaphorins are involved in axon pruning and neuronal migration in the central nervous system, they also appear to have a role in developing SGs. Semaphorin (SEMA) 3A and 3C bind co-receptors neuropilin and plexin. Neuropilin is expressed by epithelial endbuds and by activation with SEMA3A and 3C cleft formation is induced without changing proliferation and, most likely, by affecting cell movement [15]. However, additional FGF7/10 growth-promoting signals from surrounding mesenchyme were required to mediate this cleft formation. Whether additional participation of SEMAs on receptive nerves is required for cleft formation or SG development still remains unclear.

At adulthood, it remains to be determined how sensitive sympathetic nerves are to injuries such as radiation. In rodents, sympathetic nerve function was retained after radiation [52], and increased levels of *TH* as well as *NGF/NGFR* and adrenergic receptor 2 (*ADRA2B*) were detected in radiated human SMGs [49]. Whether a reestablishment of the balance between parasympathetic and sympathetic nervous system is necessary for regeneration is not known.

Furthermore, it is assumed that sensory neuronal cells are present along the sympathetic and parasympathetic nerve tracks in adult SGs [51], although they have not yet been studied in detail. While sensory neurons can be defined into multiple subtypes based on different criteria such as their origin and molecular expression patterns, they often are loosely classified as unmyelinated capsaicin-sensitive TRPV1+ receptor expressing neurons or myelinated glutamate receptor-expressing neurons. Upon activation, sensory nerves can secrete various neuropeptides, such as Substance P and CGRP. At least in the lung and pancreas, literature indicates that sensory neurons release neurotransmitters in the periphery to serve as direct mediators for recruiting and activating inflammatory cells [68, 86]. Whether a similar mechanism occurs in salivary glands is not known.

In conclusion, innervation plays an essential role for organ development, homeostasis, and repair after injury. Studies on SG biogenesis have been highly informative for defining the involvement of the PSG in branching morphogenesis, not only of SGs but also other organs such as prostate and lungs. The existence and/or loss of bidirectional communication with epithelial stem/progenitor cells have been reported to occur in rodent and human SG homeostasis and postradiation. The inhibition of parasympathetic neuronal function influences adult epithelial K5+ progenitors [49] but it is not known yet whether postradiation regeneration due to epithelial stem/progenitor transplantation repairs neuronal function, even though morphological repair has been suggested [71].

1.3.2 The Role of Blood Vessels

The vasculature in branching organs develops in close proximity to the epithelia, although its spatial pattern differs from parasympathetic nerves. Not only are endothelia important for mediating gas exchange but also as a source of endothelial secretory factors, termed angiocrines, which impact organ development (Fig. 1.4). While research on blood vessels in SGs remains limited, much can be learned from other branching organs. Overall, complex cross-communication between epithelial and endothelial cells appears to regulate both epithelial differentiation and angiogenesis. The initial cues to form an endothelial cell plexus around a condensed mesenchyme do not require epithelia. This is observed in *Fgf10*$^{-/-}$ mice that don't form initial SMG epithelia but where the placode of mesenchyme, blood vessels, and neuronal bodies are present [48]. After SG ontogenesis, however, epithelial-derived angiogenic factors, such as VEGF-A, do play a role as null mutations in *Vegf-A* or its endothelial-expressed receptor (*Vegfr*) show vascular defects in tissues, reduced epithelial budding, and ultimately embryonic lethality [13, 107]. Another epithelial-induced angiogenic mechanism may include the vitamin D pathway. The enzymes CYP27B1/24A1 that activate and catabolize vitamin D are highly upregulated just before birth and in postnatal lung. Exogenous vitamin

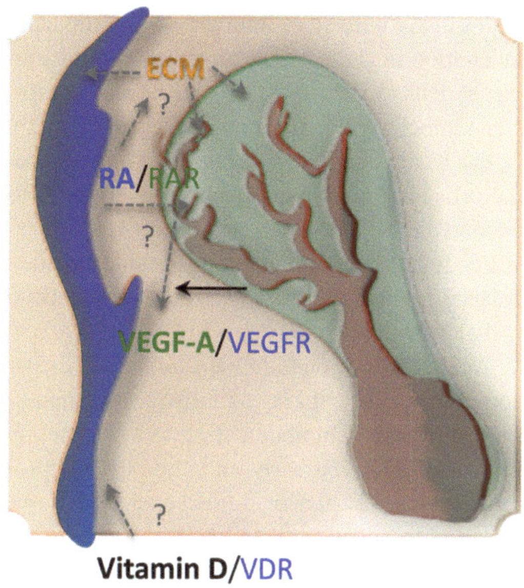

Fig. 1.4 Potential communications initiated by and to endothelial cells. Described pathways were majorly found in other branching organs. Whether they exist in SGs needs to be determined. *Green*: expressed by epithelial cells; *orange*: expressed by mesenchyme; *blue*: expressed by endothelial cells

D positively influences lung growth by inducing maturation in vitamin D receptor-expressing epithelial cells (VDR) [64]. As VDR is also present on endothelial cells, this enhanced growth might be due to direct effects of vitamin D on endothelial cells and/or indirect effects from epithelia to endothelia. Nevertheless, it is clear that epithelial-endothelial communication requires a tight balance as any hyper-vascularization inhibits epithelial growth [14].

Apart from endothelial-epithelial cross-communication, there is also endothelial-mesenchymal communication, as recently reviewed [94]. The early endothelial cells promote survival of pancreatic mesenchymal cells, which in turn have a pivotal role in organ development. A similar complex paracrine signaling network was also found in the lung. Retinoic acid (RA), which is produced by endothelial cells, induces VEGF-A expression in lung epithelia. Evidently, endothelial cells are recruited via VEGF-A, and thus angiogenesis is stimulated via this endothelial-epithelial communication loop. Furthermore, endothelial-released RA also stimulated mesenchymal cells to produce more FGF18 and ECM component elastin, thus increasing epithelial alveolar formation [108]. Other organ-specific angiocrine factors that may follow this paracrine loop include HGF, WNT, NOTCH, and BMP ligands. Mesenchymal cells also signal back to endothelial cells to stimulate survival, proliferation, migration, and autophagy via production of ECM components, such as the perlecan/heparan sulfate proteoglycan (HSPG2) fragment endorepellin, decorin, and endostatin [18, 75].

While blood vessels and nerves can independently respond to their own set of signaling factors, there also seems to be a paracrine connection via epithelial-released VEGF. Even though VEGFR is absent on nerves and not required for innervation, VEGF overexpression in pancreas not only led to hyper-vascularization but also to hyper-innervation [85]. Interestingly, endothelial cells did not produce any known neurotrophic factors, but the effect appeared to be related to their upregulated expression of basement membrane components, such as collagens and laminins. These components in turn served as scaffolds for increased axon outgrowth.

In addition to blood vessels, we must not forget the circulating cells within them. White blood cell monocyte-derived macrophages and dendritic cells arise from the bone marrow and colonize tissues via blood vessels to phagocytose cellular debris and help in the innate non-specific and specific adaptive immune defense. While macrophages normally develop in the bone marrow via granulocyte-macrophage colony-stimulating factor (GM-CSF), mesenchymal cells in tissues can also release GM-CSF to induce a similar differentiation effect on circulating monocytes. While it is clear that both macrophages and dendritic cells may be involved in organ morphogenesis, their exact functions are not always fully understood. Also in adult tissues, for example, in the lung, there is conflicting data on their specific role: antigen-sensing dendritic cells might induce different immune responses depending on their physical location in the tissue while surrounding different epithelial cell types [54]. Similarly, various macrophages invade the mesenchyme where they can interact with dendritic cells, lymphocytes, and epithelia to regulate immunity. Macrophages suppress immune responses by inhibiting both dendritic-mediated T-cell activation and inactive TGFβ production. Subsequent activation of this inactive TGFβ into bioactive TGFβ by lung epithelia is essential in order to prevent spontaneous inflammation after acute injury. Lung alveolar cells in turn secrete various ligands to receptive macrophages to ensure this prevention of inflammatory responses. Whether a similar action or disruption in this communication is occurring in adult SGs after radiation remains to be determined.

Apart from immune regulators, macrophages further shape the branching patterning of organs by remodeling the ECM around the ducts to allow outgrowth as well as survival of endbud stem/progenitor cells [11, 102]. They also regulate angiogenesis by instructing endothelial cells to undergo apoptosis via WNT signaling, counterbalancing a pro-survival factor produced by pericytes, which wrap around endothelial cells to influence functions such as blood flow [2].

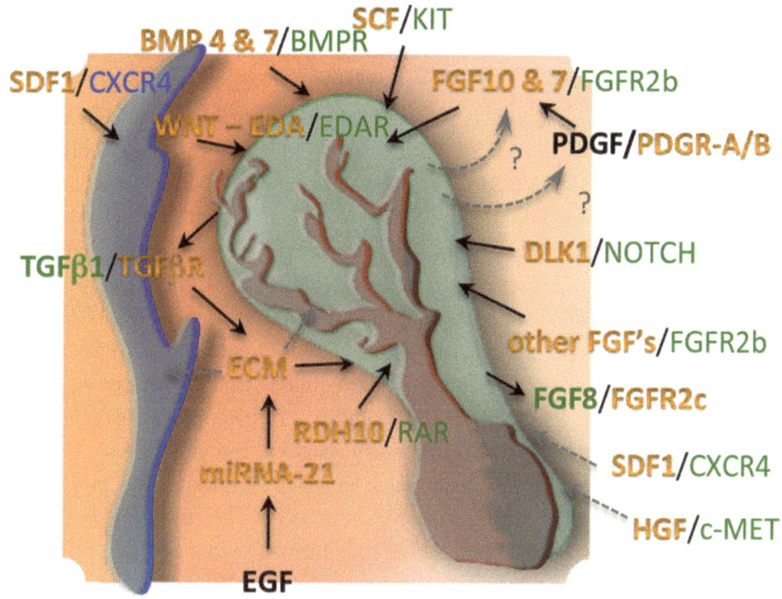

Fig. 1.5 Described signaling interactions with the SG mesenchyme. Most known communications are between the SG mesenchyme and epithelium. *Green*: expressed by epithelial cells; *orange*: expressed by mesenchyme; *blue*: expressed by endothelial cells; *black*: undefined which compartment takes part in it

In sum, blood vessels play important roles during development in other branching tissues not only for oxygen supply but also to maintain essential communication signaling pathways with epithelia and surrounding mesenchyme. The bone marrow-derived cells circulating in the blood vessels also aid in tissue branching morphogenesis and evoke or suppress immune responses after injury. Whether similar mechanisms exist in the developing and adult SGs remains to be determined.

1.3.3 Supporting Mesenchymal Cells

Embryonic SG mesenchymal cells are WNT1+ neural crest-derived cells [44, 105] and provide supportive cues such as growth factors, proteases, and ECM proteins to guide and activate epithelial, neuronal, and endothelial cells (Fig. 1.5). In vitro recombination experiments show that the SG mesenchyme induces an SG-like branching pattern in various epithelia such as a pancreatic, mammary, and pituitary gland [96]. This property, however, is not found in mesenchyme from non-SG tissues, as E11.5–13 SG epithelium does not properly branch unless it is recombined with SG mesenchyme [28, 29, 99]. This indicated that SG mesenchyme has strong and unique multicomponent instructional properties. One of them is the high production of FGF10, which is also essential for lung, lacrimal, and mammary gland initiation. Notably, non-SG epithelia only adapted SG-like branching patterns when they were placed in close vicinity to high FGF10-expressing mesenchymal tissue, confirming the importance of FGF10's spatiotemporal dosage [99]. With this in mind, it is important to understand that the SG mesenchymal component in these experimental conditions contains mesenchymal cells as well as ganglia and blood vessels albeit disconnected from the rest of the body. It can therefore not be excluded that SG-specific neuronal cells and/or blood vessels may have additional contributions to this specific SG patterning.

Interestingly, early-stage E11.5–12.5 SG epithelia, but not later stages, are able to instruct E10.5 mesenchyme from different sources to produce FGF10. However, not every mesenchyme is as competent to receive this signal as only SG and pharyngeal second arch mesenchyme responded and limb mesenchyme, for example, did not. This indicates that this initial signal is exclusively located within specific regions of the embryo, likely to restrict specific organ outgrowth to the correct location in the body. What this initial epithelial signal is remains a subject of debate. Whereas early limb and lung epithelia secrete FGF8 or FGF9 to initiate this process, it is unlikely that FGF8 serves a similar function in SGs [99]. Neither is the signal FGF4, BMP2, SHH, TGFβ1, or WNT6 [47]. One unexplored candidate is platelet-derived growth factor (PDGF). During development, PDGF-A and PDGF-B ligands are mainly produced by epithelia and mesenchyme, respectively [105], while PDGFR-A and PDGFR-B receptors are expressed in the mesenchyme. By adding exogenous PDGF, epithelial proliferation can accelerate via upregulation of mesenchymal *Fgf7*, *Fgf10*, *Fgf1*, and *Fgf3* and downregulation of growth inhibitory factor *Fgf2*. While this induction was observed during SG morphogenesis (E14), it has not been confirmed that epithelial PDGF-A is a potential FGF10 inducer at SG ontogenesis. Nevertheless, once FGF10 expression is initiated, it persists and becomes independent from epithelial cues. It is also interesting to note that mesenchymal condensation at placode initiation is independent of this FGF10 activation. As such, the mesenchyme can condense around a network of blood vessels and resting PSG cells before SG initiation and in the absence of FGF10 as seen in *Fgf10*$^{-/-}$ mice. This mesenchyme presumably awaits signal(s) from the invading oral epithelia to initiate FGF10, which in turn promotes SG-specific epithelial growth.

Even during branching morphogenesis, mesenchymal cells continue to play a part in multiple bidirectional signaling pathways. Early on, WNT/β-catenin signaling is exclusively induced in the mesenchyme before it is expressed in lumenizing ductal cells [78]. This mesenchymal WNT can activate EDA and, at least in part, influence SG morphogenesis via epithelial EDAR [31]. Branching epithelia also regulate local FGF10 expression to reduce aberrant cell proliferation. Lung epithelia release BMP ligands as well as SHH to spatially downregulate FGF10 and specifically induce secondary bud formation [12]. A recent study further points to an important role for mesenchymal retinoic acid (RA) to enhance branching. RA is a small diffusible hormone-like molecule generated by a two-step enzymatic oxidation of dietary vitamin A via RDH10 and ALDH1A. RDH10, which metabolizes vitamin A in the first step, is exclusively expressed in early SG mesenchyme, while RA activity is mainly observed in RA receptor+ epithelia. Disruption in *Rdh10* results in early embryonic lethality, often before SG epithelial invagination. However, when E13 SGs were treated with an RAR inhibitor reduced branching was observed [101]. In contrast, mesenchymal cells can also secrete signaling inhibitors to slow down branching. DLK1, a non-canonical NOTCH1 ligand produced by the mesenchyme, inhibits branching and subsequent innervation, presumably to modulate cleft formation [25]. Furthermore, DLK1 appears to regulate the epithelial balance of K14+ progenitors, although the precise mechanism is unclear [26]. The role of TGFβ1 is also unclear; there is some evidence that epithelial-secreted TGFβ1 enhances collagen production from Col11α1+ mesenchymal cells to inhibit SG acinar formation [42].

Other mesenchymal signaling pathways that may influence epithelial branching are hepatocyte growth factor (HGF)/c-MET and SDF1/CXCR4 signaling [38]. Additional cellular instruction mechanisms also include microRNAs (miRNAs). miRNAs are small noncoding RNA molecules that function to silence other mRNAs.

While most research has focused on mesenchymal-epithelial interactions, there may also be mesenchymal-endothelial interactions. When the mesenchymal factor SDF-1 was specifically inhibited from binding CXCR7+ endothelial cells, SG epithelial branching was decreased [38], thus suggesting a tri-directional loop between mesenchymal, endothelial, and epithelial cells. It is also possible that mesenchymal cells may play a role in axonal guidance. As outlined earlier, multiple studies have verified that mesenchymal cells aid in cellular migration, clefting, and differentiation via regulation of ECM production.

1.4 Translation into Future Therapies to Repair Damaged Salivary Glands

There is a tremendous need for long-term therapies to restore salivary flow. Clinical impacts of dry mouth, or xerostomia, not only include difficulty with food mastication, swallowing, taste, and speech but also increased risks for dental caries, pain, and oral infections [19]. Depending on the type of damage, different therapies could be considered. Moderate damage could be addressed by protein and/or gene activation, while cell therapies and/or bioengineered tissues to restore the entire organ may be more suitable for cases of severe radiation-induced damage. Current bioengineering, gene and mesenchymal stem cell therapies, as well as SG transfers are emphasized in the following chapters. Here, we review how different therapies could contribute to SG repair by correlating them with the developmental concepts described above.

1.4.1 Epithelial Protection and Repair

Since epithelial cells are the major component of SGs, they are the main target for any type of damage, especially radiation. Therefore, the prevention of SG cell apoptosis and/or membrane damage-induced dysfunction of acinar cells is clinically attempted by using intensity modulated radiation therapy (IMRT) rather than conventional radiotherapy. The precise delivery of radiation by IMRT reduces the amount of the SGs being targeted, still, 40 % of head and neck cancer patients experience moderate to severe oral dryness [8]. Recently, it was revealed that exclusion of a subregion of the cranial SG is essential to reduce severe loss of organ function [97]. Not surprisingly, this region appears to harbor the highest number of epithelial SG stem/progenitor cells. Superior dose distribution as delivered by proton therapy is further expected to improve dose sparing of SGs, specifically in this cranial subregion [56].

In the meantime, free radical scavengers (amifostine and tempol) and saliva-stimulating sialogogues (pilocarpine) provide relief to some patients. However, major health-related side effects induced by radical scavengers, such as vomiting and fever, need to be taken in consideration. Also the effectiveness of pilocarpine is related to the severity of tissue damage as this muscarinic agonist relies on some functional SG epithelial cells still being present in the gland. Suppressing cell apoptosis is another treatment that has had some success in animal models. Approaches shown to be effective in mouse models include pretreatment with insulin growth factor 1 (IGF1) [69], FGF2 [30, 53], Tousled kinase (TLK1B) [77], Pkcδ [1], or roscovitine. The latter is a cyclin-dependent kinase inhibitor that transiently inhibits G2/M cell cycle arrest, allowing suppression of apoptosis and DNA repair [66].

Alternatively, epithelial cell proliferation can be stimulated via posttreatment with aldehyde activator ALDH3 [5] or pre- and posttreatment with keratinocyte growth factor (KGF) or FGF7 [60, 109] to ameliorate radiation-induced hyposalivation in mice. Other signaling pathways shown to enhance adult SG regeneration postradiation or post-ductal ligation in mice are concurrent or transient activation of WNT/β-catenin [33, 34], SHH [32], and EDA [39]. Similar to developmental processes, all of these pathways affect epithelial stem/progenitors and aid in their expansion and differentiation. Notably, radiation itself does not alter endogenous WNT, EDA, or SHH pathway components, but evidently overstimulation reinforces epithelial growth mechanisms required to regenerate the damaged tissues. Another approach is to use a cocktail of chemokines, cytokines, and growth factors obtained by mobilizing or injecting bone marrow-derived cells, adipocytes, and/or mesenchymal stem cells into damaged SGs, as reviewed in [61]. These bioactive lysates are proposed to drive SG repair by reactivating signaling pathways in the epithelial and endothelial cells that remain. Whether they also stimulate post-damage neuronal repair is currently unclear.

Restoration of saliva secretion is further feasible by epithelial water channel aquaporin 1 (AQP1) protein gene therapy. After extensive animal research in mice, rats, pigs, and monkeys, the safety of an adenovirus-containing human AQP1 vector (AdhuAQP1) was studied in human Phase I clinical trials [6]. While there was some efficacy reported in some patients, follow-up studies to test the effectiveness of this therapy for long-term maintenance of increased salivary flow are ongoing. The reader is guided to a following chapter for more in-depth information.

SG cell transplantations have proven to be an effective treatment in rodent models. Several SG epithelial specific and non-SG cell types are able to integrate within the remaining epithelial compartment and contribute to its repair and subsequent homeostasis by differentiating into pools of various ductal and acinar cell types. Their potential use in the clinic has been reviewed in detail [61]. Briefly, autologous SG cell transplantation could occur in patients who still need to undergo radiation and where a SG biopsy could be taken before radiation therapy starts. Subsequent post-therapy transplantation of biopsy-isolated cells could initiate regeneration of the radiated SG. Alternatively, other cell types such as mesenchymal or adipose cells could be transplanted or mobilized post-radiotherapy to positively influence epithelial, mesenchymal, and endothelial repair.

In conclusion, the number of remaining stem/progenitor cells and functional epithelial cells will determine the probability of spontaneous regeneration of the damaged SG, as well as efficiency of sialogogues, apoptosis protectors, and signaling pathway stimulators. Restoring the epithelial compartment could elicit repair of surrounding blood vessels and nerves as well. As indicated from animal models [32, 59, 72], this can occur via paracrine communications from epithelial angiocrine and neurotrophic factor (e.g., BDNF, NRTN) release. One important clinical issue to be determined is whether these proposed treatments lead to undesirable side effects such as radioresistance and/or the acceleration of the patient's tumor cells.

1.4.2 Inducing Neuronal Survival and Reinnervation

Nerves have a limited capacity to regenerate, and irradiated human SGs have been shown to have reduced parasympathetic innervation [49]. The role of nerves and neurotransmitters during organ development and homeostasis is being elucidated, but less is known about their role during gland repair. Apart from neurotrophic release, innervation can also be influenced via cell plasma membrane-derived vesicles or exosomes, which not only mediate transmission of proteins but also mRNA and microRNAs [76]. One study proposes a participating role for mRNA and microRNAs in neuronal myelination and survival. Oligodendrocytes secrete exosomes with specific proteins and RNA toward surrounding neurons in the brain to improve their viability under stress conditions [24]. While exosome-influenced neuronal repair has yet not been investigated in SGs, neurotrophic protein deliveries via injection or gene therapy are currently being explored. In fetal SG experimental settings, radiation-induced neuronal apoptosis and denervation were reduced by postradiation delivery of exogenous neurturin (NRTN), which binds GFRα2/RET receptors on parasympathetic nerves [49]. Adult radiated SGs were shown to benefit from exogenous GDNF, which binds GFRα1 and aids in epithelial stem/progenitor cell expansion [103]. This likely occurs via neuronal communication, although its reinnervation pattern has not yet been studied.

Future efforts will also need to be directed toward determining the impact of radiation on sympathetic nerves and possible positive influences by WNT5a, NGF, or END3, which are known to be important during development of the gland.

In sum, functional neuronal repair and reinnervation may help in the release of neurotransmitters to maintain epithelial stem/progenitor cells and induce epithelial regeneration and ductal maturation. Moreover, their reactivation is necessary for proper saliva release from acinar cells and may also reestablish the communication pathways necessary for repair and subsequent homeostasis. Importantly, the major human SMG and SLG parasympathetic ganglions are located outside the gland in contrast with many rodent models [21]. The prevention or reduction of radiation to these ganglions, as well as the PAR ganglion that is located outside the tissue, could also improve the efficiency of these therapies.

1.4.3 Restoration of Blood Vessel Supply

Radiation severely impacts blood vessels to the point that capillary endothelial cells detach from the basal lamina, their density reduces, and large blood vessels dilate. Reducing damage to blood vessels will contribute to repair of damaged SGs. As iterated above, endothelial cells communicate not only with epithelial and mesenchymal but also neuronal cells. Furthermore, circulating bone marrow-derived cells migrate via blood vessels into damaged tissues to phagocytose the damaged environment. Several proposed strategies actively contribute to repair of the endothelial compartment. Pre-radiation gene delivery of angiocrine VEGF or FGF2 could ameliorate endothelial damage, resulting in reduced hyposalivation [16]. In vivo mobilization of bone marrow-derived cells is also a feasible approach to increase saliva flow as both their secretory bioactive lysate as well as the presence of endothelial progenitors can contribute to stimulation of endothelial survival and proliferation [59, 92].

1.4.4 Inflammation and Stromal Cell-Induced Fibrosis

Inflammation following radiation is an acute damage response but can persist and become chronic. While recruitment of macrophages is a necessity to phagocytose apoptotic cells [102], too much secretion of inflammatory cytokines and chemokines like IL, CCL2, and TNF can augment epithelial dysfunction. If not immediately controlled, the inflammatory response becomes pathogenic, as seen in autoimmune diseases and tissue fibrosis. In such cases, it is imperative to switch the pro-inflammatory response to an anti-inflammatory state in order to allow repair. Various potential mediators for reducing inflammation include IL-4, IL-13, T-reg, and B1 B-cells, and new evidence has also shown the ameliorating effects of stromal (mesenchymal) stem cells (MSCs) [81].

In adult tissues, the stromal cell pool harbors interstitial fibroblasts and adipocytes that continue to deposit and remodel the ECM. With aging patients, more adipose and fibrotic tissue is apparent that together with acinar cell loss leads to a 30–40 % decrease in parenchymal SG volume [7, 20]. Radiated stromal cells can enhance stress fiber formation and mature their cell matrix focal adhesions to turn into myofibroblasts, thereby producing excessive ECM. Injecting MSC/adipose-derived stem cells reduces muscular fibrosis as a consequence of radiation [93]. In the SG, these cells could presumably reduce inflammation as well as degrade the ECM and induce phagocytosis of apoptotic myofibroblasts. The timing of cell delivery, however, will be crucial as delayed therapy after radiation resulted in increased lung fibrosis due to the differentiation of mesenchymal stem cells into myofibroblasts [106]. Although TGFb has beneficial roles in wound healing and inflammatory responses, excessive levels of TGFβ1 can result in a number of serious conditions that are characterized by fibrosis, including chronic hepatitis, glomerulosclerosis, and postradiation tissue remodeling. As such, treatment with TGFβ inhibitors or genetically engineered TGFβR or HGF-expressing MSCs may be beneficial to attenuate fibrosis, as has been observed in radiated lungs [98, 104].

1.5 Future Prospects

From both this chapter and following chapters, it is clear that there are a number of different strategies to repair damaged SGs. Intravenous protein delivery, particularly of growth factors or agents that increase stem/progenitor cell activity, is often clinically disputed due to concerns regarding activation of any remaining tumor tissue. Local retrograde duct delivery of agents, such as viral vectors as part of gene therapy, provides a localized delivery and safer strategy as SGs are encapsulated and epithelial cells are easily transfected. While there is a possibility of vector diffusion into the bloodstream, this has not been a major issue in preclinical analysis. Similarly, direct cell delivery or mobilization of resident cells will always require evaluation of possible improper growth patterns, and bioengineered tissues will have to integrate with existing tissue and/or connecting ducts, nerves, and blood vessels.

Research on gland morphogenesis and repair will continue to identify targets from which new ways to approach regenerative therapies are being developed. Clearly, reestablishing crosstalk between different cell types endogenously enhances the regeneration process. Thus, by improving one compartment, one may also indirectly improve repair in another compartment. Recreating an optimal environment wherein cells can multiply and reconstruct the organ will be key. Currently, multiple approaches may be required to improve specific tissue structures within the organ and/or simultaneously influence other compartments to regenerate a damaged organ.

Acknowledgments I thank Dr. Wendy Knosp and Dr. Matthew Hoffman (National Institute for Dental and Craniofacial Research, NIH, DHHS, USA) for critical proofreading.

References

1. Arany S, Benoit DS, Dewhurst S, Ovitt CE. Nanoparticle-mediated gene silencing confers radioprotection to salivary glands in vivo. Mol Ther. 2013;21(6):1182–94. doi:10.1038/mt.2013.42.
2. Armulik A, Genove G, Betsholtz C. Pericytes: developmental, physiological, and pathological perspectives, problems, and promises. Dev Cell. 2011;21(2):193–215. doi:10.1016/j.devcel.2011.07.001.
3. Arnold K, Sarkar A, Yram MA, Polo JM, Bronson R, Sengupta S, Seandel M, Geijsen N, Hochedlinger K. Sox2(+) adult stem and progenitor cells are important for tissue regeneration and survival of mice. Cell Stem Cell. 2011;9(4):317–29. doi:10.1016/j.stem.2011.09.001.
4. Aure MH, Arany S, Ovitt CE. Salivary glands: stem cells, self-duplication, or both? J Dent Res. 2015;94(11):1502–7. doi:10.1177/0022034515599770.
5. Banh A, Xiao N, Cao H, Chen CH, Kuo P, Krakow T, Bavan B, Khong B, Yao M, Ha C, Kaplan MJ, Sirjani D, Jensen K, Kong CS, Mochly-Rosen D, Koong AC, Le QT. A novel aldehyde dehydrogenase-3 activator leads to adult salivary stem cell enrichment in vivo. Clin Cancer Res. 2011;17(23):7265–72. doi:10.1158/1078-0432.CCR-11-0179.
6. Baum BJ, Alevizos I, Zheng C, Cotrim AP, Liu S, McCullagh L, Goldsmith CM, Burbelo PD, Citrin DE, Mitchell JB, Nottingham LK, Rudy SF, Van Waes C, Whatley MA, Brahim JS, Chiorini JA, Danielides S, Turner RJ, Patronas NJ, Chen CC, Nikolov NP, Illei GG. Early responses to adenoviral-mediated transfer of the aquaporin-1 cDNA for radiation-induced salivary hypofunction. Proc Natl Acad Sci U S A. 2012;109(47):19403–7. doi:10.1073/pnas.1210662109.
7. Baum BJ, Ship JA, Wu AJ. Salivary gland function and aging: a model for studying the interaction of aging and systemic disease. Crit Rev Oral Biol Med. 1992;4(1):53–64.
8. Beetz I, Steenbakkers RJ, Chouvalova O, Leemans CR, Doornaert P, van der Laan BF, Christianen ME, Vissink A, Bijl HP, van Luijk P, Langendijk JA. The QUANTEC criteria for parotid gland dose and their efficacy to prevent moderate to severe patient-rated xerostomia. Acta Oncol. 2014;53(5):597–604. doi:10.3109/0284186X.2013.831186.
9. Bower DV, Lee HK, Lansford R, Zinn K, Warburton D, Fraser SE, Jesudason EC. Airway branching has conserved needs for local parasympathetic innervation but not neurotransmission. BMC Biol. 2014;12:92. doi:10.1186/s12915-014-0092-2.
10. Bullard T, Koek L, Roztocil E, Kingsley PD, Mirels L, Ovitt CE. Ascl3 expression marks a progenitor population of both acinar and ductal cells in mouse salivary glands. Dev Biol. 2008;320(1):72–8. doi:10.1016/j.ydbio.2008.04.018.
11. Bussard KM, Smith GH. The mammary gland microenvironment directs progenitor cell fate in vivo. Int J Cell Biol. 2011;2011:451676. doi:10.1155/2011/451676.
12. Cardoso WV, Lu J. Regulation of early lung morphogenesis: questions, facts and controversies. Development. 2006;133(9):1611–24. doi:10.1242/dev.02310.
13. Carmeliet P, Ferreira V, Breier G, Pollefeyt S, Kieckens L, Gertsenstein M, Fahrig M, Vandenhoeck A, Harpal K, Eberhardt C, Declercq C, Pawling J, Moons L, Collen D, Risau W, Nagy A. Abnormal blood vessel development and lethality in embryos lacking a single VEGF allele. Nature. 1996;380(6573):435–9. doi:10.1038/380435a0.
14. Carolan PJ, Melton DA. New findings in pancreatic and intestinal endocrine development to advance regenerative medicine. Curr Opin Endocrinol Diabetes Obes. 2013;20(1):1–7. doi:10.1097/MED.0b013e32835bc380.
15. Chung L, Yang TL, Huang HR, Hsu SM, Cheng HJ, Huang PH. Semaphorin signaling facilitates cleft formation in the developing salivary gland. Development. 2007;134(16):2935–45. doi:10.1242/dev.005066.
16. Cotrim AP, Sowers A, Mitchell JB, Baum BJ. Prevention of irradiation-induced salivary hypofunction by microvessel protection in mouse salivary glands. Mol Ther. 2007;15(12):2101–6. doi:10.1038/sj.mt.6300296.
17. Cotroneo E, Proctor GB, Carpenter GH. Regeneration of acinar cells following ligation of rat submandibular gland retraces the embryonic-perinatal pathway of cytodifferentiation. Differentiation. 2010;79(2):120–30. doi:10.1016/j.diff.2009.11.005.
18. Davis GE, Senger DR. Endothelial extracellular matrix: biosynthesis, remodeling, and functions during vascular morphogenesis and neovessel stabilization. Circ Res. 2005;97(11):1093–107. doi:10.1161/01.RES.0000191547.64391.e3.
19. Delli K, Spijkervet FK, Kroese FG, Bootsma H, Vissink A. Xerostomia. Monogr Oral Sci. 2014;24:109–25. doi:10.1159/000358792.

20. Drummond JR, Newton JP, Abel RW. Tomographic measurements of age changes in the human parotid gland. Gerodontology. 1995;12(1):26–30.
21. Ekstrom J, Khosravani N, Castagnola M, Messana I. Saliva and the control of its secretion. In: Ekberg O, editor. Dysphagia. Berlin/Heidelberg: Springer; 2012. doi:10.1007/978-3-642-17887-0.
22. Fagan AM, Zhang H, Landis S, Smeyne RJ, Silos-Santiago I, Barbacid M. TrkA, but not TrkC, receptors are essential for survival of sympathetic neurons in vivo. J Neurosci. 1996;16(19):6208–18.
23. Ferreira JN, Hoffman MP. Interactions between developing nerves and salivary glands. Organogenesis. 2013;9(3):199–205. doi:10.4161/org.25224.
24. Fruhbeis C, Frohlich D, Kuo WP, Amphornrat J, Thilemann S, Saab AS, Kirchhoff F, Mobius W, Goebbels S, Nave KA, Schneider A, Simons M, Klugmann M, Trotter J, Kramer-Albers EM. Neurotransmitter-triggered transfer of exosomes mediates oligodendrocyte-neuron communication. PLoS Biol. 2013;11(7):e1001604. doi:10.1371/journal.pbio.1001604.
25. Garcia-Gallastegui P, Ibarretxe G, Garcia-Ramirez JJ, Baladron V, Aurrekoetxea M, Nueda ML, Naranjo AI, Santaolalla F, Sanchez-del Rey A, Laborda J, Unda F. DLK1 regulates branching morphogenesis and parasympathetic innervation of salivary glands through inhibition of NOTCH signalling. Biol Cell. 2014;106(8):237–53. doi:10.1111/boc.201300086.
26. Garcia-Gallastegui P, Luzuriaga J, Aurrekoetxea M, Baladron V, Ruiz-Hidalgo MJ, Garcia-Ramirez JJ, Laborda J, Unda F, Ibarretxe G. Reduced salivary gland size and increased presence of epithelial progenitor cells in DLK1-deficient mice. Cell Tissue Res. 2015; doi:10.1007/s00441-015-2344-z.
27. Ghasemlou N, Krol KM, Macdonald DR, Kawaja MD. Comparison of target innervation by sympathetic axons in adult wild type and heterozygous mice for nerve growth factor or its receptor trkA. J Pineal Res. 2004;37(4):230–40. doi:10.1111/j.1600-079X.2004.00160.x.
28. Grobstein C. Inductive epitheliomesenchymal interaction in cultured organ rudiments of the mouse. Science. 1953a;118(3054):52–5.
29. Grobstein C. Morphogenetic interaction between embryonic mouse tissues separated by a membrane filter. Nature. 1953b;172(4384):869–70.
30. Guo L, Gao R, Xu J, Jin L, Cotrim AP, Yan X, Zheng C, Goldsmith CM, Shan Z, Hai B, Zhou J, Zhang C, Baum BJ, Wang S. AdLTR2EF1alpha-FGF2-mediated prevention of fractionated irradiation-induced salivary hypofunction in swine. Gene Ther. 2014;21(10):866–73. doi:10.1038/gt.2014.63.
31. Haara O, Fujimori S, Schmidt-Ullrich R, Hartmann C, Thesleff I, Mikkola ML. Ectodysplasin and Wnt pathways are required for salivary gland branching morphogenesis. Development. 2011;138(13):2681–91. doi:10.1242/dev.057711.
32. Hai B, Qin L, Yang Z, Zhao Q, Shangguan L, Ti X, Zhao Y, Kim S, Rangaraj D, Liu F. Transient activation of hedgehog pathway rescued irradiation-induced hyposalivation by preserving salivary stem/progenitor cells and parasympathetic innervation. Clin Cancer Res. 2014;20(1):140–50. doi:10.1158/1078-0432.CCR-13-1434.
33. Hai B, Yang Z, Millar SE, Choi YS, Taketo MM, Nagy A, Liu F. Wnt/beta-catenin signaling regulates postnatal development and regeneration of the salivary gland. Stem Cells Dev. 2010;19(11):1793–801. doi:10.1089/scd.2009.0499.
34. Hai B, Yang Z, Shangguan L, Zhao Y, Boyer A, Liu F. Concurrent transient activation of Wnt/beta-catenin pathway prevents radiation damage to salivary glands. Int J Radiat Oncol Biol Phys. 2012;83(1):e109–16. doi:10.1016/j.ijrobp.2011.11.062.
35. Hall BE, Zheng C, Swaim WD, Cho A, Nagineni CN, Eckhaus MA, Flanders KC, Ambudkar IS, Baum BJ, Kulkarni AB. Conditional overexpression of TGF-beta1 disrupts mouse salivary gland development and function. Lab Invest. 2010;90(4):543–55. doi:10.1038/labinvest.2010.5.
36. Hashizume A, Hieda Y. Hedgehog peptide promotes cell polarization and lumen formation in developing mouse submandibular gland. Biochem Biophys Res Commun. 2006;339(3):996–1000. doi:10.1016/j.bbrc.2005.11.106.
37. Hayashi T, Koyama N, Azuma Y, Kashimata M. Mesenchymal miR-21 regulates branching morphogenesis in murine submandibular gland in vitro. Dev Biol. 2011;352(2):299–307. doi:10.1016/j.ydbio.2011.01.030.
38. Hick AC, van Eyll JM, Cordi S, Forez C, Passante L, Kohara H, Nagasawa T, Vanderhaeghen P, Courtoy PJ, Rousseau GG, Lemaigre FP, Pierreux CE. Mechanism of primitive duct formation in the pancreas and submandibular glands: a role for SDF-1. BMC Dev Biol. 2009;9:66. doi:10.1186/1471-213X-9-66.
39. Hill G, Headon D, Harris ZI, Huttner K, Limesand KH. Pharmacological activation of the EDA/EDAR signaling pathway restores salivary gland function following radiation-induced damage. PLoS One. 2014;9(11):e112840. doi:10.1371/journal.pone.0112840.
40. Holmberg KV, Hoffman MP. Anatomy, biogenesis and regeneration of salivary glands. Monogr Oral Sci. 2014;24:1–13. doi:10.1159/000358776.
41. Holsinger FC, Bui DT. Anatomy, function, and evaluation of the salivary glands. In: Myers EN, Ferris RL, editors. Salivary gland disorders. Berlin/Heidelberg: Springer; 2007. p. 1–16. doi:10.1007/978-3-540-47072-4_1.
42. Janebodin K, Buranaphatthana W, Ieronimakis N, Hays AL, Reyes M. An in vitro culture system for long-term expansion of epithelial and mesenchymal salivary gland cells: role of TGF-beta1 in salivary gland epithelial and mesenchymal differentiation. Biomed Res Int. 2013;2013:815895. doi:10.1155/2013/815895.
43. Jaskoll T, Leo T, Witcher D, Ormestad M, Astorga J, Bringas Jr P, Carlsson P, Melnick M. Sonic hedgehog signaling plays an essential role during embryonic salivary gland epithelial branching morphogenesis.

Dev Dyn. 2004;229(4):722–32. doi:10.1002/dvdy.10472.
44. Jaskoll T, Zhou YM, Chai Y, Makarenkova HP, Collinson JM, West JD, Hajihosseini MK, Lee J, Melnick M. Embryonic submandibular gland morphogenesis: stage-specific protein localization of FGFs, BMPs, Pax6 and Pax9 in normal mice and abnormal SMG phenotypes in FgfR2-IIIc(+/Delta), BMP7(−/−) and Pax6(−/−) mice. Cells Tissues Organs. 2002;170(2–3):83–98.
45. Jaskoll T, Zhou YM, Trump G, Melnick M. Ectodysplasin receptor-mediated signaling is essential for embryonic submandibular salivary gland development. Anat Rec A Discov Mol Cell Evol Biol. 2003;271(2):322–31. doi:10.1002/ar.a.10045.
46. Kang JH, Kim BK, Park BI, Kim HJ, Ko HM, Yang SY, Kim MS, Jung JY, Kim WJ, Oh WM, Kim SH, Kim JH. Parasympathectomy induces morphological changes and alters gene-expression profiles in the rat submandibular gland. Arch Oral Biol. 2010;55(1):7–14. doi:10.1016/j.archoralbio.2009.11.003.
47. Kettunen P, Laurikkala J, Itaranta P, Vainio S, Itoh N, Thesleff I. Associations of FGF-3 and FGF-10 with signaling networks regulating tooth morphogenesis. Dev Dyn. 2000;219(3):322–32. doi:10.1002/1097-0177(2000)9999:9999<::AID-DVDY1062>3.0.CO;2-J.
48. Knosp WM, Knox SM, Lombaert IM, Haddox CL, Patel VN, Hoffman MP. Submandibular parasympathetic gangliogenesis requires sprouty-dependent Wnt signals from epithelial progenitors. Dev Cell. 2015;32(6):667–77. doi:10.1016/j.devcel.2015.01.023.
49. Knox SM, Lombaert IM, Haddox CL, Abrams SR, Cotrim A, Wilson AJ, Hoffman MP. Parasympathetic stimulation improves epithelial organ regeneration. Nat Commun. 2013;4:1494. doi:10.1038/ncomms2493.
50. Knox SM, Lombaert IM, Reed X, Vitale-Cross L, Gutkind JS, Hoffman MP. Parasympathetic innervation maintains epithelial progenitor cells during salivary organogenesis. Science. 2010;329(5999):1645–7. doi:10.1126/science.1192046.
51. Kobashi M, Ichikawa H, Kobashi M, Funahashi M, Mitoh Y, Matsuo R. The origin of sensory nerve fibers that innervate the submandibular salivary gland in the rat. Brain Res. 2005;1060(1–2):184–7. doi:10.1016/j.brainres.2005.08.012.
52. Kohn WG, Grossman E, Fox PC, Armando I, Goldstein DS, Baum BJ. Effect of ionizing radiation on sympathetic nerve function in rat parotid glands. J Oral Pathol Med. 1992;21(3):134–7.
53. Kojima T, Kanemaru S, Hirano S, Tateya I, Suehiro A, Kitani Y, Kishimoto Y, Ohno S, Nakamura T, Ito J. The protective efficacy of basic fibroblast growth factor in radiation-induced salivary gland dysfunction in mice. Laryngoscope. 2011;121(9):1870–5. doi:10.1002/lary.21873.
54. Kopf M, Schneider C, Nobs SP. The development and function of lung-resident macrophages and dendritic cells. Nat Immunol. 2015;16(1):36–44. doi:10.1038/ni.3052.
55. Kyriacou K, Garrett JR. Morphological changes in the rabbit submandibular gland after parasympathetic or sympathetic denervation. Arch Oral Biol. 1988;33(4):281–90.
56. Langendijk JA, Lambin P, De Ruysscher D, Widder J, Bos M, Verheij M. Selection of patients for radiotherapy with protons aiming at reduction of side effects: the model-based approach. Radiother Oncol. 2013;107(3):267–73. doi:10.1016/j.radonc.2013.05.007.
57. Lombaert IM, Abrams SR, Li L, Eswarakumar VP, Sethi AJ, Witt RL, Hoffman MP. Combined KIT and FGFR2b signaling regulates epithelial progenitor expansion during organogenesis. Stem Cell Rep. 2013;1(6):604–19. doi:10.1016/j.stemcr.2013.10.013.
58. Lombaert IM, Brunsting JF, Wierenga PK, Faber H, Stokman MA, Kok T, Visser WH, Kampinga HH, de Haan G, Coppes RP. Rescue of salivary gland function after stem cell transplantation in irradiated glands. PLoS One. 2008a;3(4):e2063. doi:10.1371/journal.pone.0002063.
59. Lombaert IM, Brunsting JF, Wierenga PK, Kampinga HH, de Haan G, Coppes RP. Cytokine treatment improves parenchymal and vascular damage of salivary glands after irradiation. Clin Cancer Res. 2008b;14(23):7741–50. doi:10.1158/1078-0432.CCR-08-1449.
60. Lombaert IM, Brunsting JF, Wierenga PK, Kampinga HH, de Haan G, Coppes RP. Keratinocyte growth factor prevents radiation damage to salivary glands by expansion of the stem/progenitor pool. Stem Cells. 2008c;26(10):2595–601. doi:10.1634/stemcells.2007-1034.
61. Lombaert IM, Movahednia MM, Adine C, Ferreira JN. Salivary gland regeneration: therapeutic approaches from stem cells to tissue organoids. Stem Cells. 2016. doi:10.1002/stem.2455
62. Maimets M, Rocchi C, Bron R, Pringle S, Kuipers J, Giepmans BN, Vries RG, Clevers H, de Haan G, van Os R, Coppes RP. Long-term in vitro expansion of salivary gland stem cells driven by Wnt signals. Stem Cell Rep. 2016;6(1):150–62. doi:10.1016/j.stemcr.2015.11.009.
63. Makita T, Sucov HM, Gariepy CE, Yanagisawa M, Ginty DD. Endothelins are vascular-derived axonal guidance cues for developing sympathetic neurons. Nature. 2008;452(7188):759–63. doi:10.1038/nature06859.
64. Mandell E, Seedorf GJ, Ryan S, Gien J, Cramer SD, Abman SH. Antenatal endotoxin disrupts lung vitamin D receptor and 25-hydroxyvitamin D 1alpha-hydroxylase expression in the developing rat. Am J Physiol Lung Cell Mol Physiol. 2015;309(9):L1018–26. doi:10.1152/ajplung.00253.2015.

65. Mandour MA, Helmi AM, El-Sheikh MM, El-Garem F, El-Ghazzawi E. Effect of tympanic neurectomy on human parotid salivary gland. Histopathologic, Histochemical, and Clinical Study. Arch Otolaryngol. 1977;103(6):338–41.
66. Martin KL, Hill GA, Klein RR, Arnett DG, Burd R, Limesand KH. Prevention of radiation-induced salivary gland dysfunction utilizing a CDK inhibitor in a mouse model. PLoS One. 2012;7(12):e51363. doi:10.1371/journal.pone.0051363.
67. Mathison R, Davison JS, Befus AD. Neuroendocrine regulation of inflammation and tissue repair by submandibular gland factors. Immunol Today. 1994;15(11):527–32.
68. McGovern AE, Mazzone SB. Neural regulation of inflammation in the airways and lungs. Auton Neurosci. 2014;182:95–101. doi:10.1016/j.autneu.2013.12.008.
69. Mitchell GC, Fillinger JL, Sittadjody S, Avila JL, Burd R, Limesand KH. IGF1 activates cell cycle arrest following irradiation by reducing binding of DeltaNp63 to the p21 promoter. Cell Death Dis. 2010;1:e50.
70. Miyazaki Y, Nakanishi Y, Hieda Y. Tissue interaction mediated by neuregulin-1 and ErbB receptors regulates epithelial morphogenesis of mouse embryonic submandibular gland. Dev Dyn. 2004;230(4):591–6. doi:10.1002/dvdy.20078.
71. Nanduri LS, Lombaert IM, van der Zwaag M, Faber H, Brunsting JF, van Os RP, Coppes RP. Salisphere derived c-Kit+ cell transplantation restores tissue homeostasis in irradiated salivary gland. Radiother Oncol. 2013;108(3):458–63. doi:10.1016/j.radonc.2013.05.020.
72. Nanduri LS, Maimets M, Pringle SA, van der Zwaag M, van Os RP, Coppes RP. Regeneration of irradiated salivary glands with stem cell marker expressing cells. Radiother Oncol. 2011;99(3):367–72. doi:10.1016/j.radonc.2011.05.085.
73. Nedvetsky PI, Emmerson E, Finley JK, Ettinger A, Cruz-Pacheco N, Prochazka J, Haddox CL, Northrup E, Hodges C, Mostov KE, Hoffman MP, Knox SM. Parasympathetic innervation regulates tubulogenesis in the developing salivary gland. Dev Cell. 2014;30(4):449–62. doi:10.1016/j.devcel.2014.06.012.
74. Nordgarden H, Johannessen S, Storhaug K, Jensen JL. Salivary gland involvement in hypohidrotic ectodermal dysplasia. Oral Dis. 1998;4(2):152–4.
75. Nussenzweig SC, Verma S, Finkel T. The role of autophagy in vascular biology. Circ Res. 2015;116(3):480–8. doi:10.1161/CIRCRESAHA.116.303805.
76. Pagella P, Jimenez-Rojo L, Mitsiadis TA. Roles of innervation in developing and regenerating orofacial tissues. Cell Mol Life Sci. 2014;71(12):2241–51. doi:10.1007/s00018-013-1549-0.
77. Palaniyandi S, Odaka Y, Green W, Abreo F, Caldito G, De Benedetti A, Sunavala-Dossabhoy G. Adenoviral delivery of Tousled kinase for the protection of salivary glands against ionizing radiation damage. Gene Ther. 2011;18(3):275–82. doi:10.1038/gt.2010.142.
78. Patel N, Sharpe PT, Miletich I. Coordination of epithelial branching and salivary gland lumen formation by Wnt and FGF signals. Dev Biol. 2011;358(1):156–67. doi:10.1016/j.ydbio.2011.07.023.
79. Patel VN, Hoffman MP. Salivary gland development: a template for regeneration. Semin Cell Dev Biol. 2014;25-26:52–60. doi:10.1016/j.semcdb.2013.12.001.
80. Patel VN, Lombaert IM, Cowherd SN, Shworak NW, Xu Y, Liu J, Hoffman MP. Hs3st3-modified heparan sulfate controls KIT+ progenitor expansion by regulating 3-O-sulfotransferases. Dev Cell. 2014;29(6):662–73. doi:10.1016/j.devcel.2014.04.024.
81. Prockop DJ. Inflammation, fibrosis, and modulation of the process by mesenchymal stem/stromal cells. Matrix Biol. 2016; doi:10.1016/j.matbio.2016.01.010.
82. Proctor GB, Asking B. A comparison between changes in rat parotid protein composition 1 and 12 weeks following surgical sympathectomy. Q J Exp Physiol. 1989;74(6):835–40.
83. Rebustini IT, Hayashi T, Reynolds AD, Dillard ML, Carpenter EM, Hoffman MP. miR-200c regulates FGFR-dependent epithelial proliferation via Vldlr during submandibular gland branching morphogenesis. Development. 2012;139(1):191–202. doi:10.1242/dev.070151.
84. Rebustini IT, Myers C, Lassiter KS, Surmak A, Szabova L, Holmbeck K, Pedchenko V, Hudson BG, Hoffman MP. MT2-MMP-dependent release of collagen IV NC1 domains regulates submandibular gland branching morphogenesis. Dev Cell. 2009;17(4):482–93. doi:10.1016/j.devcel.2009.07.016.
85. Reinert RB, Cai Q, Hong JY, Plank JL, Aamodt K, Prasad N, Aramandla R, Dai C, Levy SE, Pozzi A, Labosky PA, Wright CV, Brissova M, Powers AC. Vascular endothelial growth factor coordinates islet innervation via vascular scaffolding. Development. 2014;141(7):1480–91. doi:10.1242/dev.098657.
86. Rodriguez-Diaz R, Caicedo A. Neural control of the endocrine pancreas. Best Pract Res Clin Endocrinol Metab. 2014;28(5):745–56. doi:10.1016/j.beem.2014.05.002.
87. Rothova M, Thompson H, Lickert H, Tucker AS. Lineage tracing of the endoderm during oral development. Dev Dyn. 2012;241(7):1183–91. doi:10.1002/dvdy.23804.
88. Rugel-Stahl A, Elliott ME, Ovitt CE. Ascl3 marks adult progenitor cells of the mouse salivary gland. Stem Cell Res. 2012;8(3):379–87. doi:10.1016/j.scr.2012.01.002.
89. Ryu YK, Collins SE, Ho HY, Zhao H, Kuruvilla R. An autocrine Wnt5a-Ror signaling loop mediates sympathetic target innervation. Dev Biol. 2013;377(1):79–89. doi:10.1016/j.ydbio.2013.02.013.
90. Steinberg Z, Myers C, Heim VM, Lathrop CA, Rebustini IT, Stewart JS, Larsen M, Hoffman

MP. FGFR2b signaling regulates ex vivo submandibular gland epithelial cell proliferation and branching morphogenesis. Development. 2005;132(6):1223–34. doi:10.1242/dev.01690.
91. Sugimoto T, Taya Y, Shimazu Y, Soeno Y, Sato K, Aoba T. Three-dimensional visualization of developing neurovascular architecture in the craniofacial region of embryonic mice. Anat Rec (Hoboken). 2015;298(11):1824–35. doi:10.1002/ar.23179.
92. Sumita Y, Liu Y, Khalili S, Maria OM, Xia D, Key S, Cotrim AP, Mezey E, Tran SD. Bone marrow-derived cells rescue salivary gland function in mice with head and neck irradiation. Int J Biochem Cell Biol. 2011;43(1):80–7. doi:10.1016/j.biocel.2010.09.023.
93. Sun W, Ni X, Sun S, Cai L, Yu J, Wang J, Nie B, Sun Z, Ni X, Cao X. Adipose-derived stem cells alleviate radiation-induced muscular fibrosis by suppressing the expression of TGF-beta1. Stem Cells Int. 2016;2016:5638204. doi:10.1155/2016/5638204.
94. Talavera-Adame D, Dafoe DC. Endothelium-derived essential signals involved in pancreas organogenesis. World J Exp Med. 2015;5(2):40–9. doi:10.5493/wjem.v5.i2.40.
95. Tanida S, Kataoka H, Mizoshita T, Shimura T, Kamiya T, Joh T. Intranuclear translocation signaling of HB-EGF carboxy-terminal fragment and mucosal defense through cell proliferation and migration in digestive tracts. Digestion. 2010;82(3):145–9. doi:10.1159/000310903.
96. Tucker AS. Salivary gland development. Semin Cell Dev Biol. 2007;18(2):237–44. doi:10.1016/j.semcdb.2007.01.006.
97. van Luijk P, Pringle S, Deasy JO, Moiseenko VV, Faber H, Hovan A, Baanstra M, van der Laan HP, Kierkels RG, van der Schaaf A, Witjes MJ, Schippers JM, Brandenburg S, Langendijk JA, Wu J, Coppes RP. Sparing the region of the salivary gland containing stem cells preserves saliva production after radiotherapy for head and neck cancer. Sci Transl Med. 2015;7(305):305ra147. doi:10.1126/scitranslmed.aac4441.
98. Wang H, Yang YF, Zhao L, Xiao FJ, Zhang QW, Wen ML, Wu CT, Peng RY, Wang LS. Hepatocyte growth factor gene-modified mesenchymal stem cells reduce radiation-induced lung injury. Hum Gene Ther. 2013;24(3):343–53. doi:10.1089/hum.2012.177.
99. Wells KL, Gaete M, Matalova E, Deutsch D, Rice D, Tucker AS. Dynamic relationship of the epithelium and mesenchyme during salivary gland initiation: the role of Fgf10. Biol Open. 2013;2(10):981–9. doi:10.1242/bio.20135306.
100. Wells KL, Mou C, Headon DJ, Tucker AS. Recombinant EDA or Sonic Hedgehog rescue the branching defect in Ectodysplasin A pathway mutant salivary glands in vitro. Dev Dyn. 2010;239(10):2674–84. doi:10.1002/dvdy.22406.
101. Wright DM, Buenger DE, Abashev TM, Lindeman RP, Ding J, Sandell LL. Retinoic acid regulates embryonic development of mammalian submandibular salivary glands. Dev Biol. 2015;407(1):57–67. doi:10.1016/j.ydbio.2015.08.008.
102. Wynn TA, Chawla A, Pollard JW. Macrophage biology in development, homeostasis and disease. Nature. 2013;496(7446):445–55. doi:10.1038/nature12034.
103. Xiao N, Lin Y, Cao H, Sirjani D, Giaccia AJ, Koong AC, Kong CS, Diehn M, Le QT. Neurotrophic factor GDNF promotes survival of salivary stem cells. J Clin Invest. 2014;124(8):3364–77. doi:10.1172/JCI74096.
104. Xue J, Li X, Lu Y, Gan L, Zhou L, Wang Y, Lan J, Liu S, Sun L, Jia L, Mo X, Li J. Gene-modified mesenchymal stem cells protect against radiation-induced lung injury. Mol Ther. 2013;21(2):456–65. doi:10.1038/mt.2012.183.
105. Yamamoto S, Fukumoto E, Yoshizaki K, Iwamoto T, Yamada A, Tanaka K, Suzuki H, Aizawa S, Arakaki M, Yuasa K, Oka K, Chai Y, Nonaka K, Fukumoto S. Platelet-derived growth factor receptor regulates salivary gland morphogenesis via fibroblast growth factor expression. J Biol Chem. 2008;283(34):23139–49. doi:10.1074/jbc.M710308200.
106. Yan X, Liu Y, Han Q, Jia M, Liao L, Qi M, Zhao RC. Injured microenvironment directly guides the differentiation of engrafted Flk-1(+) mesenchymal stem cell in lung. Exp Hematol. 2007;35(9):1466–75. doi:10.1016/j.exphem.2007.05.012.
107. Yancopoulos GD, Davis S, Gale NW, Rudge JS, Wiegand SJ, Holash J. Vascular-specific growth factors and blood vessel formation. Nature. 2000;407(6801):242–8. doi:10.1038/35025215.
108. Yun EJ, Lorizio W, Seedorf GJ, Abman SH, Vu TH. VEGF and endothelium-derived retinoic acid regulate lung vascular and alveolar development. Am J Physiol Lung Cell Mol Physiol:ajplung. 2015;00229:02015. doi:10.1152/ajplung.00229.2015.
109. Zheng C, Cotrim AP, Rowzee A, Swaim W, Sowers A, Mitchell JB, Baum BJ. Prevention of radiation-induced salivary hypofunction following hKGF gene delivery to murine submandibular glands. Clin Cancer Res. 2011;17(9):2842–51. doi:10.1158/1078-0432.CCR-10-2982.

Systems Biology: Salivary Gland Development, Disease, and Regenerative Medicine

2

Melinda Larsen, Petko Bogdanov, Ravi Sood, Hae Ryong Kwon, Deirdre A. Nelson, Connor Duffy, Sarah B. Peters, and Sridar V. Chittur

Abstract

There are multiple challenges currently facing clinical salivary gland research, encompassing a wide range of topics from understanding development to understanding disease etiology and from diagnosing disease to designing more effective, personalized therapies. Systems analysis complements traditional reductionist approaches, and the integration of these two approaches is starting to provide a more comprehensive understanding of the causal relationships leading to normal and abnormal biology. Understanding normal developmental processes is critical for understanding development of disease. Morphogenesis and differentiation are complex developmental processes involving orchestrated interactions between heterotypic cell types that have proven difficult to understand through reductionist approaches alone. It has become clear that it is not possible to understand the complex molecular, cellular, and physical process integration that is required for any developmental or disease process without the use of systems biology approaches. In this chapter, we demonstrate examples in the use of systems approaches to better characterize the difference

M. Larsen, PhD (✉) • H.R. Kwon, PhD
D.A. Nelson • C. Duffy
Department of Biological Sciences, University at Albany, SUNY, State University of New York, 1400 Washington Ave LSRB 1086, Albany, NY 12222, USA

Oklahoma Medical Research Foundation, Oklahoma, OK, USA
e-mail: mlarsen@albany.edu

P. Bogdanov, PhD • R. Sood
Department of Computer Science, University at Albany, SUNY, State University of New York, 1400 Washington Ave, Albany, NY 12222, USA

S.B. Peters, PhD
Department of Biological Sciences, University at Albany, SUNY, State University of New York, 1400 Washington Ave LSRB 1086, Albany, NY 12222, USA

Department of Cell, Developmental, and Integrative Biology, University of Alabama at Birmingham, 1918 University Blvd., MCML 643, Birmingham, AL 35294, USA

S.V. Chittur, PhD
Center for Functional Genomics, University at Albany, SUNY, State University of New York, Health Sciences Campus, One Discovery Drive, Albany, NY 12144, USA

S. Cha (ed.), *Salivary Gland Development and Regeneration*, DOI 10.1007/978-3-319-43513-8_2

between embryonic submandibular salivary glands grown in vivo and ex vivo as 3D organ explants and to identify potential signaling networks in heterotypic subpopulations of cells that lead to the prediction of a function for endothelial cells in salivary gland development. We conclude with examples of how systems biology-based approaches using both tissue samples and saliva from patients are currently being used in many laboratories to make progress in salivary gland disease diagnosis, understanding disease etiology, and informing therapeutic development for cancer and regenerative medicine strategies.

2.1 Introduction

Knowledge of developmental mechanisms is critical to inform effective therapies for regeneration and repair of damaged glands [66]. During embryogenesis, the information encoded in the genome is converted into a 3D organ, where primitive clusters of cells undergo morphogenesis and differentiation to acquire mature structure and function. Embryonic development of secretory and absorptive organs including the salivary glands, kidneys, lungs, mammary glands, prostate, and lacrimal glands depends on a process known as branching morphogenesis [11]. Branching morphogenesis allows organs to maximize epithelial surface area while efficiently minimizing volume within the tissue. Differentiation begins with changes in gene expression during early development that translate into differential protein expression, allowing cells to take on specific functions. Morphogenesis and differentiation are complex processes involving orchestrated interactions between multiple cell types that have proven to be difficult to understand through reductionist approaches alone, and systems biology approaches are needed to facilitate understanding of the complex molecular, cellular, and physical process integrations required for organ development. Systems biology can be defined in many ways, but attempts to integrate molecular-level knowledge of genes, RNAs, and proteins with biophysical and cellular processes into networks of interacting components are the essence of systems biology approaches. Systems analysis complements traditional approaches, and the integration of systems biology with reductionist approaches is proving to provide a more comprehensive understanding of the causal relationships leading to normal and abnormal organogenesis.

Animal models that have provided insights relevant to human salivary gland development include the mouse, rat, and the fruit fly, *Drosophila melanogaster*, but most recent studies have used the mouse model due to its relative similarity to humans and its tractable genetics. The mouse submandibular salivary gland begins its development at embryonic day 11.5 (E11.5) (with the day of coital plug discovery defined as E0) as a thickening of the primitive oral epithelium that grows into the first branchial (mandibular) arch mesenchyme to form the solid epithelial placode (reviewed in [21, 27, 30, 34, 67, 87]). This placode protrudes into the mesenchyme by E12, forming a single, solid mass of immature epithelial cells connected to the tongue epithelium by a stalk of immature epithelial cells. By E12.5, alterations in the basement membrane form indentations on the surface of the bud termed clefts. The clefts are essential to gland development since, as they progress to separate the primary bud into multiple buds [10, 37], they define the boundary between developing proacini and ducts. As cleft progression proceeds, the epithelium proliferates, and the base of the cleft becomes the nascent ductal structure. This process of salivary gland branching morphogenesis is repeated multiple times over the course of several days, continually increasing the complexity of the organ. By late E14, the simple one-bud one-duct salivary gland has both grown and branched significantly, and the main duct begins to lumenize. The end buds undergo reorganization and begin to form acini – the main secretory units of the salivary gland. By E15–16, lumenization of the main secretory duct is nearly complete, and by E17, the forming acini lumenize, so that the gland has a continuous network of lumenized, but still immature, acini and ducts connecting to the oral cavity. Both nerves and blood vessels populate the gland in association with the branching epithelium, and the nerves are known to be critically involved in the elaboration of organ structure [32]. Cellular differentiation occurs in parallel with branching morphogenesis through pathways that are at least partially independent [9] as the glands continue to mature after birth. The developmental program is complex, with multiple processes occurring simultaneously.

Salivary gland function is equally complex in the adult as are the mechanisms leading to multiple forms of dysfunction. Multiple epithelial cell types are required to make saliva and for regulation of exocrine secretion involving both sympathetic and parasympathetic innervation [73]. Saliva production begins with the secretion of incomplete saliva by submandibular acinar cells, including contributions from both serous and mucous acinar cells in human glands. This primary fluid is modified by ductal cells as it is transported to the oral cavity, where it mixes with secretions from other salivary glands to generate whole saliva. Adult salivary glands can be affected by viral and bacterial infections, inflammation, autoimmune disease, and tumorigenesis. Infections of the salivary gland, or sialadenitis, can occur from bacterial, viral, fungal, or other causes. Sjögren's syndrome (SS) is a complex systemic autoimmune disease affecting primarily the salivary glands and the lacrimal glands with an elusive etiology [13, 40]. SS is difficult to diagnose until it is in its later stages due to many factors, including a lack of molecular markers. Up to 90 % of SS patients express antibodies targeting autoantigens [52], but these autoantibodies are not unique to SS. SS patients may also be affected by other autoimmune diseases, and such patients are classified as secondary SS patients. Around 5 % of SS patients typically progress to a non-Hodgkin's, mucosa-associated lymphoid tissue (MALT) lymphoma [33]. Independent from SS, salivary gland tumors can occur. In general, primary tumors of the major salivary glands are relatively rare, accounting for 11 % of oropharyngeal neoplasms in the USA [7]. A distinguishing feature of salivary gland neoplasms is that they are highly histologically variable due to the significant variance in their cellular origins and presumably variable etiology. As a result, histological classification of salivary gland tumors is challenging and subject to misdiagnosis by nonexperts [84, 85]. Although potentially invaluable for tumor classification, molecular signatures for salivary tumors are lacking, similar to the lack of definitive markers for SS.

Other than patients with malignant tumors, the most significant problems for patients with salivary gland disease typically result from salivary hypofunction, or a decrease in or loss of salivary function. Hyposalivation, or decreased saliva production, occurs in patients that have SS, and also as a side effect following radiation therapy for head and neck cancers. Loss of salivary flow results in "dry mouth" and affects additional processes, such as mastication and swallowing. Secondary oral disease states can subsequently develop, including sialadenitis, dental caries, periodontal disease, and persistent oral infections [61], leading to a general decline in the quality of life. The molecular basis for hyposalivation in patients suffering from SS and the mechanism of the selective destruction of salivary acinar cells leading to hyposalivation as a result of radiation therapy remain poorly understood. There are few satisfactory therapeutic options for these patients.

There are multiple challenges currently facing salivary gland research encompassing a wide range of topics from understanding development to diagnosing human disease and developing more effective therapies. Understanding disease pathogenesis and developing rational therapies will require an understanding of the emergent properties of these interactions, which is the strength of systems biology. Since this topic was last reviewed [38], systems-based approaches have been widely incorporated into many studies addressing basic biological mechanisms as well as translational studies that address salivary gland disease pathogenesis and therapeutic development. In this chapter, we will demonstrate examples of how systems-based approaches can be used to understand the complexities of salivary gland organogenesis and disease. Specifically, we will demonstrate how we have used systems-based methods to evaluate the embryonic mouse submandibular salivary gland organ explants as a model system for the study of in vivo biology and how data mining of existing databases can be used to generate novel hypothesis regarding mechanisms of development. We will also review how others have used systems-based profiling and modeling approaches to gain novel insights into mechanisms involved in glandular differentiation and salivary gland diseases and how systems-based evaluation of saliva is useful potentially not only for diagnosis of salivary diseases but of systemic conditions.

2.2 Current Reductionist Approaches to Understand Salivary Gland Development

2.2.1 Organ Explants, Organoid Culture, and Other Primary Cultures

Traditionally, cell and developmental biologists have taken a reductionist approach to understanding the development and function of salivary glands. This approach – rooted in the scientific method itself – has provided useful information regarding linear signaling pathways. First established in the 1950s [5, 6, 19, 20], the embryonic submandibular gland organ explant culture system has provided researchers with a 3D experimental system that can be used to ask in-depth, complex questions about the control of morphogenesis. Organ explants have been useful to study and identify many signaling pathways that are critical during early development since they preserve heterotypic cell-cell interactions that occur in vivo and the morphogenesis that occurs closely mimics the in vivo process. More recently, it has been demonstrated that the early differentiation of multiple epithelial cell types occurs in the developing organoids [70]. Embryonic development of the salivary glands depends upon the regulated proliferation and differentiation of progenitor cell populations. Growth of salivary gland cells in small clusters of cells known as organoids or "salispheres" is currently being used as a useful method of preserving cell phenotype through homotypic and, in some cases, heterotypic cell interactions. Although organoids are further removed from the in vivo conditions than organ explants, they offer advantages for maintaining stem/progenitor cell populations. Both organ explants and organoids offer the opportunity to manipulate multiple genes in a moderate-throughput manner via pharmacological [36], RNA-mediated knockdown [10, 22, 78], viral transduction [23, 28, 80], and other means.

2.2.2 Transgenic Mouse Models

Numerous transgenic mouse models have identified critical genetic mediators of salivary gland development (reviewed in [58]). Recent studies have additionally employed transgenic mouse models encoding fluorescent lineage reporters to characterize molecular mediators of heterotypic cell interactions and progenitor cell function that control salivary gland development and regeneration [3, 31, 32, 62], providing important insights for improved cell therapy and regenerative medicine strategies. The future looks bright for the continued use of transgenic animals in salivary gland research. The introduction of CRISPR/Cas-based methods for manipulation of the mouse genome promises to make the process of generating transgenic animals more efficient [81]. New promoters are increasingly becoming available to be able to target specific salivary gland cell populations [51]. The disadvantage of the transgenic approach is that only one gene can be manipulated at a time; however, the utilization of CRISPR/Cas methods promises to make the manipulation of multiple genes at a time [95] much more efficient and cost effective than using traditional transgenic technologies. Manipulation of genes within model organisms occurs within the context of the organismal system, and such manipulations can have widespread effects. Hence, there is a significant need to apply systems-level analysis to transgenic organisms.

2.3 Systems-Based Approaches to the Study of Gland Development

2.3.1 Salivary Gland Molecular Anatomy Project

Gene profiling is a powerful tool for researchers studying both developmental processes and human disease to identify the transcriptome. Developed by Matthew Hoffman and Kenneth Yamada in the intramural research program at the NIDCR, the Salivary Gland Molecular Anatomy Project (SGMAP) (http://sgmap.nidcr.nih.gov) provides a searchable gene expression database for multiple stages of mouse submandibular and sublingual salivary gland development that is an invaluable tool for researchers. Temporal expression patterns of genes are provided for salivary glands extracted from mice at embryonic day (E) 11.5 through adult that were obtained using Agilent mouse microarrays. The SGMAP website allows for searching of any gene symbol, gene description, or gene ontology (GO) term [2] to obtain information on its developmental expression, expression in epithelium vs mesenchyme at E13, or its spatial expression in specific regions of the gland at specific developmental time points [59].

2.3.2 Comparison of the transcriptome of embryonic salivary glands grown in vivo as compared with that of embryonic organ explants grown ex vivo

Although organ explants have been in use for many years, it has not been clear how closely the explant culture system mimics in vivo development. To address this question, we compared our own gene expression data obtained from embryonic mouse submandibular salivary glands grown on a polyacrylamide gel constructed to mimic the compliance of the in vivo environment vs gene expression data obtained from glands that were grown in vivo that is available from SGMAP. This study was founded on our prior work that demonstrated that development of the mouse submandibular salivary gland is mechanically sensitive. Both epithelial morphogenesis and differentiation proceed the most similar to in vivo when glands are grown on a substrate having a stiffness that mimics that of the in vivo environment [70, 71]. We compared the genome-wide differential expression of individual mRNAs of embryonic glands grown in vivo from E13 through E15, as reported in SGMAP,[1] with E13 embryonic organ explants cultured on a 0.5 kPa polyacrylamide gel for 24 or 72 h, as previously described [70, 71]. We quantified mRNA expression using Affymetrix Mouse Gene ST 2.0 gene arrays. Our microarray data was normalized using GeneSpring v12.6, first being quantile normalized using a PLIER16 algorithm and baseline transformed to the median of all samples. The $\log_2$ normalized signal values were filtered to remove entities that show signal in the bottom 20th percentile across all samples. We then compared gene expression levels from the organ explants with data from SGMAP.

Comparison of different datasets that were obtained from different types of samples and that were processed independently is challenging with microarray data. Ideally, the datasets should be compared starting with raw data and then processed in parallel [96]. In comparing our microarray data with the SGMAP data, both datasets were first normalized to the same scale to facilitate the comparison. Genes from the in vivo data for which no data were available were omitted from the comparison, which corresponded to 10.9 % or 2462 of 22,510 genes. Both datasets were then standardized to a zero mean and a standard deviation of one, and then the expression levels in individual developmental stages were scaled to the interval [0, 1]. The in vivo and ex vivo datasets were cross-matched, retaining only shared genes for the analysis (13,709 genes).

We first compared the corresponding in vivo and ex vivo overall distributions of differential expression between the in vivo dataset from E13 to

[1] http://sgmap.nidcr.nih.gov/sgmap/sgexp.html

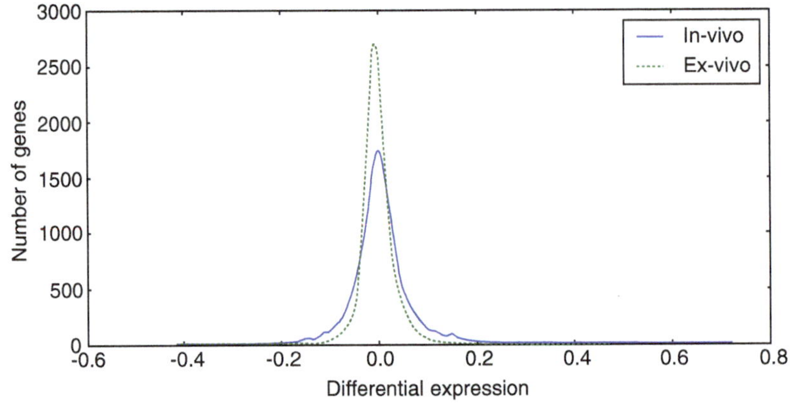

Fig. 2.1 Comparison of the differential gene expression of all genes expressed in developing salivary glands from E13 to E15 in vivo vs ex vivo. All genes showing developmental differential gene expression were plotted. Both distributions appear to have a single mode near 0, indicating that it is most common for a gene not to be significantly differentially expressed in either condition. The ex vivo mode peaks higher, indicating that there are a greater number of genes that are not significantly differentially expressed as compared to in vivo, while the in vivo distribution is broader, indicating a greater range of differential expression in comparison with ex vivo growth

E15 and the ex vivo data set grown for 24–72 h. As shown in Fig. 2.1, the overall shapes of the two distributions are similar; however, there are more genes having higher differential expression that are both upregulated and downregulated in the in vivo dataset than the ex vivo dataset. This analysis suggests that gene expression levels are generally tempered in glands cultured ex vivo compared to glands grown in vivo.

Given that there are notable differences in the global distributions of gene expression between glands grown in vivo vs ex vivo, we attempted to identify what groups of genes were different. Differential gene expression for each dataset was defined as a significant difference between the 24 h and the 72 h culture condition for the ex vivo culture and a significant difference between the E13 and E15 in vivo gene expression time points. We compared the functional agreement of the most differentially expressed genes based on their biological process annotations in the gene ontology (GO) [2]. We performed this GO term analysis for the 10 % most differentially expressed genes (i.e., those genes whose expression decreased or increased most drastically between the two time points). Agreement was most significant for the top 10 % of genes ranked by differential expression ex vivo, where 35 % of such genes were also in the top 10 % in vivo (data not shown). We summarize the annotations of the top 10 % differentially expressed genes ex vivo and the top 10 % in vivo genes (Fig. 2.2). The percentage of annotations that occur in the top 10 % ex vivo are similar to those that occur in the top 10 % in vivo across all level 1 biological process GO terms. In other words, at least at a high level, genes involved in the same processes are changing in expression levels both in vivo and ex vivo.

To understand which biological processes were the most conserved and less conserved in submandibular salivary glands grown ex vivo as compared to in vivo, we performed a Kolmogorov-Smirnov (KS) test on the genes within each of the level 1 GO categories. The KS test quantifies how dissimilar two sample distributions are, with a smaller KS indicating high similarity and a larger KS term indicating lesser similarity. Since the mean KS for all genes was 0.128, we considered biological processes with a lower KS than 0.128 and a p-value of $p < 0.05$ to be conserved. As shown in Fig. 2.3, the Kolmogorov-Smirnov test demonstrated that the most conserved biological processes in glands grown ex vivo vs in vivo were included in the GO terms for growth (KS 0.101), metabolic processes (KS 0.111), developmental processes (KS 0.118), cellular processes (KS 0.119), and biological regulation (KS 0.124). As genes included in these categories would be predicted to be essential for

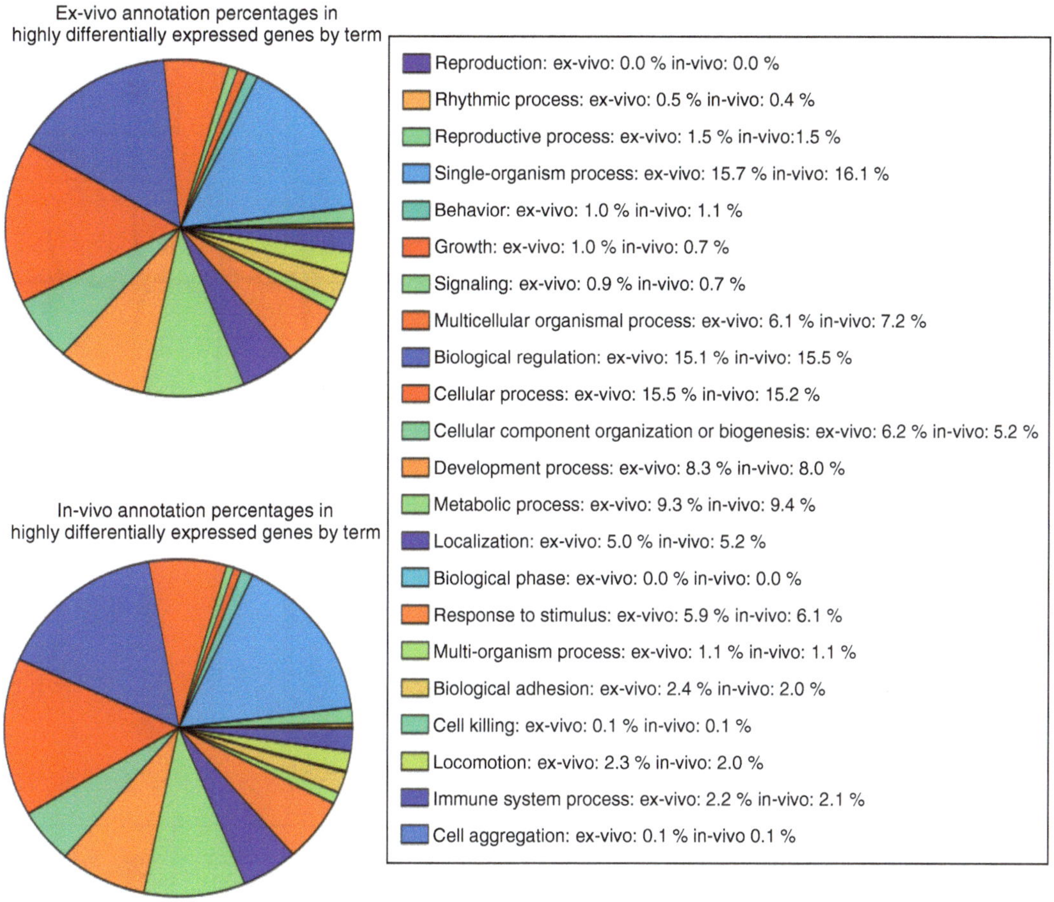

Fig. 2.2 Comparison of the biological process representation in differentially expressed in vivo and ex vivo genes. We considered the known annotations of genes with GO terms at the first level of the GO biological process ontology and show the percentage of annotations for each term in vivo and ex vivo. This analysis indicates that the relative importance of each process ex vivo is similar to the in vivo situation. Only the top 10 % of the most differentially expressed genes were considered. Genes with more specific annotations at ontology levels deeper than 1 were considered to be annotated with the corresponding ancestor biological process at level one for this analysis

development, this analysis validates the mouse embryonic submandibular salivary gland as a valid model system to evaluate developmental processes. The less-conserved biological processes, as determined by a KS value greater than the mean of 0.128 with a *p*-value of <0.05, are biological processes whose expression does not agree as well in ex vivo as in vivo (Fig. 2.4). At the top of this list were the GO categories, cell killing (KS 0.417), immune system process (KS 0.202), and behavior (KS 0.167). As the organ explants do not maintain an intact immune system, nor do they remain connected to the nervous system, vascular system, or lymphatic system, it is also not surprising that expressions of genes classified in the categories, "Immune System Process" and "Behavior," are somewhat different in vivo and ex vivo. This systems-based analysis of gland development ex vivo confirms that many general biological processes that are required for development, including cell growth, metabolic processes, and other developmental processes, occur in organ explants similarly to how they occur in vivo, while other processes that require interaction with the rest of the body system are somewhat different.

To examine how similar the differential gene expression is for specific genes in vivo vs ex vivo,

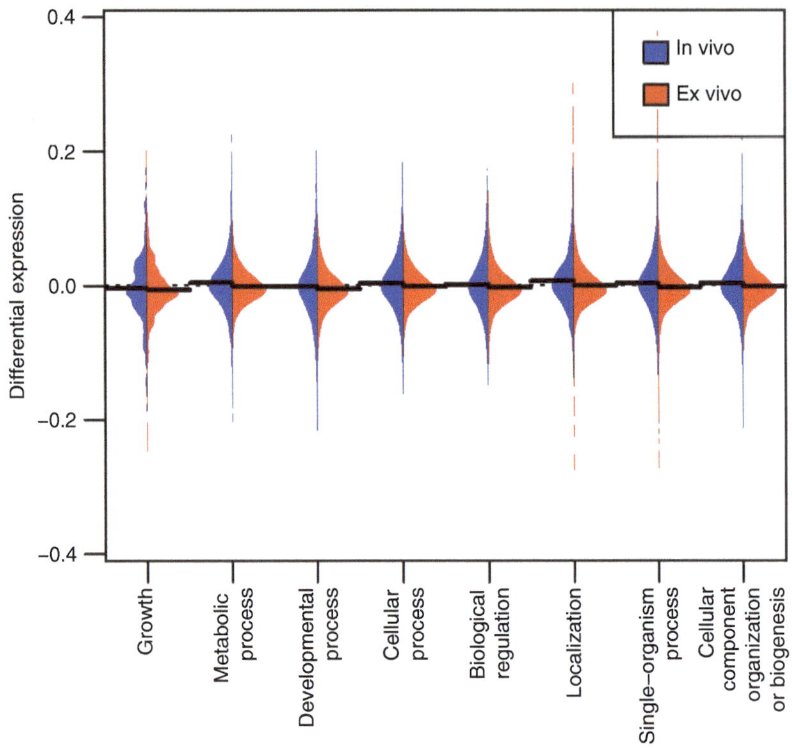

Fig. 2.3 Comparison of the in vivo and ex vivo differential gene expression in GO processes of highest agreement. Genes associated with a given biological process at the first level of the GO hierarchy are included in the comparison for each of the distributions. The pairs of distributions of all included processes have a Kolmogorov-Smirnov test distance of 0.13 or less and associated *p*-value <0.05

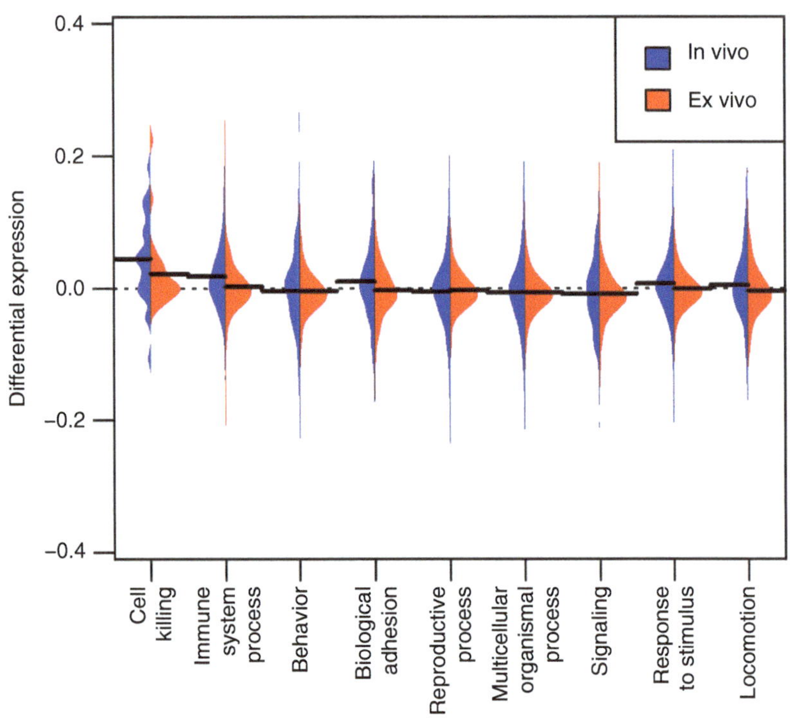

Fig. 2.4 Comparison of the in vivo and ex vivo differential gene expression in GO processes of lowest agreement. Genes associated with a given biological process at the first level of the GO hierarchy are included in the comparison for each of the distributions. The pairs of distributions of all included processes have a Kolmogorov-Smirnov test distance higher than 0.14 and associated *p*-value <0.05

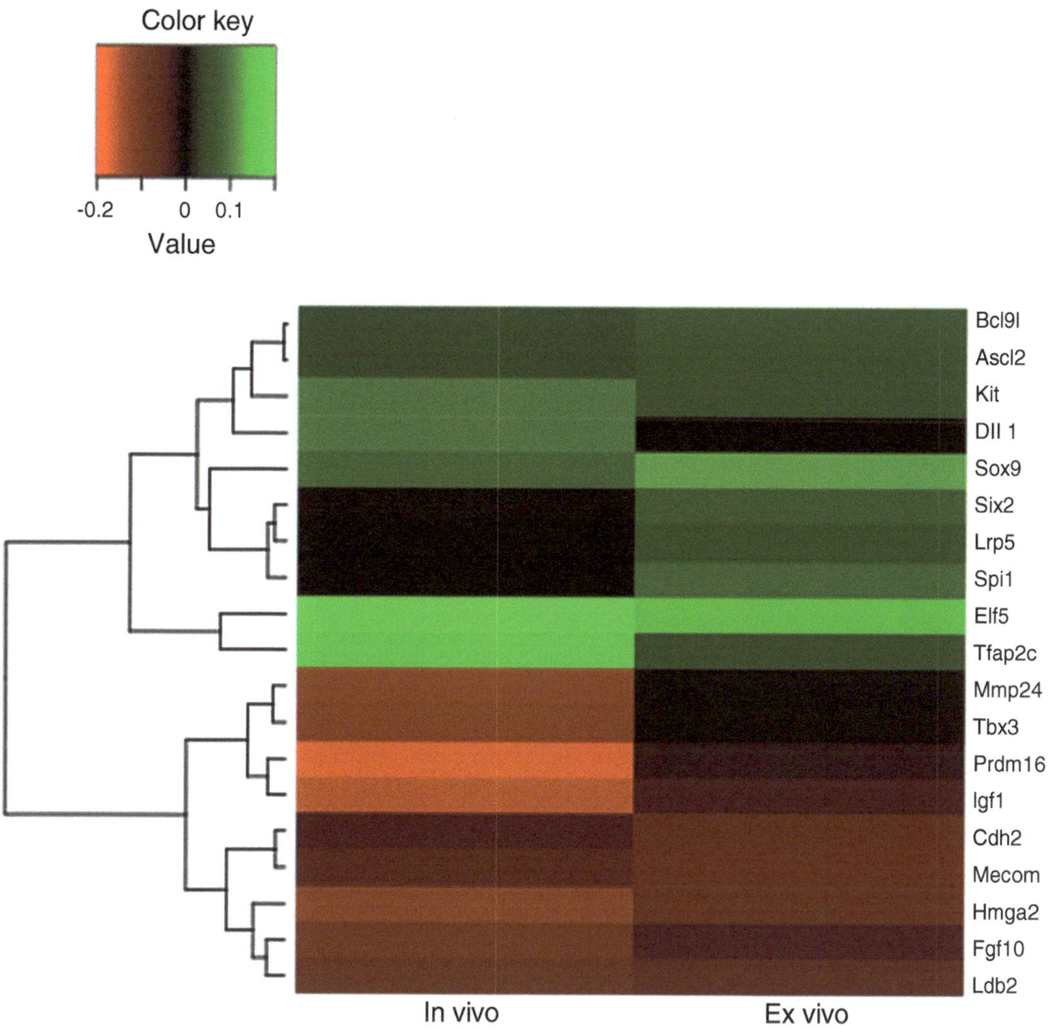

Fig. 2.5 Hierarchical clustering of differentially expressed mRNAs in "Maintenance of Cell Number" GO category. The top 10 % of differentially expressed genes were selected from the "Maintenance of Cell Number" category (GO:0098727), a subcategory of "Developmental Process" category (GO:0032502), one of the most conserved categories (Fig. 2.3), to perform hierarchical clustering, and displayed as a heatmap using the heatmap.2 function in R. The hierarchical clustering provided two distinct groups including a preferentially upregulated gene group and a downregulated gene group, suggesting that differentially expressed genes between in vivo and ex vivo are relatively well conserved. This is consistent with the high agreement of the parent GO class "Developmental Process" category (GO:0032502) in Kolmogorov-Smirnov test in Fig. 2.3. However, note that individual genes are regulated slightly differently in vivo and ex vivo

we examined the top three most conserved GO categories and divided them into subcategories based on the next tier of GO descriptors. As an example, we plotted differences in gene expression of genes annotated with GO process "Maintenance of Cell Number," a subcategory of "Developmental Process." The heat map of the subcategory, "Maintenance of Cell Number" (Fig. 2.5), demonstrates that on the individual gene level, there are some genes that are regulated more similarly than others. The analysis we report here of differences between differential gene expressions in organ explants vs glands grown in vivo is useful in understanding how the transcriptome is regulated similarly in vivo and in vitro.

2.3.3 Systems-Level Analysis to Identify Temporal-Spatial Differences in Gene Expression During Salivary Gland Development

Given that during embryonic development, heterotypic epithelial and mesenchymal cell subtypes undergo dynamic communication [63], knowledge of how specific cell subpopulations contribute to gland development and how these cell subpopulations are regulated is critical to understand gland development [34]. Cleft formation initiates the process of branching morphogenesis and subdivides buds into cells that will become nascent secondary ducts and cells that will become secretory acini. In an early systems-level analysis using serial analysis of gene expression (SAGE), fibronectin (FN) was identified as a critical regulator of cleft formation [37, 77, 78]. Cleft formation was also found to be accompanied by a concomitant loss of adjacent E-cadherin-based cell-cell junctions [78]. This conversion of cell-cell adhesions to cell-matrix adhesions was found to be regulated transcriptionally through increases in BTB (POZ) domain containing 7 (Btbd7) [64]. Btbd7, which was also identified as a cleft-enriched gene using SAGE, is reported was suggested to activate a local epithelial-to-mesenchymal transition (EMT) within the epithelial cells deep within the cleft to separate the adjacent cells to allow continuous FN assembly [64] and cleft progression.

To identify additional genes that may be required for cleft formation, we performed in silico data mining of the publically available SGMAP temporal and spatial data. Temporal and spatial mRNA expression profiles derived from laser capture microdissected regions of embryonic submandibular salivary glands were used to screen for cleft-enriched genes at developmental stages E12.5 and E13.5 [59] (Fig. 2.6a). Differentially expressed mRNAs were identified in E12.5 cleft regions compared to the main bud (528 mRNAs) or E13.5 clefts compared to the central bud/basement membrane bud (1576 genes). Of these cleft-enriched mRNAs, 317 were common to both E12.5 and E13.5 clefts (Fig. 2.6b). Among the putative cleft-enriched genes, genes were partitioned into epithelium and mesenchyme by the comparison of E13 epithelium and E13 mesenchyme gene expression profiles available in the SGMAP database. Surprisingly, over half of the cleft-enriched mRNAs (specifically, 71.3 % of the conserved cleft-enriched genes) showed mesenchymal origins, supporting a hypothesis that mesenchymal genes dynamically contribute to epithelial patterning through regulation of cleft formation (Fig. 2.6c). Additional functional prediction of the cleft-enriched genes was performed with an available integrated analysis tool, the Database for Annotation, Visualization and Integrated Discovery (DAVID) system (ver. 6.7) (https://david.ncifcrf.gov). Results of gene ontology (GO) term and Kyoto Encyclopedia of Genes and Genomes (KEGG) pathway analysis by DAVID [26] confirmed the involvement of many mRNAs known to contribute to clefts and identified some potential new cellular mechanisms to consider in cleft formation. This analysis first confirmed the enrichment of genes involved in focal adhesions, ECM-receptor interactions, and the actin cytoskeleton (Fig. 2.7 and Table 2.1), consistent with previous observations by our lab and others [28, 59, 64, 75, 79].

Interestingly, GO biological process analysis highlighted the cleft-enriched localization of specialized mesenchymal cell types, including endothelial cells, nerves, and immune cells (Fig. 2.6, Tables 2.1, and 2.2, and 2.3). The cleft endothelial/blood vessel signature predicts a possible contribution of endothelial cells to salivary gland cleft formation (Table 2.2), which is supported by the known contribution of endothelial cells to the early development of other organs, including the liver, pancreas, and lung, prior to the establishment of perfusion [35, 39, 45, 53, 74]. Recent studies have demonstrated that parasympathetic ganglia innervation regulates epithelial cytokeratin 5^+ ($K5^+$) progenitor cell expansion via muscarinic and epidermal growth factor receptor signaling and controls gland tubulogenesis through vasoactive intestinal peptide (VIP) secretion [32, 62]. However, a specific contribution for nerve-derived factors in cleft formation itself is still unclear but is implicated by this data (Table 2.3). Endothelial cells and/or nerves may be important in early patterning of the clefts

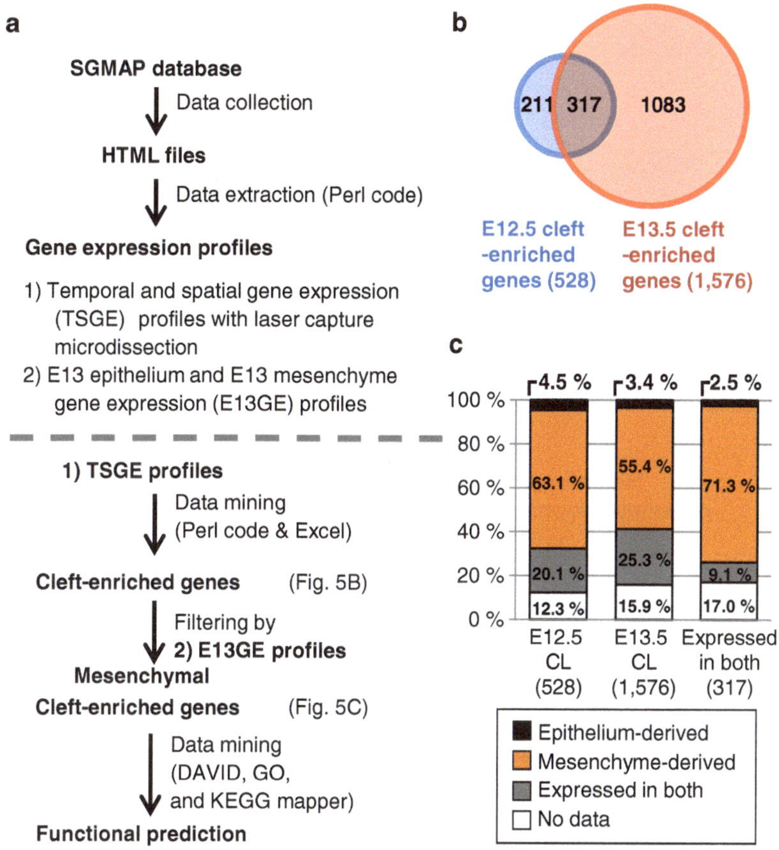

Fig. 2.6 Overview of in silico data mining of SGMAP to identify cleft-enriched genes in developing salivary glands. (**a**) Overview of data collection and analysis of SGMAP.nidcr.nih.gov mRNA expression profiles. (**b**) Cleft-enriched genes were screened from temporal and spatial mRNA expression profiles from regions of developing salivary gland that were manually collected using laser capture microdissection. E12.5 cleft-enriched genes (528 genes) were identified from the comparison between E12.5 cleft and E12.5 main bud (≥ twofold change (FC)). E13.5 cleft-enriched genes (1576 genes) were conserved genes in the two sets of comparisons between E13.5 cleft and E13.5 central bud and between E13.5 cleft and E13.5 basement membrane bud (≥2 FC). (**c**) The origin of cleft-enriched genes was predicted by the comparison of E13 mesenchyme and E13 epithelium mRNA expression profiles that were manually collected from SGMAP. The genes were classified into four origin groups including epithelium derived ($x \leq -2$ FC), mesenchyme derived (≥ 2 FC), expressed in both ($-2 < x < 2$ FC), and no data. Over half of cleft-enriched genes were predicted to be mesenchymal genes, suggesting a potential role of mesenchymal signaling in salivary gland early cleft formation and epithelial patterning

during branching morphogenesis, which merits future investigation.

Cleft-enriched mRNAs may provide insight into signaling that occurs in clefts. Mesenchymal soluble factors such as fibroblast growth factors and platelet-derived growth factors are known to provide critical contributions to salivary gland development [67, 94]. Recent work examined a role for mesenchymal gene products in cell-ECM interaction and ECM remodeling during epithelial branching [28, 71], although the specific cell subpopulations required for mesenchymal-epithelial signaling are poorly understood. Cleft enrichment of gene products that are known to stimulate cellular behaviors such as cell-cell adhesion and chemotaxis is consistent with active communication between epithelial and mesenchymal cells and/or mesenchymal cell subtypes in developing clefts (Table 2.1). Other cleft-enriched cytokines and chemokines may regulate undefined aspects of mesenchymal function in progressing clefts, such as directed cell migration

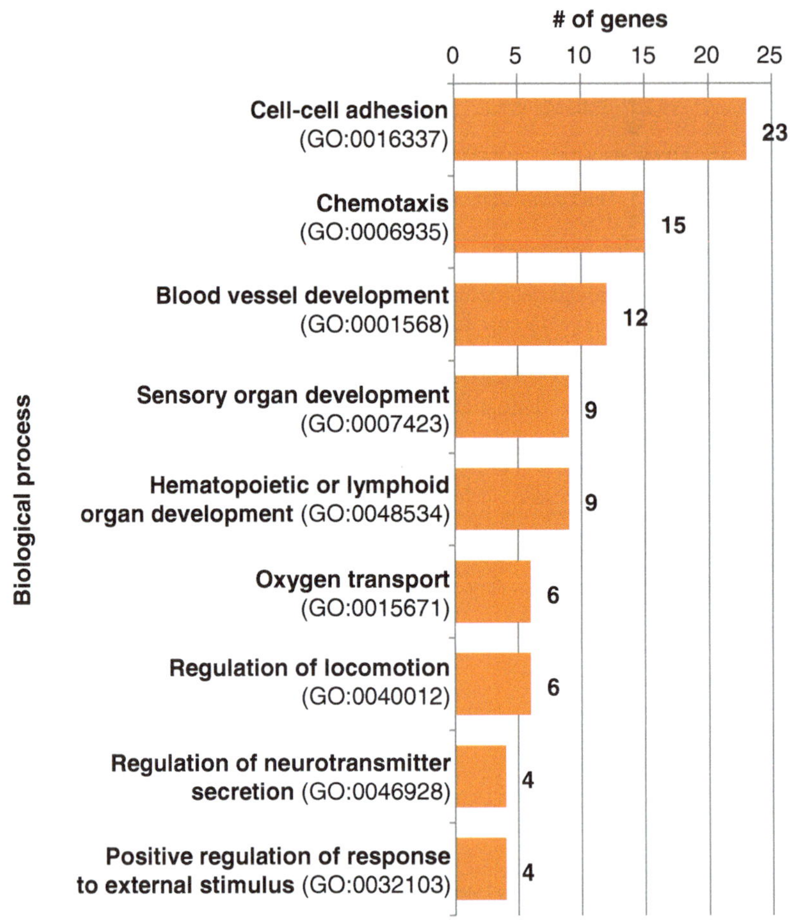

Fig. 2.7 Gene ontology of biological processes for mesenchymal cleft-enriched genes. Gene ontology (GO) of biological processes for mesenchymal cleft-enriched genes (conserved genes in both E12.5 and E13.5; *p*-value <0.01) was analyzed by Database for Annotation, Visualization and Integrated Discovery (DAVID) system (ver. 6.7)

and condensation via growth factor-mediated mechanical compaction [48, 49, 78, 90]. Taken together, our analysis reveals promising novel molecular candidates for regulators of cleft formation and salivary gland development that can be pursued in future studies to characterize their potential roles in salivary gland branching morphogenesis and tissue maintenance/regeneration.

2.3.4 Systems-Level Analysis of Salivary Gland Differentiation Networks

The initiation of differentiation occurs in developing salivary glands trailing only slightly behind the initiation of morphogenesis, but the process of differentiation continues late into development and is not complete until after sexual maturity. The signaling pathways required to induce differentiation of salivary acinar cells are poorly understood but are critical to understand for application in future regenerative medicine approaches. Although information on gene expression is available on SGMAP for late developmental stages, how these genes are coordinated to induce differentiation is not understood. Network-based modeling approaches are one way to gain insights into organization of signaling networks. Building on previous network-based modeling approaches performed by Jaskoll and Melnick [55, 88] to investigate signaling networks in late stage submandibular glands, a recent study used a combinatorial analysis of mRNA and miRNA expression profiles to suggest that the regulatory

Table 2.1 Most highly expressed mesenchymal cleft-enriched genes grouped by KEGG pathway

		KEGG pathway
Common	Both	mmu04510:Focal adhesion
		mmu04512:ECM-receptor interaction
		mmu04670:Leukocyte transendothelial migration
		mmu04062:Chemokine signaling pathway
		mmu04810:Regulation of actin cytoskeleton
		mmu04060:Cytokine-cytokine receptor interaction
Stage-specific	E12.5	mmu04010:MAPK signaling pathway
		mmu04610:Complement and coagulation cascades
	E13.5	mmu04360:Axon guidance
		mmu04610:Complement and coagulation cascades
		mmu04310:Wnt signaling pathway
		mmu04020:Calcium signaling pathway
		mmu04142:Lysosome
		mmu04650:Natural killer cell mediated cytotoxicity
		mmu04080:Neuroactive ligand-receptor interaction

Cleft (CL)-enriched genes in this table were more highly expressed in mesenchyme than epithelium (≥ twofold change). Signaling pathways were predicted by Kyoto Encyclopedia of Genes and Genomes (KEGG) pathway analysis of Database for Annotation, Visualization and Integrated Discovery (DAVID). The common category contained KEGG pathways that were conserved in both E12.5 and E13.5 (*purple*). There were also KEGG categories of genes that were highly expressed only in E12.5 (*blue*) or E13.5 clefts (*red*)

Table 2.2 Putative endothelial-expressed cleft-enriched genes

	E12.5	E13.5	Both
Blood vessel development (GO:0001568)	FGF18, FLT1, PRRX1, KDR, CITED2, ZFP697	NRP1, CSPG4, ENPEP, PRRX2, TNFSF12, GJA4, WNT2, EDNRA, HAND2, HMOX1, SOX18, LOX, NR2F2, FGF2, NKX2–5, BMP4, EFNB2, TBX1, THY1, ANXA2, GTR1A, ID1, ITGA7	RECK, ALDH1A2, MYO18B, EMCN, NTRK2, COL3A1, ROBO4, FGF10, COL1A1, ARHGAP24, CXCL12, CDH5
Gas transport (GO:0015669)		F830116E18RIK	HBA-A1, HBA-A2, HBB-BH1, HBB-B1, HBA-X, HBB-Y, CAR2

Endothelial genes shown here were predicted by gene ontology (GO) biological process analysis with DAVID (See Fig. 2.7)

Table 2.3 Putative neuronal-expressed cleft-enriched genes

		E12.5	E13.5	Both
GO	Regulation of neurotransmitter secretion (GO:0046928)		PARK2	ADORA2A, NTRK2, SNCA, CAMK2A
KEGG	mmu04360: axon guidance		SEMA5B, PLXNC1, NRP1, RAC2, EFNB3, EFNB2, PPP3CC, NFATC4, ROBO2, LRRC4C, SEMA3A, CXCL12, EPHB1, EPHA3, SLIT3, NFATC1	
	mmu04080: neuroactive ligand-receptor interaction		CALCR, GABRA2, PTGER3, GABRA1, GRIK1, ADORA2A, GABRB2, GRIA3, NTSR1, EDNRA, EDNRB, P2RY6, AGTR2, CHRM3, AGTR1A, GRM7, ADRA2A	

Neuronal genes shown here were predicted by gene ontology (GO) biological process and KEGG pathway analysis with DAVID (See Fig. 2.7 and Table 2.1)

networks driving acinar cell terminal differentiation in rat parotid salivary glands are not continuous but involve temporally distinct developmental transitions [56]. In this study, mRNA profiling of acinar cell RNA was performed using tissue collected by laser capture microdissection (LCM) at nine different stages from E18 to P25. The data grouped into four distinct clusters showing similarity of differentially expressed genes within close developmental time points and revealing that differentiation occurred in four temporally distinct phases. Clustering analysis extended these observations to suggest distinct functional modes of the temporally distinct differentiation phases, and integrated mRNA-miRNA expression analysis was used to decipher regulatory networks controlled by mRNAs and regulatory miRNAs. Interestingly, repression of a stemness-promoting pathway and stimulation of an Xbp1-stimulated pathway are required for parotid acinar differentiation. Significantly, the data implicate miRNA control of multiple mRNAs as a major driver of parotid cell differentiation. Future analyses that incorporate similar co-regulatory predictions as well as predictions of likely signaling pathways will be important to identify important pathways for experimental validation and to identify putative therapeutic targets.

2.4 Systems Approaches to the Study of Salivary Gland Disease

2.4.1 Systems-Based Analysis of Saliva

Saliva has been investigated as a surrogate for blood as a means to identify differences in normal and diseased patients. As saliva is a readily accessible bodily fluid that can be collected in a nonintrusive manner, many "-omics" studies have focused on differences in the levels of various biomolecules in saliva. Proteomic approaches have been applied extensively to saliva, resulting in a comprehensive catalog of the contents of human saliva – the salivary proteome – prepared by a team of researchers from five US universities [14]. This study identified 917 submandibular/sublingual proteins and 914 parotid proteins, of which 252 are submandibular/sublingual specific and 249 are parotid specific and are primarily extracellular or secretory in nature. The salivary proteome is available as a tool for identification and diagnosis of both systemic and salivary gland diseases. Since proteomic profiling has demonstrated its utility, it is likely that future disease studies will employ these methods as protein detection methods become increasingly sensitive. However, the processing of samples for proteomic analysis, the specifics of the instrumentation, and analysis of the data can significantly impact the results of the experiment [8]. Analysis of saliva has recently extended beyond the proteome, and has been demonstrated to provide many insights into normal and diseased functions. With the emerging field of epigenomics, evaluation of saliva promises to be an important player, which is an exciting development that has been reviewed elsewhere [91]. Saliva has also been demonstrated to contain exosomes [57], small (30–100 nm diameter) cell-secreted vesicles that carry both proteins and miRNAs, short RNAs that work on groups of expressed mRNAs to downregulate gene expression of mRNAs by interfering with their translation and targeting them for degradation [91]. Since saliva can be obtained noninvasively and only small amounts are needed for most analyses, saliva has significant potential as a diagnostic fluid for detection and diagnosis of many diseases, perhaps most obviously for oral cancers [18, 82, 92]. Evaluation of the saliva transcriptome is also currently under investigation for assessment of health and disease states in infants [50]. The future application of saliva-based diagnostics coming out of systems-level analysis of saliva has significant promise for many diseases and many types of patients [8].

2.4.2 Systems-Based Methods to Diagnose Salivary Gland Diseases: Sjögren's Syndrome

Diagnosis of SS is difficult due to a lack of specific molecular markers for the disease. Many studies have used "-omics" methods in an attempt to discover reliable diagnostic markers for SS from saliva samples. In a study comparing whole saliva from normal and Sjögren's syndrome patients, proteins, peptides, and mRNAs that are differentially expressed in these populations were identified [24]. Another study compared proteins taken from parotid salivary glands between SS patients and healthy patients through a technique known as multidimensional protein identification technology, or MudPIT [1], which is an unbiased method for rapid and large-scale proteome analysis by multidimensional liquid chromatography, followed by tandem mass spectrometry, and database searching [89]. Out of the 1246 proteins identified by MudPIT, 529 were only detected in either SS or healthy patients, 206 were significantly upregulated by more than twofold, and 34 were downregulated by more than twofold [54]. Mass spectrometry assays conducted with the same samples identified 71 proteins, 58 of which were proteins that had also been detected by MudPIT [1]. Further analysis of proteins with differential expression between SS patients and healthy controls identified 100 pathways of significance that were differentially regulated in SS vs normal patients [1]. A recent study by Deutsche et al. identified 79 peptides that were expressed in SS patients at more than a threefold rate relative to those in healthy patients. The samples, obtained from the saliva of 18 female SS patients and 18 age-matched and gender-matched controls, were subjected to protein depletion to remove high-abundance proteins that make identification of low-abundance proteins difficult, prior to semiquantitative two-dimensional gel electrophoresis and quantitative demethylation liquid chromatography tandem mass spectrometry (LC-MS/MS), resulting in a threefold increase in the ability to identify low-abundance proteins [15]. Bioinformatics analysis of the data identified proteins with a >threefold increase in SS patients, including calcium-binding proteins, defense-response proteins, proteins involved in apoptotic regulation, stress-response proteins, and cell motion-related proteins. The results borne of these proteomic studies offer the potential to create better tools to diagnose SS as well as allowing for a better understanding of the pathology of SS at the molecular level.

Similar to proteomics, metabolomics is now possible as a result of recent improvements to mass spectrometry technology. Metabolomics is an emerging field that seeks to identify the full spectrum of chemical metabolites produced during cellular metabolism since many differences in metabolism occur between normal and abnormal cells [69]. In a recent study, Kageyama et al. focused on the different levels of metabolites in saliva from primary Sjögren's syndrome (pSS) patients vs normal patients [29]. With this approach, 88 metabolites were discovered by comparing saliva taken from 12 SS patients with that obtained from 21 healthy patients of similar ages. Out of these, 41 metabolites were expressed at lower levels in pSS patients relative to healthy patients [29]. Further comparative analysis revealed a loss of metabolite diversity in SS patients, primarily resulting from decreased levels of the amino acids glycine and tyrosine, uric acid, and fucose. The discovery of these differing metabolomic profiles between healthy patients and pSS patients could prove useful in establishing more reliable diagnostic criteria for SS.

2.4.3 Sjogren's Syndrome Disease Progression

Systems biology-based approaches are beginning to be utilized as a tool to better understand complex salivary gland diseases. Hu et al. used a systems approach to identify new disease-hub genes, which they define as promising targets for therapeutic intervention and diagnosis of SS [25]. This study compared parotid tissue from three classes of patients: patients with primary SS, patients with primary SS associated with mucosa-associated lymphoid tissue (MALT) lymphoma, and patients without primary SS (non-primary SS controls). Microarray profiling to examine gene expression and proteomic analysis to examine protein expression were performed on all samples, and weighted gene co-expression network analysis was performed on these data to identify disease-hub genes. Computational analysis has also been employed in the study of animal models of SS. In one study, the widely used non-obese diabetic (NOD) mouse model of SS [47] was compared with the parent line, Balb/C, in terms of biomarker expression, autoantibody production, glandular inflammation, and saliva production [12]. Principal component analysis was used to identify significant positive and negative correlations within the data. Interestingly, in this study each biomarker typically associated exclusively with only one of the other parameters. These data indicate that SS disease progression may not follow a linear trajectory, even within this animal model. In humans, the disease etiology is likely to be significantly more complex and variable.

2.4.4 Systems Approaches to Understand Radiation Sensitivity of Salivary Glands Following Radiation Treatment for Head and Neck Cancers

A wide variety of tumors are classified as head and neck cancers; affected tissues include, but are not limited to, the mouth, nose, tongue, pharynx, larynx, lymph nodes, salivary glands, and sinuses. The typical treatment for head and neck cancer is dependent on the patient's medical history and the stage of the cancer. However, owing to the close proximity of these tumors to the salivary glands, treatments often adversely affect a patient's ability to produce saliva, which hinders the patient's capacity to eat, swallow, and/or speak. Understanding why the salivary glands are sensitive to radiation and what happens to the glands during radiation could lead to improved methods to prevent salivary gland damage. Systems approaches have been used to better understand the response of salivary glands to radiation treatment. A 2012 study by Stiubea-Cohen et al. to characterize the effects of irradiation on salivary glands, a common treatment for cancers of the head and neck, focused on the transcriptome and saliva proteome of submandibular salivary glands of rats [97]. In the rat irradiation model used in this study, saliva output fell to roughly 50 % of its output prior to IR treatment within 8–12 weeks after IR treatment. Using microarrays, which can quickly compare RNA levels in multiple subjects simultaneously, and real-time polymerase chain reaction (real-time PCR) to confirm the microarray data, 95 target genes were identified that exhibited significant differences in expression as a result of IR treatment. Out of these genes, 81 exhibited a decrease in activity following treatment and a gradual recovery to levels near to those prior to treatment. The other 14 genes, most of which are cell cycle control genes, exhibited an increase in expression and activity throughout the 12-week period. A proteomic assay performed on saliva samples taken both before and after IR treatment showed that most of the proteins encoded by the 81 genes with reduced activity after IR treatment are involved in protein secretion and saliva production. That the other 14 genes with increased activity following IR treatment are mostly associated with the cell cycle suggested that IR treatment led to DNA damage and impairment of the surviving salivary gland cells to regenerate. Taken together transcriptomic and proteomic data were used in this study to show that the loss of saliva production following IR treatment is a result of the salivary glands' loss of capacity to both produce and secrete proteins and to regenerate following cell damage.

2.4.5 Systems Approaches to Characterize Salivary Gland Tumors

Proteomics can also be used to characterize differences in protein expression in glandular tissue and has been used recently to categorize salivary tumors. Tumors of the salivary glands themselves make up roughly 8 % of head and neck cancers. Like most tumors, diagnosis of these tumors involves fine-needle aspiration biopsy and scoring of the tumor type by a pathologist on the basis of the appearance of the tumor from a thin section of the biopsy tissue that was stained with hematoxylin and eosin (H&E) to highlight tumor structure. False negatives (i.e., incorrectly identifying a malignant tumor as benign) lead to premature deaths, and false positives (i.e., incorrectly identifying a benign tumor as malignant) lead to unnecessarily stringent therapeutic regimes. As a result, biomedical scientists are turning to proteomics in their search for a more reliable diagnostic tool. In a 2013 paper, Donadio et al. sought to compare the proteome of one of the most frequent benign salivary gland tumors: Warthin's tumor (WT), a cystadenolymphoma, with that of another typically benign but more complex mixed tumor, pleomorphic adenomas (PAs) [17]. PAs are difficult to diagnose as these complex tumors involve the glandular epithelium, the myoepithelium, and the stromal compartment, and they have the potential to transition to a metastatic state. With 35 patients that had undergone parotidectomy (removal of the parotid salivary glands) as their subjects, the group ran mass spectrometry assays and gel electrophoresis on gland samples. Through a computational comparison of the data, a total of 34 proteins with different expression patterns between WT and PA were identified, which was narrowed down to 26 different proteins, of which nine were selected for Western blot analysis. Between healthy tissue and PA/WT samples, noticeably, visible increases in protein expression in all nine proteins for either PA or WT were observed. Many of the proteins whose expression is altered in WT are implicated in autoimmune disorders (e.g., rheumatoid arthritis), while those proteins with altered expression in PA are implicated in tumorigenesis events (e.g., proliferation, invasion) [17]. These results demonstrate the utility of using a proteomic approach to study characteristics of salivary gland tumors, as the differing protein expression patterns between the two types of tumors could provide the basis for a more reliable tool for differential diagnosis and subclassification. Improvement of diagnosis accuracy directly affects patient quality of life, as it can help physicians choose the best treatment regime for the patient.

Other -omics methods have been used to gain additional insights into salivary gland cancer cell function and to identify potential new biomarkers and therapeutic strategies. Salivary adenoid cystic carcinomas (SACCs) are salivary ductal-derived tumors, which account for 24 % of malignant salivary gland tumors. miRNA profiling of metastatic SACC vs less metastatic SACC identified miR-320a as a metastatic repressor that targets the integrin beta 3 mRNA (itgb3), suggesting a potential for miR-320a-based therapeutics. In addition, miR-320a levels positively correlated with prognosis for SACC patients, suggesting an additional application in disease diagnosis [86]. miRNA profiling of salivary gland adenoid cystic carcinoma (ACC) cell lines, a relatively rare malignant tumor with a poor long-term survival rate, identified reduced levels of miR-101-3p [41]. While miR-101-3p has potential utility as a biomarker, its mechanism was investigated in cell lines where it was found to inhibit cell growth and invasion by inhibiting expression of the Pim-1 serine/threonine kinase, which can function as an oncogene. Interestingly, ACC cell lines stably expressing miR-101-3p were found to have enhanced sensitivity to cisplatin, the most common treatment for ACC [41], demonstrating the possible therapeutic potential of miR-101-3p-based therapeutics for ACC.

2.5 Systems Analysis and Regenerative Therapies

2.5.1 Systems Analysis in Characterizing Cell Populations

Interest in using salivary gland stem/progenitor cell populations for cell therapy has resulted in several reports of salivary gland progenitor cell culture expansion and preclinical cell therapies [42, 46, 60, 93]. While these studies have very promising translational medicine potential, the small number of markers used to characterize cell phenotypes combined with different markers used in each study provides an incomplete understanding of the cell populations in question, and no definitive stem/progenitor cell markers have yielded long-term functional restoration of hyposalivation in an animal model as yet. This is an area that could greatly benefit from a systems analysis of functional, therapeutic progenitor cell preparations. A recent deep sequencing study of the transcriptome of SGC was corroborated and complemented with quantitative real-time PCR, immunocytochemistry, and flow cytometry analysis [72]. Postnatal hepatic stem cell markers, EpCAM, NCAM, c-kit, CD44, and CD133, and other cell markers, CD29, CD49f, K7, K19, Sca-1, and c-Met, were found to be considerably well conserved in both SGC and LPC under the isolation and culture methods performed in this study. Most of the markers above were similarly expressed in both LPCs and SGCs or more highly expressed in LPC than in SGC, while the expression levels of EpCAM, CD49f, K14, and K7 were significantly higher in SGC than heterogeneous LPC [72]. Clinical translation would benefit from further system analyses of functionally defined salivary gland stem/progenitor cell populations and benefit from similar comparisons to known stem/progenitor cell populations used for functional restoration in other organs.

2.6 Conclusions and Outlook

There is much to be gained from the application of systems biology-based approaches to understand salivary gland developmental processes, to diagnose disease, and to develop better therapeutic options for patients. The development of effective therapeutic options is typically more successful the more that is understood regarding the mechanisms through which disease develops, which requires an understanding of the normal development and homeostasis mechanisms at play in normal glands. Due to the complexity of the salivary gland and the complexity of salivary gland diseases, systems-based approaches are integral to making progress on all of these avenues.

Systems-based profiling approaches, including profiling the transcriptome, the miRNAome, the genome, the epigenome, the proteome, the metabolome, and the oral microbiome, are now in common use in the study of both salivary gland development and disease. To make progress toward a comprehensive understanding of the normal development and homeostasis of the gland and mechanisms through which homeostasis is disrupted will require moving beyond simple profiling studies. Since the enormity of systems-level datasets pose a challenge for researchers, continued development of computational tools to make the data accessible and integration possible for both researchers and clinicians is critical to facilitate the current trend toward systems thinking and incorporation of systems approaches into research. The use of pathway analysis methods and mathematical modeling, which is already at use in many diverse salivary and saliva-based research projects [4, 43, 44, 56, 65, 68, 76, 83], will become increasingly necessary to "make sense" of the enormity of data and to integrate multiple types of "-omics" data to answer specific research questions. Application of systems approaches in the future promises to deepen our understanding of the relationships between the genome and the epigenome, the transcriptome and proteome, environmental factors and disease, and the relationship between physiological states and disease. Integration of systems-level "-omics" data with the physics of tissue biology is a holy grail that promises to provide crucial insights to normal biology and to disease development [16]. Nevertheless, reductionist approaches will still be required for testing hypotheses generated from systems biology approaches to validate predictions. In the future, systems-based analysis and computational approaches will be useful in understanding disease etiology and inspiring new therapeutics. As we move into the age of personalized medicine, systems-based approaches will eventually play an integral role in assessing the patient's disease state and in defining patient-specific therapeutic regimes.

Acknowledgments The authors thank Kenneth Yamada and Matthew Hoffman for reviewing the manuscript. This work was sponsored in part by the University at Albany, SUNY, and by NIH grants R01DE022467 and C06 RR015464.

References

1. Ambatipudi KS, Swatkoski S, Moresco JJ, Tu PG, Coca A, Anolik JH, Gucek M, Sanz I, Yates 3rd JR, Melvin JE. Quantitative proteomics of parotid saliva in primary Sjogren's syndrome. Proteomics. 2012;12(19–20):3113–20. doi:10.1002/pmic.201200208.

2. Ashburner M, Ball CA, Blake JA, Botstein D, Butler H, Cherry JM, Davis AP, Dolinski K, Dwight SS, Eppig JT, Harris MA, Hill DP, Issel-Tarver L, Kasarskis A, Lewis S, Matese JC, Richardson JE, Ringwald M, Rubin GM, Sherlock G. Gene ontology: tool for the unification of biology. The gene ontology consortium. Nat Genet. 2000;25(1):25–9. doi:10.1038/75556.
3. Aure MH, Konieczny SF, Ovitt CE. Salivary gland homeostasis is maintained through acinar cell self-duplication. Dev Cell. 2015;33(2):231–7. doi:10.1016/j.devcel.2015.02.013.
4. Bilgin CC, Ray S, Baydil B, Daley WP, Larsen M, Yener B. Multiscale feature analysis of salivary gland branching morphogenesis. PLoS One. 2012;7(3):e32906. doi:10.1371/journal.pone.0032906.
5. Borghese E. The development in vitro of the submandibular and sublingual glands of Mus musculus. J Anat. 1950a;84(3):287–302.
6. Borghese E. Explanation experiments on the influence of the connective tissue capsule on the development of the epithelial part of the submandibular gland of Mus musculus. J Anat. 1950b;84(3):303–18.
7. Boukheris H, Curtis RE, Land CE, Dores GM. Incidence of carcinoma of the major salivary glands according to the WHO classification, 1992 to 2006: a population-based study in the United States. Cancer Epidemiol Biomarkers Prev Publ Am Assoc Cancer Res cosponsored Am Soc Prev Oncol. 2009;18(11):2899–2906.
8. Castagnola M, Cabras T, Iavarone F, Fanali C, Nemolato S, Peluso G, Bosello SL, Faa G, Ferraccioli G, Messana I. The human salivary proteome: a critical overview of the results obtained by different proteomic platforms. Expert Rev Proteomics. 2012;9(1):33–46. doi:10.1586/epr.11.77.
9. Cutler LS. The dependent and independent relationships between cytodifferentiation and morphogenesis in developing salivary gland secretory cells. Anat Rec. 1980;196(3):341–7.
10. Daley WP, Gulfo KM, Sequeira SJ, Larsen M. Identification of a mechanochemical checkpoint and negative feedback loop regulating branching morphogenesis. Dev Biol. 2009;336(2):169–82. doi:10.1016/j.ydbio.2009.09.037.
11. Davies J. Branching morphogenesis. Molecular biology intelligence unit, 1st ed. Boston: Springer; 2005.
12. Delaleu N, Immervoll H, Cornelius J, Jonsson R. Biomarker profiles in serum and saliva of experimental Sjogren's syndrome: associations with specific autoimmune manifestations. Arthritis Res Ther. 2008;10(1):R22. doi:10.1186/ar2375.
13. Delaleu N, Jonsson R, Koller MM. Sjogren's syndrome. Eur J Oral Sci. 2005;113(2):101–113.
14. Denny P, Hagen FK, Hardt M, Liao L, Yan W, Arellanno M, Bassilian S, Bedi GS, Boontheung P, Cociorva D, Delahunty CM, Denny T, Dunsmore J, Faull KF, Gilligan J, Gonzalez-Begne M, Halgand F, Hall SC, Han X, Henson B, Hewel J, S H, Jeffrey S, Jiang J, Loo JA, Ogorzalek Loo RR, Malamud D, Melvin JE, Miroshnychenko O, Navazesh M, Niles R, Park SK, Prakobphol A, Ramachandran P, Richert M, Robinson S, Sondej M, Souda P, Sullivan MA, Takashima J, Than S, Wang J, Whitelegge JP, Witkowska HE, Wolinsky L, Xie Y, Xu T, Yu W, Ytterberg J, Wong DT, Yates 3rd JR, Fisher SJ. The proteomes of human parotid and submandibular/sublingual gland salivas collected as the ductal secretions. J Proteome Res. 2008;7(5):1994–2006.
15. Deutsch O, Krief G, Konttinen YT, Zaks B, Wong DT, Aframian DJ, Palmon A. Identification of Sjogren's syndrome oral fluid biomarker candidates following high-abundance protein depletion. Rheumatology (Oxford). 2015;54(5):884–90. doi:10.1093/rheumatology/keu405.
16. Discher D, Dong C, Fredberg JJ, Guilak F, Ingber D, Janmey P, Kamm RD, Schmid-Schonbein GW, Weinbaum S. Biomechanics: cell research and applications for the next decade. Ann Biomed Eng. 2009;37(5):847–59. doi:10.1007/s10439-009-9661-x.
17. Donadio E, Giusti L, Seccia V, Ciregia F, da Valle Y, Dallan I, Ventroni T, Giannaccini G, Sellari-Franceschini S, Lucacchini A. New insight into benign tumours of major salivary glands by proteomic approach. PLoS One. 2013;8(8):e71874. doi:10.1371/journal.pone.0071874.
18. Dumache R, Rogobete AF, Andreescu N, Puiu M. Genetic and epigenetic biomarkers of molecular alterations in oral carcinogenesis. Clin Lab. 2015;61(10):1373–81.
19. Grobstein C. Inductive epitheliomesenchymal interaction in cultured organ rudiments of the mouse. Science. 1953a;118(3054):52–5.
20. Grobstein C. Morphogenetic interaction between embryonic mouse tissues separated by a membrane filter. Nature. 1953b;172(4384):869–70.
21. Harunaga J, Hsu JC, Yamada KM. Dynamics of salivary gland morphogenesis. J Dent Res. 2011;90(9):1070–7. doi:10.1177/0022034511405330.
22. Hoffman MP, Kidder BL, Steinberg ZL, Lakhani S, Ho S, Kleinman HK, Larsen M. Gene expression profiles of mouse submandibular gland development: FGFR1 regulates branching morphogenesis in vitro through BMP- and FGF-dependent mechanisms. Development. 2002;129(24):5767–78.
23. Hsu JC, Di Pasquale G, Harunaga JS, Onodera T, Hoffman MP, Chiorini JA, Yamada KM. Viral gene transfer to developing mouse salivary glands. J Dent Res. 2012;91(2):197–202. doi:10.1177/0022034511429346.
24. Hu S, Wang J, Meijer J, Ieong S, Xie Y, Yu T, Zhou H, Henry S, Vissink A, Pijpe J, Kallenberg C, Elashoff D, Loo JA, Wong DT. Salivary proteomic and genomic biomarkers for primary Sjogren's syndrome. Arthritis Rheum. 2007;56(11):3588–600. doi:10.1002/art.22954.
25. Hu S, Zhou M, Jiang J, Wang J, Elashoff D, Gorr S, Michie SA, Spijkervet FK, Bootsma H, Kallenberg CG, Vissink A, Horvath S, Wong DT. Systems biology analysis of Sjogren's syndrome and mucosa-associated lymphoid tissue lymphoma in parotid glands. Arthritis Rheum. 2009;60(1):81–92.
26. Huang DW, Sherman BT, Tan Q, Kir J, Liu D, Bryant D, Guo Y, Stephens R, Baseler MW, Lane

HC, Lempicki RA. DAVID bioinformatics resources: expanded annotation database and novel algorithms to better extract biology from large gene lists. Nucleic Acids Res. 2007;35(Web Server issue):W169–75. doi:10.1093/nar/gkm415.
27. Jaskoll T, Melnick M. Embryonic salivary gland branching morphogenesis. In: Davies JA, editor. Branching morphogenesis. New York: Landes Bioscience/Springer; 2005. p. 160–75.
28. Joo EE, Yamada KM. MYPT1 regulates contractility and microtubule acetylation to modulate integrin adhesions and matrix assembly. Nat Commun. 2014;5:3510. doi:10.1038/ncomms4510.
29. Kageyama G, Saegusa J, Irino Y, Tanaka S, Tsuda K, Takahashi S, Sendo S, Morinobu A. Metabolomics analysis of saliva from patients with primary Sjogren's syndrome. Clin Exp Immunol. 2015;182(2):149–53. doi:10.1111/cei.12683.
30. Knosp WM, Knox SM, Hoffman MP. Salivary gland organogenesis. Wiley Interdiscip Rev Dev Biol. 2012;1(1):69–82. doi:10.1002/wdev.4.
31. Knosp WM, Knox SM, Lombaert IM, Haddox CL, Patel VN, Hoffman MP. Submandibular parasympathetic gangliogenesis requires sprouty-dependent Wnt signals from epithelial progenitors. Dev Cell. 2015;32(6):667–77. doi:10.1016/j.devcel.2015.01.023.
32. Knox SM, Lombaert IM, Reed X, Vitale-Cross L, Gutkind JS, Hoffman MP. Parasympathetic innervation maintains epithelial progenitor cells during salivary organogenesis. Science. 2010;329(5999):1645–7. doi:10.1126/science.1192046.
33. Kovacs L, Szodoray P, Kiss E. Secondary tumours in Sjogren's syndrome. Autoimmunity reviews; 2009.
34. Kwon HR, Larsen M. The contribution of specific cell subpopulations to submandibular salivary gland branching morphogenesis. Curr Opin Gen Dev (in press). 2015. doi:10.1016/j.gde.2015.01.007.
35. Lammert E, Cleaver O, Melton D. Role of endothelial cells in early pancreas and liver development. Mech Dev. 2003;120(1):59–64.
36. Larsen M, Hoffman MP, Sakai T, Neibaur JC, Mitchell JM, Yamada KM. Role of PI 3-kinase and PIP3 in submandibular gland branching morphogenesis. Dev Biol. 2003;255(1):178–91.
37. Larsen M, Wei C, Yamada KM. Cell and fibronectin dynamics during branching morphogenesis. J Cell Sci. 2006;119(Pt 16):3376–84.
38. Larsen M, Yamada KM, Musselmann K. Systems analysis of salivary gland development and disease. Wiley Interdiscip Rev Syst Biol Med. 2010;2(6):670–82. doi:10.1002/wsbm.94.
39. Lazarus A, Del-Moral PM, Ilovich O, Mishani E, Warburton D, Keshet E. A perfusion-independent role of blood vessels in determining branching stereotypy of lung airways. Development. 2011;138(11):2359–68. doi:10.1242/dev.060723.
40. Lee BH, Tudares MA, Nguyen CQ. Sjogren's syndrome: an old tale with a new twist. Arch Immunol Ther Exp. 2009;57(1):57–66. doi:10.1007/s00005-009-0002-4.
41. Liu XY, Liu ZJ, He H, Zhang C, Wang YL. MicroRNA-101-3p suppresses cell proliferation, invasion and enhances chemotherapeutic sensitivity in salivary gland adenoid cystic carcinoma by targeting Pim-1. Am J Cancer Res. 2015;5(10):3015–29.
42. Lombaert IM, Brunsting JF, Wierenga PK, Faber H, Stokman MA, Kok T, Visser WH, Kampinga HH, de Haan G, Coppes RP. Rescue of salivary gland function after stem cell transplantation in irradiated glands. PLoS One. 2008;3(4):e2063. doi:10.1371/journal.pone.0002063.
43. Lubkin SR, Li Z. Force and deformation on branching rudiments: cleaving between hypotheses. Biomech Model Mechanobiol. 2002;1(1):5–16.
44. Lubkin SR, Murray JD. A mechanism for early branching in lung morphogenesis. J Math Biol. 1995;34(1):77–94.
45. Magenheim J, Ilovich O, Lazarus A, Klochendler A, Ziv O, Werman R, Hija A, Cleaver O, Mishani E, Keshet E, Dor Y. Blood vessels restrain pancreas branching, differentiation and growth. Development. 2011;138(21):4743–52. doi:10.1242/dev.066548.
46. Maimets M, Rocchi C, Bron R, Pringle S, Kuipers J, Giepmans BN, Vries RG, Clevers H, de Haan G, van Os R, Coppes RP. Long-term in vitro expansion of salivary gland stem cells driven by Wnt signals. Stem Cell Rep. 2016;6(1):150–62. doi:10.1016/j.stemcr.2015.11.009.
47. Makino S, Kunimoto K, Muraoka Y, Mizushima Y, Katagiri K, Tochino Y. Breeding of a non-obese, diabetic strain of mice. Jikken Dobutsu. 1980;29(1):1–13.
48. Mammoto T, Mammoto A, Jiang A, Jiang E, Hashmi B, Ingber DE. Mesenchymal condensation-dependent accumulation of collagen VI stabilizes organ-specific cell fates during embryonic tooth formation. Dev Dyn Off Publ Am Assoc Anatomists. 2015;244(6):713–23. doi:10.1002/dvdy.24264.
49. Mammoto T, Mammoto A, Torisawa YS, Tat T, Gibbs A, Derda R, Mannix R, de Bruijn M, Yung CW, Huh D, Ingber DE. Mechanochemical control of mesenchymal condensation and embryonic tooth organ formation. Dev Cell. 2011;21(4):758–69. doi:10.1016/j.devcel.2011.07.006.
50. Maron JL. The neonatal salivary transcriptome. Cold Spring Harb Perspect Med. 2015;6(3):a026369. doi:10.1101/cshperspect.a026369.
51. Maruyama EO, Aure MH, Xie X, Myal Y, Gan L, Ovitt CE. Cell-specific cre strains for genetic manipulation in salivary glands. PLoS One. 2016;11(1):e0146711. doi:10.1371/journal.pone.0146711.
52. Mathews SA, Kurien BT, Scofield RH. Oral manifestations of Sjogren's syndrome. J Dent Res. 2008;87(4):308–18. doi:87/4/308 [pii].
53. Matsumoto K, Yoshitomi H, Rossant J, Zaret KS. Liver organogenesis promoted by endothelial cells prior to vascular function. Science. 2001;294(5542):559–63. doi:10.1126/science.1063889.
54. Mauri P, Scarpa A, Nascimbeni AC, Benazzi L, Parmagnani E, Mafficini A, Della Peruta M, Bassi

C, Miyazaki K, Sorio C. Identification of proteins released by pancreatic cancer cells by multidimensional protein identification technology: a strategy for identification of novel cancer markers. FASEB J. 2005;19(9):1125–7.
55. Melnick M, Jaskoll T. Mouse submandibular gland morphogenesis: a paradigm for embryonic signal processing. Crit Rev Oral Biol Med Off Publ Am Assoc Oral Biologists. 2000;11(2):199–215.
56. Metzler MA, Venkatesh SG, Lakshmanan J, Carenbauer AL, Perez SM, Andres SA, Appana S, Brock GN, Wittliff JL, Darling DS. A systems biology approach identifies a regulatory network in parotid acinar cell terminal differentiation. PLoS One. 2015;10(4):e0125153. doi:10.1371/journal.pone.0125153.
57. Michael A, Bajracharya SD, Yuen PS, Zhou H, Star RA, Illei GG, Alevizos I. Exosomes from human saliva as a source of microRNA biomarkers. Oral Dis. 2010;16(1):34–8. doi:10.1111/j.1601-0825.2009.01604.x.
58. Miletich I. Introduction to salivary glands: structure, function and embryonic development. Front Oral Biol. 2010;14:1–20. doi:10.1159/000313703.
59. Musselmann K, Green JA, Sone K, Hsu JC, Bothwell IR, Johnson SA, Harunaga JS, Wei Z, Yamada KM. Salivary gland gene expression atlas identifies a new regulator of branching morphogenesis. J Dent Res. 2011;90(9):1078–84. doi:10.1177/0022034511413131.
60. Nanduri LS, Lombaert IM, van der Zwaag M, Faber H, Brunsting JF, van Os RP, Coppes RP. Salisphere derived c-Kit+ cell transplantation restores tissue homeostasis in irradiated salivary gland. Radiother Oncol. 2013;108(3):458–63. doi:10.1016/j.radonc.2013.05.020.
61. Napenas JJ, Brennan MT, Fox PC. Diagnosis and treatment of xerostomia (dry mouth). Odontol Soc Nippon Dent Univ. 2009;97(2):76–83.
62. Nedvetsky PI, Emmerson E, Finley JK, Ettinger A, Cruz-Pacheco N, Prochazka J, Haddox CL, Northrup E, Hodges C, Mostov KE, Hoffman MP, Knox SM. Parasympathetic innervation regulates tubulogenesis in the developing salivary gland. Dev Cell. 2014;30(4):449–62. doi:10.1016/j.devcel.2014.06.012.
63. Nelson DA, Larsen M. Heterotypic control of basement membrane dynamics during branching morphogenesis. Dev Biol. 2015;401(1):103–9. doi:10.1016/j.ydbio.2014.12.011.
64. Onodera T, Sakai T, Hsu JC, Matsumoto K, Chiorini JA, Yamada KM. Btbd7 regulates epithelial cell dynamics and branching morphogenesis. Science. 2010;329(5991):562–5. doi:10.1126/science.1191880.
65. Palk L, Sneyd J, Patterson K, Shuttleworth TJ, Yule DI, Maclaren O, Crampin EJ. Modelling the effects of calcium waves and oscillations on saliva secretion. J Theor Biol. 2012;305:45–53. doi:10.1016/j.jtbi.2012.04.009.
66. Patel VN, Hoffman MP. Salivary gland development: a template for regeneration. Semin Cell Dev Biol. 2014;25-26:52–60. doi:10.1016/j.semcdb.2013.12.001.
67. Patel VN, Rebustini IT, Hoffman MP. Salivary gland branching morphogenesis. Differ Res Biol Divers. 2006;74(7):349–64. doi:10.1111/j.1432-0436.2006.00088.x.
68. Patterson K, Catalan MA, Melvin JE, Yule DI, Crampin EJ, Sneyd J. A quantitative analysis of electrolyte exchange in the salivary duct. Am J Physiol Gastrointest Liver Physiol. 2012;303(10):G1153–63. doi:10.1152/ajpgi.00364.2011.
69. Patti GJ, Yanes O, Siuzdak G. Innovation: metabolomics: the apogee of the omics trilogy. Nature Rev. 2012;13(4):263–9. doi:10.1038/nrm3314.
70. Peters SB, Naim N, Nelson DA, Mosier AP, Cady NC, Larsen M. Biocompatible tissue scaffold compliance promotes salivary gland morphogenesis and differentiation. Tissue Eng A. 2014;20(11–12):1632–42. doi:10.1089/ten.TEA.2013.0515.
71. Peters SB, Nelson DA, Kwon HR, Koslow M, DeSantis KA, Larsen M. TGFbeta signaling promotes matrix assembly during mechanosensitive embryonic salivary gland restoration. Matrix Biol. 2015;43:109–24. doi:10.1016/j.matbio.2015.01.020.
72. Petrakova OS, Terskikh VV, Chernioglo ES, Ashapkin VV, Bragin EY, Shtratnikova VY, Gvazava IG, Sukhanov YV, Vasiliev AV. Comparative analysis reveals similarities between cultured submandibular salivary gland cells and liver progenitor cells. Springerplus. 2014;3:183. doi:10.1186/2193-1801-3-183.
73. Proctor GB, Carpenter GH. Regulation of salivary gland function by autonomic nerves. Auton Neurosci Basic Clin. 2007;133(1):3–18. doi:10.1016/j.autneu.2006.10.006.
74. Rafii S, Butler JM, Ding BS. Angiocrine functions of organ-specific endothelial cells. Nature. 2016;529(7586):316–25. doi:10.1038/nature17040.
75. Ray S, Fanti JA, Macedo DP, Larsen M. LIM kinase regulation of cytoskeletal dynamics is required for salivary gland branching morphogenesis. Mol Biol Cell. 2014;25(16):2393–407. doi:10.1091/mbc.E14-02-0705.
76. Ray S, Yuan D, Dhulekar N, Oztan B, Yener B, Larsen M. Cell-based multi-parametric model of cleft progression during submandibular salivary gland branching morphogenesis. PLOS Computational Biology (in press). 2013.
77. Sakai T, Larsen M, Yamada KM. Microanalysis of gene expression in tissues using T7-SAGE: Serial Analysis of Gene Expression after high-fidelity T7-Based RNA amplification. In: Current protocols in cell biology. New York: John Wiley & Sons ; 2002. pp. 19.13.11–19.14.10.
78. Sakai T, Larsen M, Yamada KM. Fibronectin requirement in branching morphogenesis. Nature. 2003;423(6942):876–81.
79. Sakai T, Larsen M, Yamada KM. Morphogenesis and branching of salivary glands: characterization of new

matrix and signaling regulators. Oral Biosci Med. 2005;2(2/3):105–13.
80. Sequeira SJ, Gervais EM, Ray S, Larsen M. Genetic modification and recombination of salivary gland organ cultures. J Vis Exp JoVE. 2013;71:e50060. doi:10.3791/50060.
81. Singh P, Schimenti JC, Bolcun-Filas E. A mouse geneticist's practical guide to CRISPR applications. Genetics. 2015;199(1):1–15. doi:10.1534/genetics.114.169771.
82. Sivadasan P, Kumar Gupta M, Sathe GJ, Balakrishnan L, Palit P, Gowda H, Suresh A, Abraham Kuriakose M, Sirdeshmukh R. Data from human salivary proteome – A resource of potential biomarkers for oral cancer. Data Brief. 2015;4:374–8. doi:10.1016/j.dib.2015.06.014.
83. Sneyd J, Crampin E, Yule D. Multiscale modelling of saliva secretion. Math Biosci. 2014;257:69–79. doi:10.1016/j.mbs.2014.06.017.
84. Speight PM, Barrett AW. Salivary gland tumours. Oral Dis. 2002;8(5):229–40.
85. Speight PM, Barrett AW. Prognostic factors in malignant tumours of the salivary glands. Br J Oral Maxillofac Surg. 2009;47(8):587–93.
86. Sun L, Liu B, Lin Z, Yao Y, Chen Y, Li Y, Chen J, Yu D, Tang Z, Wang B, Zeng S, Fan S, Wang Y, Li Y, Song E, Li J. MiR-320a acts as a prognostic factor and Inhibits metastasis of salivary adenoid cystic carcinoma by targeting ITGB3. Mol Cancer. 2015;14:96. doi:10.1186/s12943-015-0344-y.
87. Tucker AS. Salivary gland development. Semin Cell Dev Biol. 2007;18(2):237–44. doi:10.1016/j.semcdb.2007.01.006.
88. Waddington CH. The strategy of the genes. London: George Allen & Unwin; 1957. p. 11–58.
89. Washburn MP, Wolters D, Yates 3rd JR. Large-scale analysis of the yeast proteome by multidimensional protein identification technology. Nat Biotechnol. 2001;19(3):242–7. doi:10.1038/85686.
90. Wells KL, Gaete M, Matalova E, Deutsch D, Rice D, Tucker AS. Dynamic relationship of the epithelium and mesenchyme during salivary gland initiation: the role of Fgf10. Biol Open. 2013;2(10):981–9. doi:10.1242/bio.20135306.
91. Wren ME, Shirtcliff EA, Drury SS. Not all biofluids are created equal: chewing over salivary diagnostics and the epigenome. Clin Ther. 2015;37(3):529–39. doi:10.1016/j.clinthera.2015.02.022.
92. Wu CC, Chu HW, Hsu CW, Chang KP, Liu HP. Saliva proteome profiling reveals potential salivary biomarkers for detection of oral cavity squamous cell carcinoma. Proteomics. 2015;15(19):3394–404. doi:10.1002/pmic.201500157.
93. Xiao N, Lin Y, Cao H, Sirjani D, Giaccia AJ, Koong AC, Kong CS, Diehn M, Le QT. Neurotrophic factor GDNF promotes survival of salivary stem cells. J Clin Invest. 2014;124(8):3364–77. doi:10.1172/JCI74096.
94. Yamamoto S, Fukumoto E, Yoshizaki K, Iwamoto T, Yamada A, Tanaka K, Suzuki H, Aizawa S, Arakaki M, Yuasa K, Oka K, Chai Y, Nonaka K, Fukumoto S. Platelet-derived growth factor receptor regulates salivary gland morphogenesis via fibroblast growth factor expression. J Biol Chem. 2008;283(34):23139–49. doi:10.1074/jbc.M710308200.
95. Yang H, Wang H, Shivalila CS, Cheng AW, Shi L, Jaenisch R. One-step generation of mice carrying reporter and conditional alleles by CRISPR/Cas-mediated genome engineering. Cell. 2013;154(6):1370–9. doi:10.1016/j.cell.2013.08.022.
96. Zhang A. Advanced analysis of gene expression microarray data. Singapore: World Scientific; 2006.
97. Stiubea-Cohen R, David R, Neumann Y, Krief G, Deutsch O, Zacks B, Aframian DJ, Palmon A. Effect of irradiation on cell transcriptome and proteome of rat submandibular salivary glands. PloS One. 2012;7:e40636

Mucins in Salivary Gland Development, Regeneration, and Disease

3

Isabel Castro, María-José Barrera, Sergio González, Sergio Aguilera, Ulises Urzúa, Juan Cortés, and María-Julieta González

Abstract

Mucins are large glycoproteins that can be grouped as membrane-bound or secreted. Membrane-bound mucins are essential contributors of the glycocalyx of mucosal surfaces where they play important biological roles in cell interactions and signaling. Secreted mucins are the main structural components of the mucus gel that covers the epithelium and contribute to the protection of the mucosa surface against allergens, debris, pathogens, drying, injury, and abrasive stress. MUC1 and MUC4 are plasma membrane-anchored mucins expressed in oral epithelium and salivary glands and are ubiquitously located in normal epithelia. MUC5B is the major secreted polymeric mucin present in human saliva and contains negatively charged glycans allowing the formation of a hydrophilic gel that hydrates and protects the oral epithelium. MUC7 is a secreted mucin present in saliva and has low viscoelasticity and high bacteria-agglutinating properties allowing clearance of microorganisms from the mouth by swallowing.

The roles of mucins during development of salivary glands remain unknown. Few studies on mucin expression in human salivary glands' development and in murine models have been described to date. In salivary gland regeneration, mucins have been evaluated only as markers of functionality of the glandular acini after damage and/or regeneration therapy. Interestingly, changes in quality and quantity of salivary mucins have been observed in pathological conditions. Over-expression of specific mucins induced by pro-inflammatory cytokines and ectopic secretion of

I. Castro, PhD • M.-J. Barrera, PhD • U. Urzúa, PhD
J. Cortés, PhD • M.-J. González, MSc (✉)
Institute of Biomedical Sciences, Faculty of Medicine, University of Chile, Santiago, Chile
e-mail: jgonzale@med.uchile.cl

S. González, MSc
Mayor University, Santiago, Chile

S. Aguilera, MD
INDISA Clinic, Santiago, Chile

S. Cha (ed.), *Salivary Gland Development and Regeneration*, DOI 10.1007/978-3-319-43513-8_3

mucins to extracellular matrix (ECM) support evidence on a self-perpetuating mucin-cytokine signaling loop that may facilitate the maintenance of an inflammatory environment in chronic inflammatory diseases, such as Sjögren's syndrome (SS). However, the loss of a mucin member could predispose to infections and inflammatory diseases in a mucosa. A reduced sulfation of MUC5B has been observed in salivary glands of SS patients, aggravating their oral dryness. Tumoral cells of salivary glands express aberrant forms or high amounts of mucins, being proposed as markers of malignant transformation. In this chapter, we summarize the main structural and functional characteristics of salivary mucins, their expression during salivary gland development, and regeneration and alterations in mucin quantity and quality in pathological processes affecting the salivary gland.

3.1 Structure, Biosynthesis and Functions of Salivary Mucins

Mucins are O-glycoproteins composing the mucus layer that protects mucosal surfaces from external insult and desiccation. Human salivary mucins are synthesized by the submandibular, sublingual and minor salivary glands. Although their contribution to the total volume of saliva is low, the minor salivary glands contribute up to 70 % of total mucin found in saliva [1]. Mucins are high-molecular-weight O-glycoproteins containing at least one region of repeated sequences that in some cases include variable numbers of tandem repeats (VNTR) polymorphisms [2, 3] (Fig. 3.1). These regions are rich in serine and threonine, residues that covalently bond with a variety of O-glycans. Importantly, oligosaccharides contribute up to 80 % of the mucin mass [9], and many of these glycans are sialylated and/or sulfated, conferring hydrophilic and polyanionic properties to mucins [10–12]. The N- and C-terminal regions of mucins are rich in cysteine and are non-glycosylated or sparsely N- and O-glycosylated [13] (Fig. 3.1).

A total of 20 mucins in humans, numbered in the order of their discovery, have been described to date [14]; they can be structurally categorized in two major classes. High-molecular-mass secreted mucins that form polymeric structures and include MUC2, MUC5AC, MUC5B, MUC6, and MUC19. MUC7 and MUC8 are also secreted but consist of low-molecular-mass non-polymeric mucins. The second class consists of membrane-tethered mucins and includes MUC1, MUC3A, MUC3B, MUC4, MUC12, MUC13, MUC15, MUC16, MUC17, MUC20, MUC21 and MUC22 [14–17]. Table 3.1 summarizes the tissue distribution of human mucins.

The biosynthesis of polymeric mucins starts at the endoplasmic reticulum (ER), where the apomucin or polypeptide backbone is translated. The N-glycosylation and the formation of intramolecular disulfide bonds within the cysteine-rich N- and C-termini and the internal cys domains occur co-translationally. Dimerization takes place at the ER by intermolecular disulfide linkage between C-terminal domains of mucin monomers [18, 19]. The O-glycosylation is carried out in the Golgi apparatus, where UDP-GalNAc:polypeptide N-acetylgalactosaminyl transferases transfer N-acetylgalactosamine to a serine or threonine residue of the mucin protein backbone [20]. The O-glycosylated products are further elongated by sequential action of glycosyltransferases to produce a wide array of oligosaccharides [21]. Intermolecular disulfide linkage between N-terminal domains mediates polymerization of mucin dimers in the trans-Golgi network or in secretory granules. High concentrations of Ca^{+2} and H^+ inside the secretory granules contribute to mucin aggregation, reducing repulsive forces among oligosaccharide negative charges [22]. Polymers are packaged and stored in dehydrated form within secretory granules until regulated exocytosis [23].

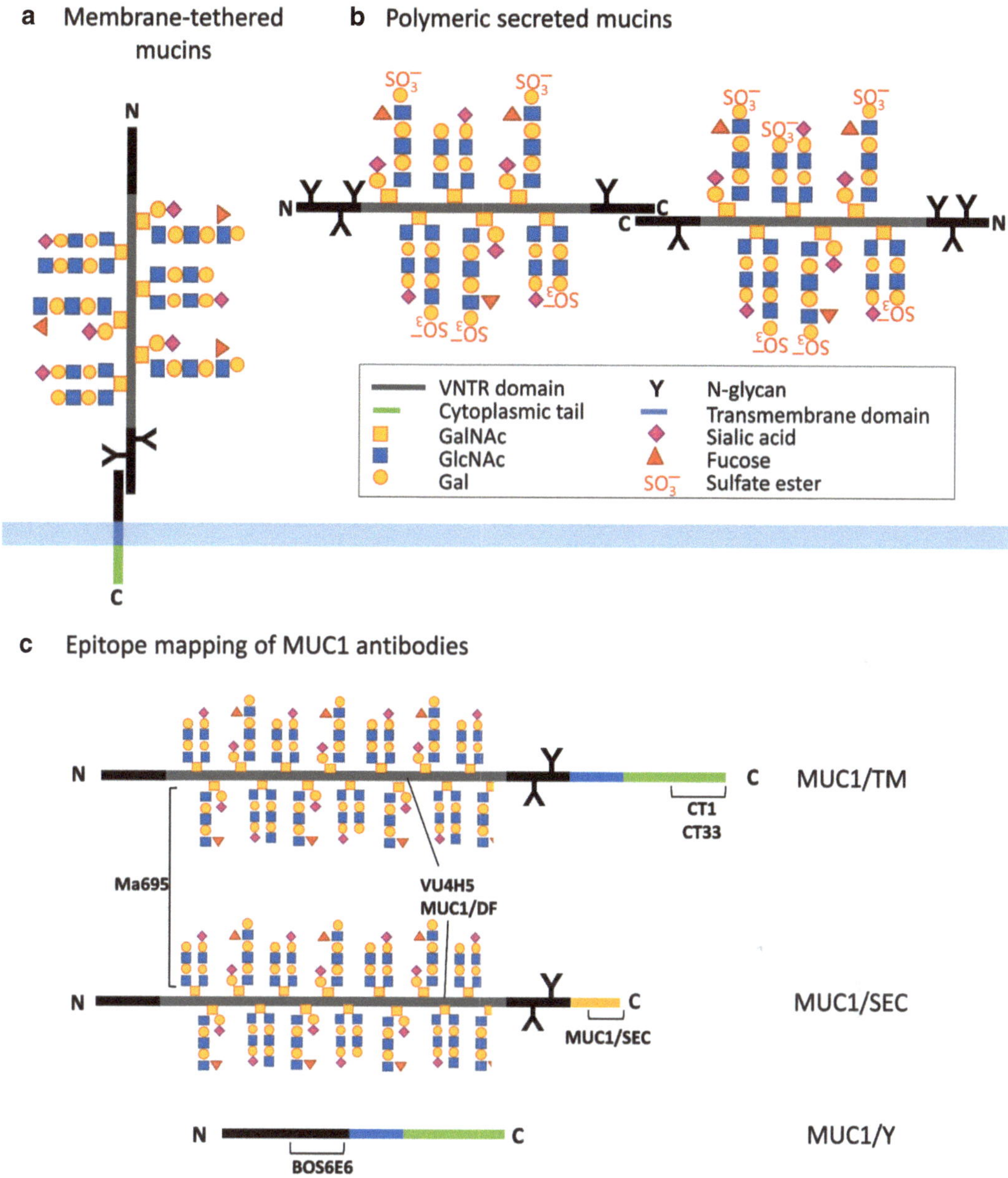

Fig. 3.1 Scheme of the structure of membrane-tethered mucins (**a**) and polymeric secreted mucins (**b**). (**c**) The epitope mapping of several MUC1 antibodies: CT1 [4], CT33 [5], Ma695 [6], VU4H5 [7], MUC1/DF [8], MUC1/SEC [7] and BOS6E6 [7]

The major polymeric mucin found in human saliva is MUC5B (>1 MDa), formerly named MG1 [24, 25]. MUC5B comprises 15 % protein, 78 % carbohydrate, and 7 % sulfate by weight [26]. MUC5B-linked carbohydrates constitute a highly heterogeneous set of neutral, sulfated, and sialylated oligosaccharides [27, 28]. Anion exchange chromatography and electrophoretic mobility analysis of MUC5B revealed two products of high- and low-charge glycoforms [29, 30]. Studies of salivary mucin preparations have shown that the highly charged MUC5B glycoform reacts with the F2 monoclonal antibody, which specifically recognizes the sulfated carbohydrate epitope SO_3Galβ1-3GlcNAc- of sulfo-Lewis[a] [31]. Negatively charged glycans of

Table 3.1 Classification and tissue expression of mucins

Mucin	Class	Chromosome	Tandem repeat length	Tissue distribution
MUC1	Membrane anchored with secreted variants	1q21	20	Most epithelia
MUC2	Large polymeric mucin	11p15.5	23	Endometrium, jejunum, ileon, colon
MUC3A	Membrane anchored	7q22	17	Small intestine, colon, gallbladder
MUC3B	Membrane anchored	7q22	17	Small intestine, colon, gallbladder
MUC4	Membrane anchored	3q29	16	Most epithelia
MUC5AC	Large polymeric mucin	11p15.5	8	Conjunctiva, respiratory tract, stomach, endocervix, endometrium
MUC5B	Large polymeric mucin	11p15.5	29	Submandibular, sublingual and minor salivary glands, respiratory tract, endocervix
MUC6	Large polymeric mucin	11p15.5	169	Stomach, ileum, gallbladder, endocervix, endometrium
MUC7	Non-polymeric secreted mucin	4q13-q21	23	Submandibular, sublingual and minor salivary glands
MUC8	Non-polymeric secreted mucin	12q24.3	13/41	Respiratory tract, uterus, endocervix, endometrium
MUC9	Non-polymeric secreted mucin	1p13	15	Fallopian tubes
MUC12	Membrane anchored	7q22	28	Colon, pancreas, prostate, uterus
MUC13	Membrane anchored	3q21.2	27	Colon, trachea, kidney, small intestine
MUC15	Membrane anchored	11p14.3	None	Colon, respiratory tract, small intestine, prostate
MUC16	Membrane anchored	19p13.2	156	Ovary, cornea, conjunctiva, respiratory tract, endometrium
MUC17	Membrane anchored	7q22	59	Stomach, duodenum, colon
MUC19	Large polymeric mucin	12q12	19	Submandibular, sublingual and minor salivary glands
MUC20	Membrane anchored	3q29	18	Placenta, colon, respiratory tract, prostate, liver
MUC21	Membrane anchored	6p21	15	Respiratory tract, thymus, colon
MUC22	Membrane anchored	6p21.33	10	Respiratory tract

MUC5B confer hygroscopic properties, allowing formation of a hydrophilic gel that hydrates and protects the oral epithelium [10, 12]. The visco-elastic properties of salivary mucins provide local protection against mechanical forces during eating and speaking [32]. Lubricant properties, adhesiveness and elasticity of salivary MUC5B reduce the friction of tooth surfaces during chewing and contribute to the formation and swallowing of a food bolus [33]. Adsorption of MUC5B to tooth surfaces, given its high affinity for hydroxyapatite, contributes to the formation of the acquired enamel pellicle that protects hard tissues against the demineralization caused by acids produced by microorganisms [34, 35]. MUC5B glycans serve as adhesion receptors allowing binding of microorganisms including *Haemophilus parainfluenzae* [36] and *Helicobacter pylori* [37].

Another mucin found in saliva is MUC7, formerly named MG2 [38–40]. MUC7 has a molecular weight of ~200 kDa and contains about 30 % protein, 68 % carbohydrate, and 1.6 % sulfate [40–42]. The oligosaccharides linked to the MUC7 peptide backbone are mainly fucosylated and sialylated trisaccharides [43]. MUC7 has low viscoelastic and high bacteria-agglutinating properties [44]. MUC7 binds to microorganisms such as *Pseudomona aeruginosa* [45], *Agregatibacter actinomycetemcomitans* [46], *Escherichia coli* [47], and *Streptococcus mutans* [48], among others. As a result, bacterial aggregates are formed and are easily cleared out of the mouth by swallowing. In addition to MUC7 glycans, for example, sialic acid [45, 46], peptide domains in MUC7 are involved in the interaction with bacteria, and some regions show broad-spectrum antimicrobial activity[49–51]. The non-glycosylated regions of mucins have many structural motifs and domains that allow interaction with various salivary mucin-interacting proteins (e.g., histatin, statherin, acidic proline-rich proteins, α-amylase and lactoferrin, among others) [52]. Such interactions might help to enhance protein stability and function in the maintenance of oral physiology [52, 53].

Membrane-anchored mucins are initially translated as a single polypeptide chain that is cleaved to generate two sub-units, one containing the extracellular domain and the other containing the transmembrane (TM) and C-terminal cytoplasmic domains of the molecule [54, 55]. These fragments produced in the ER are non-covalently associated and remain linked until insertion into the plasma membrane [54, 55] (Fig. 3.1).

Cell-tethered mucins expressed by the oral epithelium and salivary glands include MUC1 and MUC4, which are almost ubiquitously found in normal epithelia [13]. These plasma membrane–anchored mucins have a rigid conformation, which extends up to 1.5 μm from the cell surface providing a protective barrier against microorganisms and other cytotoxic agents [56]. It has been proposed that MUC1 and MUC4 serve as a scaffold that contributes to retention of secretory mucins and other salivary proteins in the oral cavity [13]. Many alternative splice variants encode several MUC1 isoforms [55]. Some MUC1 isoforms have TM and C-terminal cytosolic domains, while others lack the TM domain and are soluble, e.g., MUC1/SEC [57] (Fig. 3.1). The extracellular domain of human MUC1 comprises the N-terminal signal sequence and the VNTR domain of 20–100 repeats of the GSTAPPAHGVTSAPDTRPAP sequence [54] (Fig. 3.1). This domain can be released from the cell surface by proteolytic shedding [58]. The C-terminal cytoplasmic domain of cell-anchored mucins has been implicated in cell signaling in physiological and pathological conditions [59].

3.2 Expression of Mucins During Salivary Gland Development

The specific roles played by mucins in human salivary gland development have not been fully studied, mostly due to ethical and technical restrictions. The submandibular gland starts to develop around the 6th week of human gestation, followed by the parotid gland around the 7th week, and the sublingual gland around the 8th [60].

A single study on mucins expression in human salivary gland development was carried out by Lourenço et al. (2011). It consisted of an immunohistochemical analysis of minor salivary glands collected from 20 post mortem human fetuses ranging from 4 to 24 weeks of gestation after natural miscarriage. Using the monoclonal Ma695 antibody directed against a glycosylated epitope of the human MUC1 VNTR region (Fig. 3.1), these authors observed MUC1 in clusters of epithelial cells at the bud stage, showing that MUC1 expression begins early in salivary gland development. At the pseudoglandular stage, MUC1 was observed in the rudimentary luminal space and appeared in the luminal space of all glandular ductal systems in the canalicular stage. During the terminal bud stage, MUC1 retained high expression along the luminal pole of epithelial cells throughout the entire ductal system. In certain areas, MUC1 was also observed in the basal pole of larger excretory ductal cells. In fully developed salivary glands, MUC1 was observed in the cytoplasm of acinar cells and in the luminal space border of the ducts [6]. Using the same antibody, MUC1 was detected in striated ducts and in basal cells of excretory ducts in parotid and submandibular glands of adult individuals. In acinar cells, MUC1 was observed in the apical plasma membrane of serous cells, while mucous cells were negative [61]. A similar MUC1 expression pattern was recently observed in human labial salivary glands from adult subjects [7]. Using the VU4H5 antibody directed against the VNTR region of MUC1 (Fig. 3.1), a staining in the apical pole of serous acini and duct cells was observed. Using the BOS6E6 antibody that recognizes the MUC1/Y isoform, staining was observed mainly in the apical region of serous acinar cells and throughout the cytoplasm of ductal cells. Using an antibody directed against the C-terminal region of the MUC1/SEC isoform, a staining localized mainly at the apical region of serous acinar cells was observed. Mucous cells were negative for MUC1 with several antibodies used [7].

Analysis of mucin transcripts in homogenates of adult human salivary glands have shown expression of MUC1 (MUC1A, MUC1B, MUC1/Y, MUC1/SEC), MUC2, MUC3, MUC4, MUC5B, MUC5AC, MUC6, MUC7 and MUC19 [7, 13, 29, 62, 63]. In human salivary gland development, MUC3, MUC4, MUC5B, and MUC16 were observed, while MUC2, MUC5AC and MUC6 were not detected [6]. At the initial bud stage, MUC5B and MUC16 were expressed weakly in some cells. At the pseudoglandular stage, MUC3 was observed in small cytoplasmic deposits in epithelial cells and MUC4 was strongly detected in the luminal region of developing ductal structures. MUC4 was also detected in the canalicular stage, in the luminal space of well-developed ducts and in the wall of blood vessels. At the terminal bud stage, MUC5B and MUC16 were detected in mucous cells [6]. In adult human salivary glands, MUC5B and MUC16 were detected in the luminal border of excretory ducts, and MUC16 was also observed in acinar cells [6]. However, several studies using antibodies directed against different epitopes of MUC5B showed expression in mucous acinar cells of fully developed submandibular, sublingual, and labial salivary glands [28, 64–66]

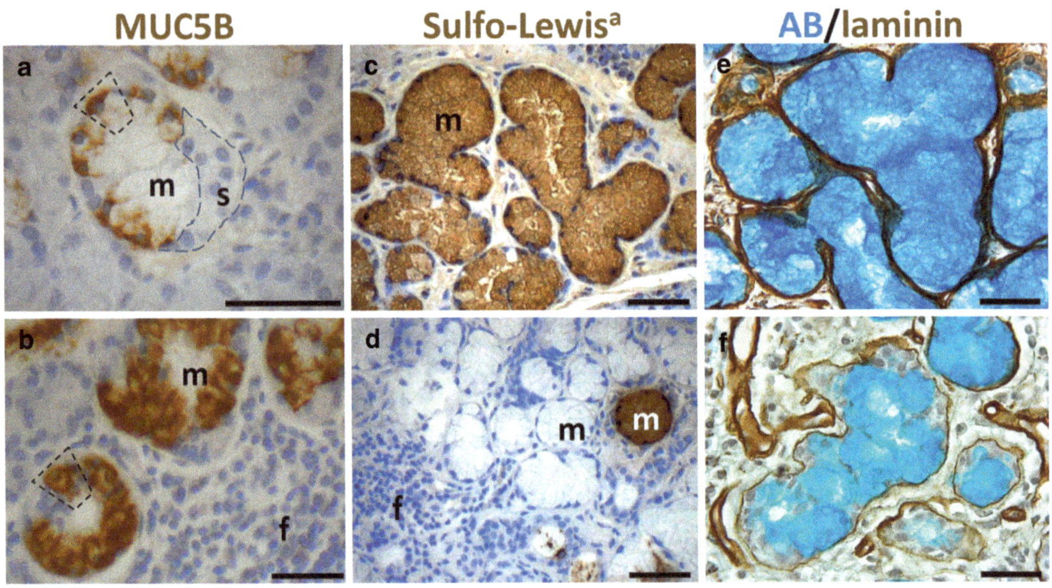

Fig. 3.2 Sections of labial salivary glands from control individuals (**a**, **c**, **e**) and SS patients (**b**, **d**, **f**). (**a**, **b**) Immunohistochemistry of MUC5B. Gland section showing MUC5B in the basal region of mucous acinar cells (**a**). In SS patients, MUC5B was observed both in basal and apical cytoplasms of mucous acinar cells (**b**). (**c**, **d**) Sulfo-Lewis^a antigen immuno-detection. Mucous acini showed sulfo-Lewis^a immunoreactivity in the whole cytoplasm (**c**). Low abundance of mucous acini showing immunoreaction for sulfo-Lewis^a in SS patients was observed (**d**). (**e**, **f**) Double staining with Alcian blue (**a**, **b**) and immunohistochemistry for laminin. Positive (**a**, **b**) staining was stronger in control individuals (**e**) than in SS patients (**f**). Decreased (**a**, **b**) staining correlated with decreased laminin immunoreactivity in SS patients (**f**). *Bars* 50 μm, *m* mucous acini, *s* serousacini; f: inflammatory focus. Images reproduced with permission of Castro et al. 137].

(Fig. 3.2 a, c and e). Mucous acini expressing MUC5B showed a mosaic pattern of sulfo-Lewis[a] stain, indicating that one and the same salivary gland synthesizes different MUC5B glycoforms [28, 65]. When and how these phenotypes originate during gland differentiation is yet unknown. Are the mucous acini formed by different types of acinar cells? Or is there asynchrony in the secretory response, among others? The simultaneous detection of MUC5B and MUC7 demonstrated that the staining patterns were non-overlapping [64]. While MUC5B is observed in mucous cells, MUC7 is expressed in serous acinar cells of fully developed human submandibular, sublingual, and labial salivary glands [28, 64, 66, 67].

Most of the studies on mucin expression during development have been performed on murine models, with the submandibular gland the most studied. In mouse submandibular gland development, the initial bud (E12.5) starts out as a solid chord that then develops into a network of solid stalks and end buds. Branching morphogenesis occurs in the pseudoglandular stage (E13.5). During the canalicular (E15.5) and terminal bud (E18.5) stages, branches and buds hollow out in their center by apoptosis of cells that do not make contact with the basal lamina [68]. Mucin is the initial marker of epithelial differentiation in mouse submandibular gland. Jaskoll et al. (1998) demonstrated differences between mRNA and protein expression of mucin in the embryonic, neonatal, and adult mouse submandibular gland. By northern blot assays, E17 and 1-day-old neonates exhibited two mucin transcripts (1.2 and 0.85 kb) which are different in size than the single (1.01 kb) adult transcript. Two embryonic mucin isoforms of ~110 and 152 kDa were detected in comparison to ~136 kDa adult mucin. The ~152 kDa embryonic isoform persisted in neonatal glands. *In situ* hybridization showed mucin transcripts localized in the branching epithelia by E14. The hybridization signal increased with age in terminal bud and proacinar cells. Immunofluorescence assays showed mucin protein from E17 in plasma membranes of terminal bud and proacinar cells [69]. Mucin expression in submandibular development is modulated by different factors. Melnick and Jaskoll demonstrated that glucocorticoids are required for mucin expression, and treatment *in utero* with exogenous glucocorticoids induces a significant increase of embryonic mucin mRNA and protein [70]. In addition, they also find that IL-6, but not TNF-α signaling, modulates embryonic mucin expression *in vitro* [71].

Denny et al. (1982) prepared a rabbit anti-mucin serum using purified mouse-submandibular sialomucin from 80 to 90-day–old female mice[72]. In newborn animals, the sialomucin was detected in secretory terminal-tubule cells and in proacinar cells, neither of which is morphologically identical to the mature acinar cell [73]. The sialomucin was also detected in acinar cells and in granular intercalating-duct cells of the submandibular gland of adult mice [73]. Moreover, a strong Alcian blue (AB) staining was observed in proacinar cell granules. Radioimmunoassays were carried out for mucin quantitation in the homogenates of submandibular glands. Mucin concentration was lowest in the newborn group and highest in 20-day-old mice. This increase in mucin concentration was associated with changes in acinar cell size and may reflect their maturation stage. The mucin detected in all age groups was apparently antigenically identical to the mucin purified from adult mice. Together these results showed that mucin expression begins prior to the final acinar cell differentiation and that sialomucin is present in the submandibular gland from birth to adulthood [72, 74, 75].

The expression of Muc1 was analyzed during mouse post-implantation development. Muc1 protein localization was determined by immunohistochemistry using CT1, a polyclonal antiserum directed against the 17 C-terminal amino acids of the cytoplasmic tail of human transmembrane Muc1 isoforms (Fig. 3.1). This epitope is highly conserved in a variety of tissues and species [4]. In salivary glands, Muc1 immunodetection with CT1 showed signal in ducts of 15-day-old embryos and, subsequently, Muc1 was detected in terminal tubules. Muc1 protein expression increased with time during epithelial branching and highest levels were observed in lumen, days 15–18 [76]. From these data, a role for MUC1 in glandular morphogenesis was suggested due to the coincident onset of glandular differentiation with changes of MUC1 expression pattern during embryonic development [76]. Hudson et al. [77]evaluated the potential role of MUC1 on morphogenesis using cell lines cultured on type I collagen gels. MUC1 altered the three-dimensional growth pattern of cells and induced tubular and branching morphogenesis. MUC1-expressing cells showed altered morphogenesis characterized by reduced cellular adhesion and enhanced migration [77]. These results suggest that MUC1 may induce changes in tissue architecture in development and also in cancer, where it is frequently overexpressed, misglycosylated and redistributed [78–80].

3.3 Mucin and the Regeneration of Salivary Glands

Antecedents on mucin in salivary gland regeneration processes are extremely scarce. Mucins have been only evaluated as markers to address the functionality of the glandular acini after damage and/or regeneration therapy. In these studies, mucins have been indirectly detected using staining with AB (unspecified pH) or Periodic Acid Schiffs (PAS). Table 3.2 summarizes the main changes observed in murine salivary glands after damage induced by ligation and radiation.

3.3.1 Regeneration of Salivary Glands Damaged by Radiation

Radiation therapy for head and neck cancer results in significant side effects in normal salivary glands, provoking a decreased quality of life for these patients. The salivary gland is extremely sensitive to radiation, mainly acinar cells [101] and manifests acute and chronic responses to radiotherapy. During the first week of treatment, patients may lose up to 50–60 % of salivary flow [81–83] due to high levels of apoptosis and glandular shrinkage [85, 104, 105], which affects saliva flow and composition [81, 83]. Chronic responses persist months or years after radiotherapy, where a great number of patients continue to show a significant decrease in both stimulated and non-stimulated salivary flow [82, 83, 106].

Saliva of the submandibular gland from irradiated patients was evaluated in order to determine a relationship between oral dryness and MUC5B concentration [91]. Patients with severe xerostomia 12 months after radiation therapy showed a tendency to decreased levels of MUC5B in saliva, compared to patients with mild xerostomia. Interestingly, half of the patients (8 individuals) with severe xerostomia did not have detectable MUC5B levels by 12 months post-radiotherapy; however, the significance of these data should be confirmed with a greater number of cases [91]. Both groups of patients, with severe and mild

Table 3.2 Summary of the effects of radiotherapy and duct ligation on the salivary gland

	Radiotherapy	References	Ligation	References
Glandular function	Decrease salivary flow	[81–83]	Decrease salivary flow, Secretory dysfunction	[84]
	Glandular shrinkage	[85]	Inflammatory cell infiltration	[84, 86–90]
	Tendency to decreased levels of MUC5B in submandibular gland saliva	[91]	Loss of glandular weight	[88, 92]
Ductal cells			Proliferation of ductal cells	[90, 93, 94]
Acinar cells	Apoptosis and replacement by fibrotic tissue	[95]	Apoptosis	[93, 94, 96]
	Decreased glycoproteins in acinar cells	[97–100]	Decreased glycoproteins in acinar cells	[88, 90, 92]
	Dilated lumens, loss of cell polarity and discharge of mucins into the stroma	[101]	Loss of secretory granules	[102, 103]

xerostomia, showed comparable volumes of saliva, but they differed in the severity of their symptoms. These differences may be due to features directly related to the mucins, such as the amount of MUC5B bound to the oral mucosa, which is a better predictor of dry mouth compared to levels of free or soluble MUC5B in saliva [91]. Apparently, mucin quality is a relevant point, as it has been reported that non-irradiated patients with dry mouth and low salivary flow still display MUC5B on their mucosal surfaces [107]. This finding suggests that mucins retained in the oral mucosa of patients with dry mouth may be less hydrated than normal subjects [91]. Alliende et al. described how changes of MUC5B post-translational processing in labial salivary glands of SS patients, had specifically reduced levels of sulfation and were able to decrease water uptake of mucins, thereby explaining the dry mouth sensation [65]. It is noteworthy that saliva contains heterogeneous MUC5B glycoforms; therefore, it would be necessary to perform studies to determine the greatest number of MUC5B glycoforms. Thus, changes in the quantity and quality of MUC5B will provide insight on why patients who recover submandibular gland salivary flow after radiotherapy still have oral dryness.

Glandular hypofunction after radiotherapy has been attributed to a loss of acinar cells followed by replacement of fibrotic tissue [95]. Based on this evidence, studies on mice and rats revealed that post-irradiation transplant with salivary stem cells can restore normal salivary function evaluated by an improvement in glandular morphology, increased number of acinar cells, salivary flow, and glycoproteins [97, 98]. Moreover, intraglandular transplantation with bone marrow-derived clonal mesenchymal stem cells (BM-cMSCs) also preserved acinar cells and salivary gland morphology in murine models with radiation damage [99]. In these studies, the presence of mucin was used as a functional marker of salivary gland acini, observing an increase of positive acini (AB+) number in BM-cMSC-treated salivary glands compared to controls [99]. This result suggests that BM-cMSCs can counteract salivary gland damage post irradiation and can be used as a source of cell-based therapy for the restoration of induced salivary hypofunction [99]. Similar results have been observed with transplants using adipose tissue-derived human mesenchymal stem cells (hAdMSC) for regenerating salivary glands damaged by radiation in mice. Systemic administration of hAdMSCs improves salivary flow rate at 12 weeks post-radiation, and salivary glands show less damage and higher levels of mucin production than untreated irradiated salivary glands [100]. However, these studies require molecular analysis on the quality of secretory products, especially of mucins.

Interestingly, salivary gland acini of irradiated rats showed dilated lumens, loss of cell polarity, and discharge of mucins into the stroma and not to the acinar lumen [101]. Comparable changes have been reported in salivary glands from SS patients. In this case, loss of cell polarity is triggered by the alteration of tight junctions [108] inducing apico-basal relocation of protein secretion machinery (e.g., SNARE proteins [95] and Rab3D [67]), accounting for ectopic mucin exocytosis into the extracellular matrix (ECM) [109]. Recently it was revealed that such anomalously located mucins in the ECM can induce a pro-inflammatory response, which may participate in the pathogenesis of this disease [110].

3.3.2 Bioengineered Germ Transplants for Salivary Gland Regeneration

The organs are generated from reciprocal interactions between the epithelium and mesenchyme of germ layers. Ogawa and Tsuji developed a bioengineering method termed "organ germ" that can reproduce salivary gland organogenesis through the epithelial-mesenchymal interactions [111]. The structure of the bioengineered salivary glands, including the location of myoepithelial cells, aquaporin 5 water channel, and neuronal connections was analogous to control mouse submandibular glands [111]. At 30 days after transplantation, the bioengineered submandibular gland showed a mucous gland phenotype with strong PAS-positive staining, suggesting the presence of sialomucins [111]. In the future, it is expected that studies of these regeneration models could provide information about the quality and functionality of the secretory products that are synthesized rather than their mere detection.

3.3.3 Regeneration of Salivary Glands Atrophied by Ligation

The obstruction of major excretory ducts (rat, mouse, rabbit, and cat) generates salivary gland atrophy and inflammation and severely affects the secretory function of remnant parenchyma [86, 87, 112–115]. The development of animal models involving the ligation of the major excretory ducts of salivary glands has contributed to the understanding of inflammation and atrophy [102, 115, 116]. There are two models of excretory duct ligation: extra-oral duct ligation (including lingual nerve cord) and intra-oral duct ligation, the latter being the most appropriate model to study obstructive diseases of the salivary gland since it does not damage the nerve involved in the normal secretion of saliva [88, 117]. The atrophy followed by extra-oral duct ligation is characterized by inflammation, loss of acini, proliferation of ductal cells, and secretory dysfunction [89, 93, 112, 115].

Some studies showed that the secretory function of acinar cells decreases significantly within 24 h of intra-oral duct ligation [84], and that all secretory granules disappear after 2–3 days [87], the time coinciding with the apoptosis peak of acinar cells [94]. In rat submandibular glands, loss of acini and inflammation was observed from 7–14 days of ligation [87, 88, 90]. By the fourth week of ligation, the atrophy of the glands was evident with a loss of over half of their normal weight, while AB/PAS staining showed a large decrease in glycoprotein levels in acinar cells [87, 90, 92]. Unlike acini, ductal cells begin to proliferate 2–3 days after ligation [94], forming long undifferentiated ductal structures not discernible as striated, granular, or intercalated. Cotroneo et al. reported that both ligated and deligated glands showed an increase in the proportion of ducts and presence of abnormal branched entities characterized by short, tubular-shaped structures terminating with small acini [90]. These structures were more common in the deligated glands and resembled those structures that arise during branching morphogenesis in embryonic development (day 18) of the submandibular gland. In deligated glands, some of the acini at the ends of branched structures showed positive AB/PAS staining, which suggest an increase of glycoproteins [90].

Morphological studies have established that after intra- and extra-oral duct ligation, the deligation allows regenerating new salivary gland tissue. After 3 days of deligation in rats, an increase in submandibular gland weight, acini size, AQP5 expression, and glycoprotein content (staining with AB/PAS) were observed in comparison to ligated submandibular gland [90]. Following 8 weeks of deligation, the submandibular glands recovered half of their weight, normal morphology, differentiated ductal and acinar structures, and the presence of glycoproteins (AB/PAS stain) in their acini [92]. The parasympathetic nerves were able to re-associate with new target cells to form functional neuro-effector junctions [92]. Moreover, after a long period of deligation (24 weeks), the rat submandibular gland is able to recover 92 % of its normal size, salivary flow, secretion of total protein, normal acinar morphology, and content of granules suggestive of mucin (AB/PAS positive) [88].

The atrophy of the salivary gland is observed in different circumstances, such as SS, irradiation therapy and obstructive sialadenitis, among others. In severe atrophy of the rat submandibular gland induced by ligation of the excretory duct, most acinar cells disappeared – mainly through apoptosis – while the ductal cells proliferated and dedifferentiated early [96]. Moreover, the gland can survive in the atrophic state almost indefinitely, with the capability of full recovery if deligated. Silver et al. reported that approximately 10 % of the acinar cells survive in atrophy induced by ligation, where activation of mTOR and autophagosomal pathways can help preserve acinar cells during salivary gland atrophy after injury [96]. This study also showed an apparent contradiction, because gene expression analyses by quantitative real-time PCR and microarray of ligated glands revealed sustained transcription of genes specific to acinar cells, while genes specific to ductal cells decreased to baseline levels [96]. Taking into account the large loss of acinar cells, this would suggest that the remaining acinar cells still have a robust transcription of secretory proteins. However, evidence of mucins expression has not been generated from this type of analysis.

3.4 Mucins and Sjögren's Syndrome

Overview of SS Primary SS is an autoimmune exocrinopathy characterized by chronic mononuclear cell infiltration into salivary and lachrymal glands [118, 119]. Ro/SS-A and La/SS-B autoantibodies are frequently found in sera of primary SS patients. Their presence is associated with prolonged disease duration, increased frequency of non-exocrine manifestations, and a higher intensity of lymphocytic infiltrates invading minor salivary glands [118, 119].

3.4.1 Factors Related to the Secretory Activity of Salivary Glands of SS Patients

The secretory activity of salivary glands of SS patients is compromised by diverse factors leading to severe dryness of the mouth (xerostomia) and eyes (keratoconjunctivitis sicca) [120]. Autoantibodies against muscarinic M3 receptors [121], imbalances of cytokine levels [122], glandular denervation [123], acinar atrophy and decrease of glandular parenchyma [124], redistribution of aquaporin-5 in the acinar cells [125], and increased levels of cholinesterase [126] are the most frequent changes associated with dryness symptoms. However, SS patients present only some, if any, of these signs, and so this topic is still a controversial issue in the field [126, 127]. Additionally, disorganization of the basal lamina (BL) of acini and ducts, which correlates with an altered secretory pole of acinar cells and morphological changes in the secretory granules, has been consistently found in all patients evaluated [128, 129]. The BL establishes molecular interactions with plasma membrane receptors and, through them, activates signaling pathways involved in proliferation, differentiation, and survival [130, 131]. Moreover, the disruption of tight junctions (TJ) induces relocalization of secretory machinery proteins. Altered localization of TJ proteins in the salivary glands of SS patients promotes the redistribution of both apico-basal Rab3D and SNARE proteins [67, 108, 109]. The presence of these proteins in the basolateral plasma membrane may explain the ectopic presence of mucins in the ECM [109]. As these mucins are normally exocyted by the apical pole of acinar cells, its altered localization may be an inflammatory trigger [110]. A relevant function derived from these interactions is the maintenance of the differentiated state of the salivary gland [132]. In other words, cell polarity is important for the organization and function of all epithelia, including secretory epithelium. Interestingly, an animal model of SS – the NOD-strain of mice – shows some alterations resembling SS patients in salivary glands prior to lymphocytic infiltration [133].

3.4.2 Mucins in SS Patients

As mentioned above, acinar cells synthesize, modify and secrete proteins [134], glycoproteins [53] and are also implicated in the transport and release of electrolytes and water [135]. The mechanisms involved in acinar cell differentiation are fundamental to the turnover and activation of the molecular machinery that participates in the synthesis, post-translational processing of proteins, formation of the secretory granules and exocytosis events [134]. The acini of human labial salivary glands, which are used as a relevant marker of SS diagnostic criteria, are of the seromucous type, with both mucous and serous acinar cells able to synthesize mucins (MG1 and MG2) [28]. Previous studies showed that a high MG1 concentration in the resting saliva of SS patients could be due to a low water content or to a low water-retaining capacity. It was then suggested that this might explain xerostomia [136]. More recently, it has been shown that post-translational modifications of mucins, rather than their net amount, can strongly affect their function, thus providing an alternative explanation for xerostomia [11, 137].

Indeed, MUC5B mRNA and protein levels were similar between controls and SS patients, while sulfo-Lewis[a] antigen levels were lower in glandular extracts from SS patients (Fig. 3.2) [65]. This finding correlated with a dramatically lower number of sulfo-Lewis[a] antigen-positive mucous acini (Fig. 3.2) [65]. In SS patients, the MUC5B electrophoretic pattern was heterogeneous, with at least two protein variants present in all analyzed samples. Both MUC5B and sulfated MUC5B showed similar electrophoretic patterns with two broad bands larger than 200 kd and migrating within the stacking gel [65]. These findings are consistent with an earlier description of MUC5B in whole saliva and salivary glands where different glycoforms of MUC5B were shown [28]. Interestingly, no correlation between whole unstimulated salivary flow (USF) and the percentage of mucous acini with sulfo-Lewis[a] antigen was found [65]. As mentioned above, mucins, in particular MUC5B, play an important role in lubrication, since they preserve mucosa hydration by the interaction of water molecules with mucin hydrophilic moieties, including sulfate, sialyl acid, and hydroxyl groups. Thus, independent of normal or decreased USF found in SS patients, the dry mouth sensation observed in all cases could be explained by a low sulfation degree of MUC5B and other mucins present in these glands, as determined by the monoclonal antibody F2 and by microdensitometric analysis of AB staining in mucous acini, respectively (Fig. 3.2) [65]. A microdensitometric analysis confirmed a decrease in the total sulfated oligosaccharides in mucous acini of SS patients, which correlated with a strong BL disorganization (Fig. 3.2) [65]. Altogether, these results suggest that a loss of function of mucous acinar cells occurs in SS patients. Given the high water-retaining capacity of sulfated mucins, these severe and generalized changes in the quantity of sulfated oligosaccharides of MUC5B could account for the xerostomia found in SS patients. Moreover, Saari et al. have shown a high MG1 concentration in the resting whole saliva of SS patients, and our results rather support their hypothesis that low water retaining capacity might explain xerostomia [136].

In mucins, sulfated and sialic acid residues interact with Ca^{2+} and H^{+}, leading to inter-strand cross-links displacing water molecules and compacting the mucin granule. These events take place during the biogenesis of such granules and thus the low sulfation of mucins will affect the organization of secretory granules [22, 23]. Later, during exocytosis, these ions are replaced again by water; however, in the presence of low-sulfated mucins, such an exchange will not occur [22, 23]. It is noteworthy that the presence of big, pleomorphic and low electron density secretory granules in acini of labial salivary glands from SS patients has been previously reported [128]. Initially, these changes were interpreted as the result of the fusion of granules. However, considering that sulfation also plays an important role in the biogenesis of mucin granules, the enlarged structures observed could also be a result of the low mucin sulfation detected [65]. In brief, the mucous acini of labial salivary glands of SS patients experience mucin desulfation, in particular of MUC5B.

This change is not related to the volumes of saliva produced by SS patients, but are actually linked to the dry mouth sensation that could be due to the decreased water binding ability of under-sulfated mucins. Thus, post-translational modifications of MUC5B play a role in the salivary hypofunction observed in SS patients and might contribute significantly to xerostomia. An important corollary of these studies is that future treatments for SS patients should target not only enhanced water production but rather an improved capacity of water retention by modulating the synthesis of mucins with adequate post-translational processing. The major therapeutic approach to reducing mouth dryness in SS patients is using secretagogues, such as cholinergic agonists, which bind to muscarinic receptors increasing the salivary flow, mainly by increasing water transport [138]. However, these treatments neither consider the quantity nor quality of the secretory products present in saliva, including mucins, which are essential for lubrication of the oral epithelium.

3.4.3 Evaluation of the Sulfo-Lewis[a] Biosynthesis Pathway

If a low degree of MUC5B sulfate content were not related to decreased levels of MUC5B polypeptides, the question would emerge as to whether the metabolic pathway of sulfo-Lewis[a] synthesis is altered. To address this, mRNA and protein levels were determined, as was the activity of enzymes involved [139]. Synthesis and elongation of mucin oligosaccharides initiates with the transfer of *N*-acetylgalactosamine to serine or threonine residues of the peptidic mucin core [20]. The oligosaccharide may be extended with galactose (Gal), *N*-acetylglucosamine (GlcNAc), fucose, or sialic acid [21, 140]. Each sequential step is catalyzed by a distinct glycosyltransferase [21] (Fig. 3.3). Modifications of mucin oligosaccharides include sulfation of Gal and GlcNAc, reactions catalyzed by Gal3-O-sulfotransferases (Gal3ST), and GlcNAc-6-sulfotransferases (GlcNAc6ST), respectively [140]. In the labial salivary glands of SS patients, levels of Gal3ST activity were significantly decreased, without changes of mRNA and protein levels [139]. Importantly, glycosyltransferases

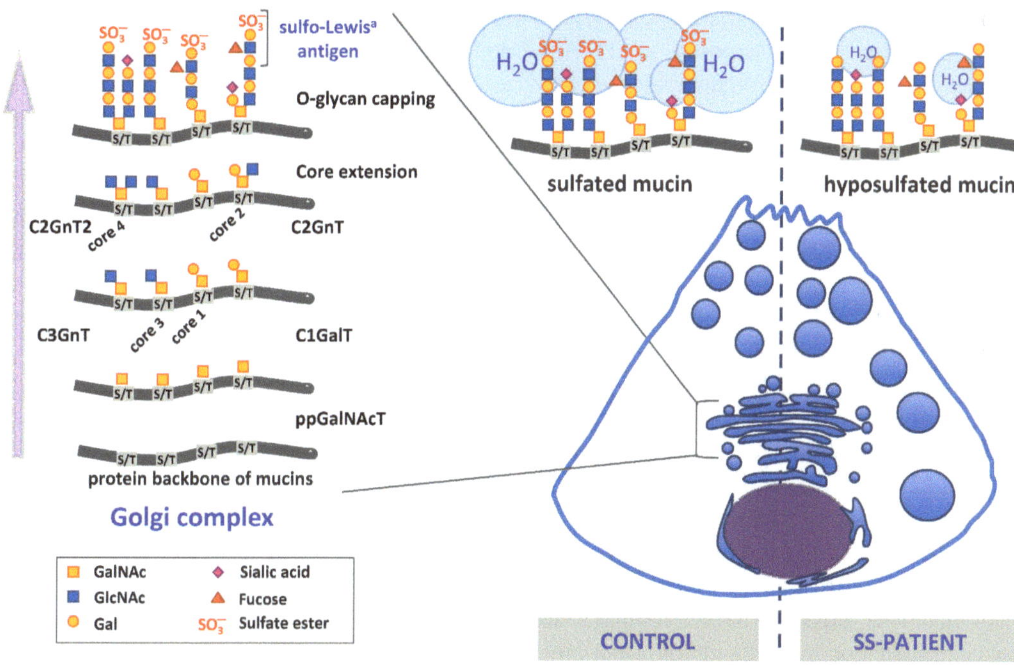

Fig. 3.3 O-glycosylation of mucins in salivary acinar cells. In the Golgi complex, ppGalNAcT initiates O-glycosylation by adding GalNAc to the hydroxyl group of either serine (S) or threonine (T) residues of the mucin protein backbone. C1GalT adds a Gal residue to synthesize the Core 1 structure that can be branched by C2GnT, forming Core 2. The Core 3 structure is synthesized by C3GnT that adds GlcNAc to GalNAc. Core 3 can be branched by C2GnT2 to form Core 4. All core structures can be further extended, branched, and modified to form complex O-glycans. Terminal GlcNAc residues can be used as the basis for the attachment of Lewis determinants. The Sulfo-Lewis[a] antigen is shown. Sulfated mucins retain large amounts of water on the epithelial surface. In SS patients, the hyposulfated mucins lose hygroscopic properties contributing to mouth dryness sensation (images Reproduced with permission of Castro et al. [137])

enzymatic activities involved in the glycosylation pathway of mucins were similar in controls and SS patients [139]. An inverse correlation was observed between Gal3ST activity and glandular function measured by scintigraphy, but not with USF. An inverse correlation between Gal3ST activity and focus score, as well as with the autoantibodies Ro/SS-A and La/SS-B were also detected. The decrease in sulfotransferase activity may explain the observed mucin hyposulfation in labial salivary glands from SS patients. Since no difference was found either in Gal3STs mRNA or protein levels, decreased activity was not a consequence of gene down-regulation. Interestingly, the sulfotransferase activity correlated with secretory function, inflammation, and autoimmunity [139].

These results suggest that pro-inflammatory cytokines may modulate Gal3ST activity, thereby altering mucin quality and leading to mouth dryness. Data on this topic in other cellular types, like bovine synoviocites exposed to TNF-α, showed a decrease of sulfotransferase activity [141], demonstrating that elevated levels of pro-inflammatory cytokines, as occurring in rheumatoid arthritis and SS, could modulate the activity and expression of glycosyltransferases [139, 141]. Reduced sulfation of mucins has been also described in inflammatory and neoplastic intestinal diseases [142]. Mucins in ulcerative colitis have shorter oligosaccharide chains and lower sulfate content than normal colon mucosa [142]. Sulfomucins in colon adenocarcinoma are notably lower than those of the adjacent normal mucosa [142–144]. The synthesis of these sulfomucins involves β3Galactosyltransferase-5 (β3GalT-5) and Gal3ST-2 [145, 146]. Lower activity and reduced expression of these enzymes in non-mucinous adenocarcinoma compared to adjacent normal mucosa, is thought to contribute to mucin hyposulfation in this pathology [147].

3.4.4 MUC5B as an Inducer of Inflammation

Although the underlying cause of SS-pathogenesis is not fully understood, among several characteristics described, the loss of apico-basal polarity of salivary acinar cells is relevant [67, 108, 109, 148, 149]. Based on this observation, a working hypothesis attributes a significant role of the salivary gland epithelium itself to the initiation and perpetuation of local autoimmune responses. According to this idea, molecular changes in the epithelial cells result in recruitment, homing, activation, proliferation, and differentiation of inflammatory cells [148, 150]. As mentioned, the loss of apico-basal polarity in salivary acinar cells of SS patients is associated with the redistribution of the molecular machinery involved in the exocytosis of secretory granules [67, 108, 109, 128, 149]. Proteins involved in membrane fusion (SNARE proteins) relocate from the apical to the baso-lateral region of acinar cells [109], and redistribution of mature secretory granules in the cytoplasm is observed [67, 109]. In addition, exocytic fusion complexes formed by SNAREs, usually present in the apical plasma membrane, and secretory granules, are found in the basolateral plasma membrane of salivary acinar cells from SS patients. Concomitant with these changes, mucins MUC5B and MUC7 are aberrantly secreted to the ECM [109].

It seems reasonable to hypothesize that mucins present in the ECM may trigger an inflammatory response by acting as ligands that activate potential receptors of epithelial or immune cells. A key question arising in this scenario concerns the type of receptors that could mediate such a response. Toll-like receptors (TLRs) are a class of pattern recognition receptors (PRRs) that play a key role in innate immunity and trigger a specific immune response [151]. TLRs are stimulated by a variety of structural signatures found in pathogens, referred to as pathogen-associated molecular patterns, and TLR activation induces the production of pro-inflammatory cytokines [151]. TLRs expressed on the cell surface, such as TLR1,2,4,5,6 and 10, recognize outer cell wall components of bacteria and fungi, whereas TLRs expressed in the intracellular compartments (TLR3,7,8 and 9) are involved in recognition of nucleic acid components [152]. In addition, TLRs may also be activated by damage-associated molecular patterns (DAMPs) produced by the cell or the ECM, endogenous molecules released, activated, or secreted by host cells and tissues undergoing stress, damage, and non-physiological cell death [152]. Interestingly, TLR4 recognizes DAMPs that contain oligossacharides and are present in the ECM, such as fibronectin, hyaluronic acid (HA), and heparan sulfate [152–154]. Termeer et al. (2002) demonstrated that hyaluronan oligosaccharides activate dendritic cells via TLR4 [155]. Furthermore, the polysaccharide portion has been shown to play a role in salmonella lipopolysaccharide-induced activation through human TLR4 [156]. After ligand binding, the cytoplasmic Toll/IL-1 receptor (TIR) domain of the TLRs associates with the TIR domain of adaptor proteins MyD88, TIRAP, TRIF and TRAM [157, 158]. Moreover, and depending on the ligand, TLR4 activation may occur in a manner either dependent or independent of MyD88. In the first case, MyD88 and TIRAP protein adapters are recruited to the TIR domain and induce a signaling pathway leading to the expression of pro-inflammatory cytokines and chemokines. Alternatively, the MyD88-independent pathway engages TRAM and TRIF as protein adapters to induce type-I interferon [157].

Normal salivary acinar cells express TLR4, which is significantly increased in SS patients [159, 160]. Moreover, chronic inflammation is evident in the salivary glands of SS patients, although the mechanisms that trigger these processes are not known. In this context, the hypothesis that was tested was whether or not the salivary mucins are involved in the expression of pro-inflammatory cytokines, exploring the molecular sensor involved in this response. Furthermore, it was investigated whether mucin oligosaccharides might act as DAMPs that are recognized by TLR4, and activate the innate immune response. Human salivary epithelial cells (HSG) were stimulated with purified MUC5B or Sulfo-Lewis^a, both inducing a significant increase of CXCL8, TNF-α, IFN-α, IFN-β, IL-6, IL-1β, but not BAFF [110]. Cytokine induction was mediated by TLR4 as shown by using the TBX2-peptide, an inhibitor of TIRAP, which is a signaling molecule downstream of TLR4 and TLR2 [161], or, alternatively, using a specific blocking antibody raised against TLR4 [110]. In summary, alterations of acinar cell polarity led to the loss of innate epithelial barrier function, triggering a series of changes that result in the anomalous release of mucins to the ECM. Human salivary MUC5B and Sulfo-Lewis^a are recognized by epithelial cell TLR4 and initiate a pro-inflammatory response. These signals originally produced by epithelial cells could attract inflammatory cells, thus perpetuating inflammation and the development of chronic disease. The findings highlight the importance of salivary gland epithelial cell organization in controlling innate immunity, and the etiopathogenesis of SS [110].

3.4.5 MUC1/SEC AND MUC1/Y Over-Expression Is Associated with Inflammation in SS

The secreted MUC1 isoform (MUC1/SEC), which lacks the cytoplasmic and transmembrane domains, contains a unique 11 amino-acid peptide at the COOH terminus that is not found in other isoforms (Fig. 3.1) [57, 162]. This sequence is referred to as the immuno-enhancing peptide (IEP) due to its ability to stimulate the immune response [163] probably by STAT-1 up-regulation [164]. IEP has been proposed to modulate both the innate and adaptive immune responses [163, 165]. MUC1/SEC may induce over-expression of cytokines through its IEP and/or via formation of a MUC1/SEC-MUC1/Y complex [163]. MUC1/Y is a MUC1 transmembrane protein without VNTR (Fig. 3.1) [166–168]. The interaction between MUC1/Y and MUC1/SEC can be compared to a receptor-ligand interaction that might trigger cytokine production and thereby modulate the immune response [166]. In addition, the formation of a receptor-ligand complex between MUC1/SEC and MUC1/Y in mammary tumors initiates a cell-signaling response that alters cell morphology [163, 164]. On the other hand, MUC1/Y has been associated with transcriptional induction of pro-inflammatory cytokine genes via NF-κB [169]. Considering these previous findings and the relevance of MUC1 in several pathologies, it was interesting to determine whether MUC1/SEC and MUC1/Y are expressed in the salivary glands of SS patients and which cytokines are able to induce their expression [7].

Significantly higher mRNA and protein levels of both these variants were found in SS patients [7]. The MUC1 gene is subject to several control mechanisms by cytokines (IFN-γ, TNF-α, IL-7) and epigenetic factors (methylation of promoter's CpG islands, histone modifications, miRNA effects) [170]. However, regulatory mechanisms involved in differential expression of MUC1 splice variants, specifically MUC1/SEC and MUC1/Y, have not yet been reported. MUC1/SEC has been associated with progressive tumoral development inhibition and anti-tumoral immune responses supported by increased STAT-1 expression [164]. Although this mechanism has not been fully elucidated, a plausible explanation is that such effects might be mediated by STAT-1 activation with: (1) over-expression of IFN-γ responsive signal transducer and/or (2) activation of pro-apoptotic and pro-inflammatory genes [164, 171]. Thus, higher levels of MUC1/SEC and MUC1/Y mRNA and protein observed in the salivary glands of SS patients may induce the synthesis of cytokines. Interestingly, we have recently demonstrated that MUC1/SEC and MUC1/Y mRNA levels are induced by TNF-α and IFN-γ in HSG cells [7], supporting previous evidence of a self-perpetuating mucin-cytokine signaling loop in inflammatory conditions [110]. Studies evaluating MUC1 function in salivary glands are not available; however, in other tissues, such as lung, bowel, and brain, studies using MUC1 knockout mice suggest an anti-inflammatory role [172–174]. In some tissues, a complete loss of all MUC1 isoforms was reported, and in the bowel, the anti-inflammatory function of MUC1 was linked to the mucosa protection provided by MUC1 anchored to the membrane [173]. These results have been reproduced by MUC1 knockdown with a siRNA targeted to a sequence shared by all known MUC1 variants[173]. However, it is important to emphasize that these studies do not provide light on the specific function of particular MUC1 isoforms, such as MUC1/SEC and MUC1/Y.

The immunohistochemical analysis of salivary glands revealed a significantly higher proportion of acini with MUC1/SEC in the cytoplasm of SS patients, while for controls, a significantly higher percentage of acini with MUC1/SEC in the apical region and low presence in the cytoplasm was observed [7]. This cytoplasmic distribution of MUC1/SEC observed in the acini of SS patients was associated with the loss of cell polarity, where the increased intensity of cytoplasmic MUC1/SEC staining coincided with increased acinar alterations[7]. These results confirmed observations showing altered distribution and accumulation of MUC1/VNTR in the cytoplasm of acinar and ductal cells from labial salivary glands of SS patients [7]. MUC1/Y is redistributed from the apical region of acini in labial salivary glands of controls to the cytoplasm and nuclei in labial salivary glands from SS patients [7]. Nuclear localization of some MUC1 isoforms has been previously described, but the mechanism involved is still unknown. Such nuclear function has been related with increased transcription of pro-inflammatory cytokines [169]. In summary, the over-expression and aberrant localization of MUC1/SEC and MUC1/Y observed in the labial salivary glands of SS patients and their over-expression induced *in vitro* in HSG cells by pro-inflammatory cytokines support previous evidence of a self-perpetuating mucin-cytokine signaling loop that may facilitate the maintenance of an inflammatory environment leading to the disruption of salivary glandular homeostasis in SS patients [110].

3.4.6 MUC7 in SS Patients

MUC7 studies in normal human labial salivary glands were previously reported by Veerman et al., showing cytoplasmic localization in acinar cells of serous acini [28]. Preliminary studies showed that MUC7mRNA levels are similar in the salivary glands of SS patients and in controls;

however, protein levels detected by western blot and immunohistochemistry showed a significant increase in SS patients [67]. MUC7 cytoplasmic localization in serous acini supported previous findings [28]. More studies are needed to collect further information on MUC7 changes in SS patients and to evaluate whether this mucin contributes to the xerostomia that afflicts these patients.

3.5 Mucins in Salivary Gland Tumors

Cancer cells express aberrant forms or high amounts of mucins, which have been suggested as molecular markers of malignant transformation in several organs and tissues [175]. Altered MUC1 expression in malignancies occurs in various modes, including up-regulation, mis-localization, and aberrant glycosylation. MUC1 expression is related to aggressive tumor behavior and a poor prognosis for patients with human neoplasms [176]. Secreted mucin expression profiles of adenocarcinomas have been associated with etiology [177], tumor progression [8], prognosis [178] and histologic characteristics [61].

Salivary gland neoplasms are characterized by morphological variability. These tumors imitate the histology, and may arise from epithelial, mesenchymal, and/or lymphoid components. Several benign and malignant salivary gland tumors showed abundant extracellular or intracellular mucins [179].

3.5.1 Mucins in Benign Salivary Gland Neoplasms

Overall, MUC1 is the mucin most related to cancer and it is transcribed as multiple [80] alternatively spliced variants; some of these have been detected in salivary gland tumors. As mentioned, MUC1 can be found as plasma membrane or secreted isoforms, and its over-expression has been associated with biochemical events that occur in carcinogenesis and/or tumor invasion [8, 61, 178, 180].

Most salivary gland tumors – both benign and malignant – express MUC1, where positive cells range from 67 to 100 %. However, it has been difficult to establish whether MUC1 over-expression is higher in malignant than in benign salivary tumors, probably due to the use of different MUC1 mAb (Fig. 3.1). Sensitivity of antibodies depends on differential glycosylation degrees between MUC1 isoforms, thus affecting epitope recognition in immunohistochemical assays. Additionally, factors such as small sample sizes and varied scoring systems in various studies make it difficult to analyze the data in order to estimate MUC1 over-expression.

Pleomorphic adenoma (PA), or benign mixed tumors of salivary glands, are the most common benign neoplasms, representing 40–70 % of cases [181]. Etiology and mechanisms involved in PA growth are not fully understood [180]. PA might display a wide range of recurrence rates (2.5–32.5 %).

Immunohistochemical markers have been used to explain this behavior and predict recurrence. Primary PA shows scarce MUC1 expression, while recurrent PAs exhibit higher MUC1 levels, suggesting the association of this mucin with recurrence [181] (Table 3.3). Using a MUC1/DF Ab (Fig. 3.1), Hamada et al. found high MUC1 expression in patients with PA recurrence (RPA) and proposed that MUC1 would be an independent risk factor of recurrence [8]. RPA also showed malignant transformation in areas where MUC1 expression was higher. However, Brieger et al. found that MUC1 did not correlate with PA recurrence in parotid glands, suggesting that MUC1 might actually play a role in cellular dysfunction [180]. It is accepted that the biological behavior of tumors during their progression is influenced by changes in structure and the distribution of cell-surface glycoproteins. Soares et al. showed differential MUC1 expression during malignant transformation of PAs toward widely invasive carcinomas [178]. Therefore, low MUC1 expression in primary PA may indicate a reduced invasive potential and aggressiveness, while a higher expression may be indicative of an aggressive biological behavior, such as can be observed in recurrent PAs and carcinoma ex-pleomorphic

Table 3.3 Immunoreactivity of MUC1, MUC2, MUC4, MUC5AC, MUC5B and MUC6 in salivary gland tumors

	MUC1	MUC2	MUC4	MUC5AC	MUC5B	MUC6
ACA	19/21(90)	12/13(92)	0/8	0/11	NT	0/11
ACC	60/60(100)	2/26(8)	1/2(50)	0/20	NT	0/20
CA	1/1(100)	0/1	1/1(100)	0/1	0/1	1/1(100)
Ca-ex-PA	11/11(100)	NT	NT	NT	NT	NT
MEC	91/114(84)	24/112(21)	109/131(83)	33/40(83)	33/40(83)	15/60(25)
PLGA	NT	NT	1/1(100)	NT	NT	NT
SDC	2/2(100)	6/6(100)	1/1(100)	4/5(80)	4/5(80)	6/6(100)
PA	106/158(67)	27/70(31)	7/51(14)	NT	NT	19/49(39)
WT	28/29(97)	22/22(100)	3/3(100)	NT	NT	NT

In parentheses are the percentages of positive tumors
ACA Acinic cell adenocarcinoma, *ACC* Adenoid cystic carcinoma, *CA* Cystadenoma, Ca-*ex-PA* Carcinoma ex-pleomorphic adenoma, *MEC* Mucoepidermoid carcinoma, *PLGA* Plymorphous low-grade adenocarcinoma, *SDC* Salivary duct carcinoma, *PA* Pleomorphic adenoma, *WT* Warthin's tumor, *NT* Not tested

adenoma [8, 61, 178]. Moreover, Soares et al. showed an augmented MUC1 expression in RPA relative to primary PA. Altogether, these findings suggest that MUC1 may be a useful indicator of a potential malignancy [178].

MUC2 has also been studied in benign salivary gland tumors and its expression associated with indolent behavior in both human and animal neoplasms [182]. PAs showed a weak cytoplasmic MUC2 signal in single cells; these findings had no relation to the clinical behavior [61]. Furthermore, MUC1 and MUC2 have been detected in Warthin's tumors (WT), a benign, well-encapsulated neoplasm usually originating in the caudal pole of parotid glands. Mannweiler et al. found that WTs were characterized by strong MUC1 and MUC2 expression with 90 % of tumor cells showing a diffuse cytoplasmic staining. In addition, membranes of luminal cells exhibited higher MUC1 expression compared to MUC2 [61]. Conversely, Yamada et al. found that MUC1 was restricted to basal tumor cells [183].

3.5.2 Mucins in Malignant Salivary Gland Neoplasms

Mucoepidermoid carcinoma (MEC) is the most common malignant neoplasm of the salivary glands, accounting for about 30–40 % of all salivary carcinomas. MEC may produce high levels of extracellular mucin and possess morphologic diversity with primary MEC displaying a variety of biologic behaviors. While the low- and intermediate-grade MECs are of a benign-like nature with high survival rates, the high-grade MEC usually has a poor prognosis [184]. Other less common mucin-producing tumors are colloid (mucinous) carcinoma (CC), mucinous cystadenocarcinoma (MCA), salivary duct carcinoma (SDC), signet ring cell carcinoma, adenocarcinoma (NOS), and, occasionally, a metastatic tumor [8]. Among membrane-bound mucins, MUC1 is frequently overexpressed in carcinomas, particularly adenocarcinomas, which in most tumor types are correlated to an adverse effect on prognosis [175], increased metastatic potential, and poor survival rates [175]. MUC1 expression is elevated in MECs, and is positively correlated with lymph node metastasis in the clinical stage; it is also a strong independent prognostic factor [184]. Besides localizing in the apical membranes of luminal tumoral cells, MUC1 was also detected in the cytoplasm and sub-cellular membranes of the epidermoid, intermediate, mucous, and clear MEC tumor cells. Studies on salivary MECs have shown a relationship between MUC1 expression in tumor cells and outcome [66, 185]. MUC1 expression in 5–10 % of MEC tumor cells is enough to indicate an adverse prognosis, such as recurrence, metastasis, and/or cancer-related death. Alos et al. [66] and Handra-Luca et al. [185] detected MUC1 expression in MEC using Ma695 monoclonal Ab (Fig. 3.1), but it remains to be determined whether other MUC1 glycoforms behave similarly and what their prognostic potential in this type of tumor is.

MUC2 is highly expressed in mucinous carcinomas, such as colon, breast, pancreas, ovary, and stomach [186]. However, its expression is low in malignant salivary gland neoplasms. Alos et al. detected low MUC2 expression in MEC (5 % of tumors analyzed with 5–10 % having positive cells). Furthermore, MUC2 content was not related to MEC histologic grade and prognosis for the patients [66]. In other research, Mannweiler et al. observed MUC2 positivity in all cases, but only 5–25 % of the tumor cells were stained [61]. In both studies, the same Ab (NCL-clone Ccp58) was used, and the differences observed may be due to sample processing, e.g., antigen retrieval. Muc2 knockout mice frequently developed adenomas in the small intestine that progressed to invasive adenocarcinomas[187]. Also, MUC2 was highly expressed in non-invasive tumors of the pancreas and intrahepatic bile duct, which show more favorable outcome than invasive carcinomas of the pancreas and intrahepatic bile duct [176].

MUC4, MUC5AC, MUC5B, and MUC6 are also expressed in MEC [66, 185]. Handra-Luca et al. found that MUC4 is redistributed from an apical surface to a basolateral surface in duct cells in intermediate and epidermoid tumor cells. However, MUC4 was not related to prognosis and seems to be associated with MEC grades [185]. MUC5AC was expressed in intermediate cells of the tumor, considered undifferentiated tumor cells, and high grade MEC showed reactive cells [185]. MUC5B was similarly distributed to MUC5AC, mainly in low-grade tumors, but not related with tumoral progression and patient outcome. MUC6 was detected in MEC predominantly in the cytoplasm of mucous cells and was not related to the histological grade of the tumor, the tumor progression, or the patient's outcome [66] (Table 3.3). In adenoid cystic carcinoma (ACC), Mannweiler et al. observed MUC1 staining in all tumors, although immunoreactivity was heterogeneous with tumor areas strongly positive and others negative [61]. MUC1 localization was preferentially in glandular structures with apical predominance. Regarding MUC2, only 2 of 9 ACC cases were positive [61]. Another malignant, mucin-producing neoplasm is the acinic cell adenocarcinoma (ACA) that showed cytoplasmic MUC1 positivity in all studied tumors. Conversely, ACC immunoreactivity was rarely observed in glandular structures [61]. This difference could be explained by alterations in processing and targeting pathways, allowing the intracellular accumulation of MUC1 in ACA [61]. On the other hand, the difference observed in MUC2 immunoreactivity between ACA and ACC, could be partially explained by different cell populations found in these tumors. Particularly in ACA, tumor cells laden with secretory granules are frequent, while ACC showed a large number of basaloid cells and few cells with secretory characteristics [61].

Conclusions

The specific roles played by mucins in human salivary gland development have not been fully studied, mostly due to ethical and technical restrictions. Most of the studies on mucin expression during development have been performed on murine models, showing that mucin is the initial marker of epithelial differentiation in the mouse submandibular gland. A role for MUC1 in glandular morphogenesis was suggested due to the coincident onset of glandular differentiation with changes of MUC1 expression pattern during embryonic development. It has been suggested that MUC1 may induce changes in tissue architecture in development and also in cancer, where it is frequently overexpressed, misglycosylated, and redistributed. Mucin expression and glycosylation have been associated with etiology, tumor progression, prognosis, and histologic characteristics. The altered glycosylation of mucins confers a wide range of potential ligands on tumor cells for interaction with other receptors at the cell surface, contributing to the survival of these tumor cells during invasion and metastasis. Mucins are used as diagnostic markers in cancer, and are being researched as therapeutic targets for this disease. In addition, mucins have been evaluated as markers to address the functionality of the glandular acini after damage and/or regeneration therapy. In the future, it is expected that studies using regeneration models as bioengineered submandibular glands could provide information about the quality and functionality of the mucins that are rather than their mere detection.

A decreased quality of salivary mucins and reduced salivary flow lead to xerostomia. These changes produce a variety of oral and dental disorders that affect the quality of life of SS patients. The main therapeutic approach to reduce mouth dryness in SS patients involves the use of secretagogues. These cholinergic agonists bind to muscarinic receptors and increase salivary flow, mainly by enhancing water release; the treatments neither consider the quantity nor quality of the secretory products present in saliva, such as mucins, which are complex O-linked glycoproteins with sialylated and/or sulfated oligosaccharides attached to their protein backbone. This characteristic allows mucins to bind large amounts of water and lubricate the oral epithelium. In salivary glands of SS patients, altered trafficking and maturation of salivary mucins were observed. These alterations are likely to contribute to the dryness sensation in SS patients. A better characterization of the molecular mechanisms that cause these alterations will favor the development of more effective therapies to treat xerostomia in SS patients [137].

In SS patients, the alterations in cell polarity lead to the loss of the innate epithelial barrier function, triggering a series of changes that result in the release of mucins to the ECM. Human salivary MUC5B and Sulfo-Lewis[a] (just to be consistent with Sulfo-Lewis in other sections of the book chapter). are recognized by epithelial cell TLR4 and initiate a pro-inflammatory response. These signals, initially produced by epithelial cells, could attract inflammatory cells, which perpetuate inflammation and the development of chronic disease. These findings highlight the importance of salivary gland epithelial cell organization in controlling innate immunity, and in the etiopathogenesis of SS [110, 150].

The over-expression and aberrant localization of MUC1/SEC and MUC1/Y observed in the LSG of SS patients and their over-expression induced *in vitro* in HSG cells by pro-inflammatory cytokines are consistent with previous evidence that points toward the existence of a self-perpetuating mucin-cytokine signaling loop that may facilitate the maintenance of an inflammatory environment leading to disruption of salivary glandular homeostasis in SS patients [7].

In short, although mucins have been studied for many years, their exact roles and mechanisms in salivary gland development and diseases are still in the process of being discovered. As mentioned, the reasons are various and diverse, such as structural complexity, glycosylation processing and the intricate and complex processes in which they participate, among others. Fortunately, today there are more methodological tools to study them, which augur a more promising future in this field.

Acknowledgements Supported by Fondecyt-Chile [#1120062] (MJG, SG and SA), Fondecyt [#1160015] (MJG, SG, SA and IC), PhD fellowship Conicyt-Chile (MJB and JC).

References

1. Milne RW, Dawes C. The relative contributions of different salivary glands to the blood group activity of whole saliva in humans. Vox Sang. 1973;25(4): 298–307.
2. Gendler SJ, Burchell JM, Duhig T, Lamport D, White R, Parker M, et al. Cloning of partial cDNA encoding differentiation and tumor-associated mucin glycoproteins expressed by human mammary epithelium. Proc Natl Acad Sci U S A. 1987;84(17): 6060–4.
3. Swallow DM, Gendler S, Griffiths B, Corney G, Taylor-Papadimitriou J, Bramwell ME. The human tumour-associated epithelial mucins are coded by an expressed hypervariable gene locus PUM. Nature. 1987;328(6125):82–4.
4. Pemberton L, Taylor-Papadimitriou J, Gendler SJ. Antibodies to the cytoplasmic domain of the MUC1 mucin show conservation throughout mammals. Biochem Biophys Res Commun. 1992;185(1): 167–75.
5. Croce MV, Isla-Larrain M, Remes-Lenicov F, Colussi AG, Lacunza E, Kim KC, et al. MUC1

cytoplasmic tail detection using CT33 polyclonal and CT2 monoclonal antibodies in breast and colorectal tissue. Histol Histopathol. 2006;21(8):849–55.
6. Teshima TH, Ianez RF, Coutinho-Camillo CM, Buim ME, Soares FA, Lourenco SV. Development of human minor salivary glands: expression of mucins according to stage of morphogenesis. J Anat. 2011;219(3):410–7.
7. Sung HH, Castro I, Gonzalez S, Aguilera S, Smorodinsky NI, Quest A, et al. MUC1/SEC and MUC1/Y over-expression is associated with inflammation in Sjogren's syndrome. Oral Dis. 2015;21(6):730–8.
8. Hamada T, Matsukita S, Goto M, Kitajima S, Batra SK, Irimura T, et al. Mucin expression in pleomorphic adenoma of salivary gland: a potential role for MUC1 as a marker to predict recurrence. J Clin Pathol. 2004;57(8):813–21.
9. Gendler SJ, Spicer AP. Epithelial mucin genes. Annu Rev Physiol. 1995;57:607–34.
10. Tabak LA, Levine MJ, Mandel ID, Ellison SA. Role of salivary mucins in the protection of the oral cavity. J Oral Pathol. 1982;11(1):–17.
11. Chaudhury NM, Shirlaw P, Pramanik R, Carpenter GH, Proctor GB. Changes in saliva rheological properties and mucin glycosylation in dry mouth. J Dent Res. 2015;94(12):1660–7.
12. van der Reijden WA, Veerman EC, Amerongen AV. Shear rate dependent viscoelastic behavior of human glandular salivas. Biorheology. 1993;30(2):141–52.
13. Offner GD, Troxler RF. Heterogeneity of high-molecular-weight human salivary mucins. Adv Dent Res. 2000;14:69–75.
14. Corfield AP. Mucins: a biologically relevant glycan barrier in mucosal protection. Biochim Biophys Acta. 2015;1850(1):236–52.
15. Hijikata M, Matsushita I, Tanaka G, Tsuchiya T, Ito H, Tokunaga K, et al. Molecular cloning of two novel mucin-like genes in the disease-susceptibility locus for diffuse panbronchiolitis. Hum Genet. 2011;129(2):117–28.
16. Itoh Y, Kamata-Sakurai M, Denda-Nagai K, Nagai S, Tsuiji M, Ishii-Schrade K, et al. Identification and expression of human epiglycanin/MUC21: a novel transmembrane mucin. Glycobiology. 2008;18(1):74–83.
17. Thornton DJ, Rousseau K, McGuckin MA. Structure and function of the polymeric mucins in airways mucus. Annu Rev Physiol. 2008;70:459–86.
18. Asker N, Axelsson MA, Olofsson SO, Hansson GC. Human MUC5AC mucin dimerizes in the rough endoplasmic reticulum, similarly to the MUC2 mucin. Biochem J. 1998;335(Pt 2):381–7.
19. Asker N, Axelsson MA, Olofsson SO, Hansson GC. Dimerization of the human MUC2 mucin in the endoplasmic reticulum is followed by a N-glycosylation-dependent transfer of the mono- and dimers to the Golgi apparatus. J Biol Chem. 1998;273(30):18857–63.
20. Ten Hagen KG, Fritz TA, Tabak LA. All in the family: the UDP-GalNAc:polypeptide N-acetylgalactosaminyltransferases. Glycobiology. 2003;13(1):1R–16R.
21. Brockhausen I. Glycodynamics of mucin biosynthesis in gastrointestinal tumor cells. Adv Exp Med Biol. 2003;535:163–88.
22. Verdugo P. Goblet cells secretion and mucogenesis. Annu Rev Physiol. 1990;52:157–76.
23. Perez-Vilar J, Hill RL. The structure and assembly of secreted mucins. J Biol Chem. 1999;274(45):31751–4.
24. Nielsen PA, Bennett EP, Wandall HH, Therkildsen MH, Hannibal J, Clausen H. Identification of a major human high molecular weight salivary mucin (MG1) as tracheobronchial mucin MUC5B. Glycobiology. 1997;7(3):413–9.
25. Wickstrom C, Davies JR, Eriksen GV, Veerman EC, Carlstedt I. MUC5B is a major gel-forming, oligomeric mucin from human salivary gland, respiratory tract and endocervix: identification of glycoforms and C-terminal cleavage. Biochem J. 1998;334(Pt 3):685–93.
26. Levine MJ, Reddy MS, Tabak LA, Loomis RE, Bergey EJ, Jones PC, et al. Structural aspects of salivary glycoproteins. J Dent Res. 1987;66(2):436–41.
27. Thomsson KA, Prakobphol A, Leffler H, Reddy MS, Levine MJ, Fisher SJ, et al. The salivary mucin MG1 (MUC5B) carries a repertoire of unique oligosaccharides that is large and diverse. Glycobiology. 2002;12(1):1–14.
28. Veerman EC, van den Keijbus PA, Nazmi K, Vos W, van der Wal JE, Bloemena E, et al. Distinct localization of MUC5B glycoforms in the human salivary glands. Glycobiology. 2003;13(5):363–6.
29. Thornton DJ, Khan N, Mehrotra R, Howard M, Veerman E, Packer NH, et al. Salivary mucin MG1 is comprised almost entirely of different glycosylated forms of the MUC5B gene product. Glycobiology. 1999;9(3):293–302.
30. Sheehan JK, Howard M, Richardson PS, Longwill T, Thornton DJ. Physical characterization of a low-charge glycoform of the MUC5B mucin comprising the gel-phase of an asthmatic respiratory mucous plug. Biochem J. 1999;338(Pt 2):507–13.
31. Veerman EC, Bolscher JG, Appelmelk BJ, Bloemena E, van den Berg TK, Nieuw Amerongen AV. A monoclonal antibody directed against high M(r) salivary mucins recognizes the SO3-3Gal beta 1-3GlcNAc moiety of sulfo-Lewis(a): a histochemical survey of human and rat tissue. Glycobiology. 1997;7(1):37–43.
32. Amerongen AV, Bolscher JG, Veerman EC. Salivary mucins: protective functions in relation to their diversity. Glycobiology. 1995;5(8):733–40.
33. Tabak LA. In defense of the oral cavity: structure, biosynthesis, and function of salivary mucins. Annu Rev Physiol. 1995;57:547–64.
34. Tabak LA, Levine MJ, Jain NK, Bryan AR, Cohen RE, Monte LD, et al. Adsorption of human salivary

mucins to hydroxyapatite. Arch Oral Biol. 1985;30(5):423–7.
35. Nieuw Amerongen AV, Oderkerk CH, Veerman EC. Influence of phytate on the adsorption of human salivary mucins onto hydroxyapatite. J Biol Buccale. 1988;16(4):203–8.
36. Veerman EC, Ligtenberg AJ, Schenkels LC, Walgreen-Weterings E, Nieuw Amerongen AV. Binding of human high-molecular-weight salivary mucins (MG1) to Hemophilus parainfluenzae. J Dent Res. 1995;74(1):351–7.
37. Veerman EC, Bank CM, Namavar F, Appelmelk BJ, Bolscher JG, Nieuw Amerongen AV. Sulfated glycans on oral mucin as receptors for Helicobacter pylori. Glycobiology. 1997;7(6):737–43.
38. Bobek LA, Tsai H, Biesbrock AR, Levine MJ. Molecular cloning, sequence, and specificity of expression of the gene encoding the low molecular weight human salivary mucin (MUC7). J Biol Chem. 1993;268(27):20563–9.
39. Mehrotra R, Thornton DJ, Sheehan JK. Isolation and physical characterization of the MUC7 (MG2) mucin from saliva: evidence for self-association. Biochem J. 1998;334(Pt 2):415–22.
40. Prakobphol A, Levine MJ, Tabak LA, Reddy MS. Purification of a low-molecular-weight, mucin-type glycoprotein from human submandibular-sublingual saliva. Carbohydr Res. 1982;108(1): 111–22.
41. Ramasubbu N, Reddy MS, Bergey EJ, Haraszthy GG, Soni SD, Levine MJ. Large-scale purification and characterization of the major phosphoproteins and mucins of human submandibular-sublingual saliva. Biochem J. 1991;280(Pt 2):341–52.
42. Reddy MS, Bobek LA, Haraszthy GG, Biesbrock AR, Levine MJ. Structural features of the low-molecular-mass human salivary mucin. Biochem J. 1992;287(Pt 2):639–43.
43. Reddy MS, Levine MJ, Prakobphol A. Oligosaccharide structures of the low-molecular-weight salivary mucin from a normal individual and one with cystic fibrosis. J Dent Res. 1985;64(1):33–6.
44. Amerongen AV, Veerman EC. Saliva--the defender of the oral cavity. Oral Dis. 2002;8(1):12–22.
45. Reddy MS, Levine MJ, Paranchych W. Low-molecular-mass human salivary mucin, MG2: structure and binding of Pseudomonas aeruginosa. Crit Rev Oral Biol Med. 1993;4(3–4):315–23.
46. Groenink J, Ligtenberg AJ, Veerman EC, Bolscher JG, Nieuw Amerongen AV. Interaction of the salivary low-molecular-weight mucin (MG2) with Actinobacillus actinomycetemcomitans. Antonie Van Leeuwenhoek. 1996;70(1):79–87.
47. Moshier A, Reddy MS, Scannapieco FA. Role of type 1 fimbriae in the adhesion of *Escherichia coli* to salivary mucin and secretory immunoglobulin A. Curr Microbiol. 1996;33(3):200–8.
48. Liu B, Rayment SA, Gyurko C, Oppenheim FG, Offner GD, Troxler RF. The recombinant N-terminal region of human salivary mucin MG2 (MUC7) contains a binding domain for oral Streptococci and exhibits candidacidal activity. Biochem J. 2000;345(Pt 3):557–64.
49. Liu B, Rayment S, Oppenheim FG, Troxler RF. Isolation of human salivary mucin MG2 by a novel method and characterization of its interactions with oral bacteria. Arch Biochem Biophys. 1999;364(2):286–93.
50. Groenink J, Walgreen-Weterings E, Nazmi K, Bolscher JG, Veerman EC, van Winkelhoff AJ, et al. Salivary lactoferrin and low-Mr mucin MG2 in Actinobacillus actinomycetemcomitans-associated periodontitis. J Clin Periodontol. 1999;26(5):269–75.
51. Bobek LA, Situ H. MUC7 20-Mer: investigation of antimicrobial activity, secondary structure, and possible mechanism of antifungal action. Antimicrob Agents Chemother. 2003;47(2):643–52.
52. Senapati S, Das S, Batra SK. Mucin-interacting proteins: from function to therapeutics. Trends Biochem Sci. 2010;35(4):236–45.
53. Iontcheva I, Oppenheim FG, Troxler RF. Human salivary mucin MG1 selectively forms heterotypic complexes with amylase, proline-rich proteins, statherin, and histatins. J Dent Res. 1997;76(3):734–43.
54. Gendler SJ, Lancaster CA, Taylor-Papadimitriou J, Duhig T, Peat N, Burchell J, et al. Molecular cloning and expression of human tumor-associated polymorphic epithelial mucin. J Biol Chem. 1990;265(25): 15286–93.
55. Ligtenberg MJ, Kruijshaar L, Buijs F, van Meijer M, Litvinov SV, Hilkens J. Cell-associated episialin is a complex containing two proteins derived from a common precursor. J Biol Chem. 1992;267(9):6171–7.
56. Hattrup CL, Gendler SJ. Structure and function of the cell surface (tethered) mucins. Annu Rev Physiol. 2008;70:431–57.
57. Smorodinsky N, Weiss M, Hartmann ML, Baruch A, Harness E, Yaakobovitz M, et al. Detection of a secreted MUC1/SEC protein by MUC1 isoform specific monoclonal antibodies. Biochem Biophys Res Commun. 1996;228(1):115–21.
58. Thathiah A, Blobel CP, Carson DD. Tumor necrosis factor-alpha converting enzyme/ADAM 17 mediates MUC1 shedding. J Biol Chem. 2003;278(5): 3386–94.
59. Singh PK, Hollingsworth MA. Cell surface-associated mucins in signal transduction. Trends Cell Biol. 2006;16(9):467–76.
60. Miletich I. Introduction to salivary glands: structure, function and embryonic development. In: Tucker AS, Miletich I, editors. Salivary glands development, adaptations and disease. Basel/New York: Karger; 2010.
61. Mannweiler S, Beham A, Langner C. MUC1 and MUC2 expression in salivary gland tumors and in non-neoplastic salivary gland tissue. APMIS. 2003;111(10):978–84.
62. Chen Y, Zhao YH, Kalaslavadi TB, Hamati E, Nehrke K, Le AD, et al. Genome-wide search and identification of a novel gel-forming mucin MUC19/

Muc19 in glandular tissues. Am J Respir Cell Mol Biol. 2004;30(2):155–65.

63. Culp DJ, Robinson B, Cash MN, Bhattacharyya I, Stewart C, Cuadra-Saenz G. Salivary mucin 19 glycoproteins: innate immune functions in Streptococcus mutans-induced caries in mice and evidence for expression in human saliva. J Biol Chem. 2015;290(5):2993–3008.
64. Nielsen PA, Mandel U, Therkildsen MH, Clausen H. Differential expression of human high-molecular-weight salivary mucin (MG1) and low-molecular-weight salivary mucin (MG2). J Dent Res. 1996;75(11):1820–6.
65. Alliende C, Kwon YJ, Brito M, Molina C, Aguilera S, Perez P, et al. Reduced sulfation of muc5b is linked to xerostomia in patients with Sjogren syndrome. Ann Rheum Dis. 2008;67(10):1480–7.
66. Alos L, Lujan B, Castillo M, Nadal A, Carreras M, Caballero M, et al. Expression of membrane-bound mucins (MUC1 and MUC4) and secreted mucins (MUC2, MUC5AC, MUC5B, MUC6 and MUC7) in mucoepidermoid carcinomas of salivary glands. Am J Surg Pathol. 2005;29(6):806–13.
67. Bahamondes V, Albornoz A, Aguilera S, Alliende C, Molina C, Castro I, et al. Changes in Rab3D expression and distribution in the acini of Sjogren's syndrome patients are associated with loss of cell polarity and secretory dysfunction. Arthritis Rheum. 2011;63(10):3126–35.
68. Tucker AS. Salivary gland development. Semin Cell Dev Biol. 2007;18(2):237–44.
69. Jaskoll T, Chen H, Denny PC, Denny PA, Melnick M. Mouse submandibular gland mucin: embryo-specific mRNA and protein species. Mech Dev. 1998;74(1–2):179–83.
70. Melnick M, Jaskoll T. Mouse submandibular gland morphogenesis: a paradigm for embryonic signal processing. Crit Rev Oral Biol Med. 2000;11(2):199–215.
71. Melnick M, Chen H, Zhou Y, Jaskoll T. Embryonic mouse submandibular salivary gland morphogenesis and the TNF/TNF-R1 signal transduction pathway. Anat Rec. 2001;262(3):318–30.
72. Denny PA, Denny PC. Localization of a mouse submandibular sialomucin by indirect immunofluorescence. Histochem J. 1982;14(3):403–8.
73. Denny PC, Ball WD, Redman RS. Salivary glands: a paradigm for diversity of gland development. Crit Rev Oral Biol Med. 1997;8(1):51–75.
74. Denny PA, Denny PC, Jenkins K. Purification and biochemical characterization of a mouse submandibular sialomucin. Carbohydr Res. 1980;87(2):265–74.
75. Denny PA, Pimprapaiporn W, Kim MS, Denny PC. Quantitation and localization of acinar cell-specific mucin in submandibular glands of mice during postnatal development. Cell Tissue Res. 1988;251(2):381–6.
76. Braga VM, Pemberton LF, Duhig T, Gendler SJ. Spatial and temporal expression of an epithelial mucin, Muc-1, during mouse development. Development. 1992;115(2):427–37.
77. Hudson MJ, Stamp GW, Chaudhary KS, Hewitt R, Stubbs AP, Abel PD, et al. Human MUC1 mucin: a potent glandular morphogen. J Pathol. 2001;194(3):373–83.
78. Miles DW, Taylor-Papadimitriou J. Therapeutic aspects of polymorphic epithelial mucin in adenocarcinoma. Pharmacol Ther. 1999;82(1):97–106.
79. Gendler SJ, Spicer AP, Lalani EN, Duhig T, Peat N, Burchell J, et al. Structure and biology of a carcinoma-associated mucin, MUC1. Am Rev Respir Dis. 1991;144(3 Pt 2):S42–7.
80. Brockhausen I, Yang JM, Burchell J, Whitehouse C, Taylor-Papadimitriou J. Mechanisms underlying aberrant glycosylation of MUC1 mucin in breast cancer cells. Eur J Biochem. 1995;233(2):607–17.
81. Henson BS, Eisbruch A, D'Hondt E, Ship JA. Two-year longitudinal study of parotid salivary flow rates in head and neck cancer patients receiving unilateral neck parotid-sparing radiotherapy treatment. Oral Oncol. 1999;35(3):234–41.
82. Eisbruch A, Kim HM, Terrell JE, Marsh LH, Dawson LA, Ship JA. Xerostomia and its predictors following parotid-sparing irradiation of head-and-neck cancer. Int J Radiat Oncol Biol Phys. 2001;50(3):695–704.
83. Dirix P, Nuyts S, Van den Bogaert W. Radiation-induced xerostomia in patients with head and neck cancer: a literature review. Cancer. 2006;107(11): 2525–34.
84. Carpenter GH, Osailan SM, Correia P, Paterson KP, Proctor GB. Rat salivary gland ligation causes reversible secretory hypofunction. Acta Physiol (Oxf). 2007;189(3):241–9.
85. Robar JL, Day A, Clancey J, Kelly R, Yewondwossen M, Hollenhorst H, et al. Spatial and dosimetric variability of organs at risk in head-and-neck intensity-modulated radiotherapy. Int J Radiat Oncol Biol Phys. 2007;68(4):1121–30.
86. Tamarin A. Submaxillary gland recovery from obstruction. I. Overall changes and electron microscopic alterations of granular duct cells. J Ultrastruct Res. 1971;34(3):276–87.
87. Tamarin A. Submaxillary gland recovery from obstruction. II. Electron microscopic alterations of acinar cells. J Ultrastruct Res. 1971;34(3):288–302.
88. Osailan SM, Proctor GB, McGurk M, Paterson KL. Intraoral duct ligation without inclusion of the parasympathetic nerve supply induces rat submandibular gland atrophy. Int J Exp Pathol. 2006;87(1):41–8.
89. Norberg LE, Abok K, Lundquist PG. Effects of ligation and irradiation on the submaxillary glands in rats. Acta Otolaryngol. 1988;105(1–2):181–92.
90. Cotroneo E, Proctor GB, Paterson KL, Carpenter GH. Early markers of regeneration following ductal ligation in rat submandibular gland. Cell Tissue Res. 2008;332(2):227–35.
91. Dijkema T, Terhaard CH, Roesink JM, Raaijmakers CP, van den Keijbus PA, Brand HS, et al. MUC5B levels in submandibular gland saliva of patients

treated with radiotherapy for head-and-neck cancer: a pilot study. Radiat Oncol. 2012;7:91.
92. Carpenter GH, Khosravani N, Ekstrom J, Osailan SM, Paterson KP, Proctor GB. Altered plasticity of the parasympathetic innervation in the recovering rat submandibular gland following extensive atrophy. Exp Physiol. 2009;94(2):213–9.
93. Walker NI, Gobe GC. Cell death and cell proliferation during atrophy of the rat parotid gland induced by duct obstruction. J Pathol. 1987;153(4):333–44.
94. Takahashi S, Nakamura S, Suzuki R, Islam N, Domon T, Yamamoto T, et al. Apoptosis and mitosis of parenchymal cells in the duct-ligated rat submandibular gland. Tissue Cell. 2000;32(6):457–63.
95. Radfar L, Sirois DA. Structural and functional injury in minipig salivary glands following fractionated exposure to 70 Gy of ionizing radiation: an animal model for human radiation-induced salivary gland injury. Oral Surg Oral Med Oral Pathol Oral Radiol Endod. 2003;96(3):267–74.
96. Silver N, Proctor GB, Arno M, Carpenter GH. Activation of mTOR coincides with autophagy during ligation-induced atrophy in the rat submandibular gland. Cell Death Dis. 2010;1:e14.
97. Lombaert IM, Brunsting JF, Wierenga PK, Faber H, Stokman MA, Kok T, et al. Rescue of salivary gland function after stem cell transplantation in irradiated glands. PLoS One. 2008;3(4):e2063.
98. Jeong J, Baek H, Kim YJ, Choi Y, Lee H, Lee E, et al. Human salivary gland stem cells ameliorate hyposalivation of radiation-damaged rat salivary glands. Exp Mol Med. 2013;45:e58.
99. Lim JY, Yi T, Choi JS, Jang YH, Lee S, Kim HJ, et al. Intraglandular transplantation of bone marrow-derived clonal mesenchymal stem cells for amelioration of post-irradiation salivary gland damage. Oral Oncol. 2013;49(2):136–43.
100. Lim JY, Ra JC, Shin IS, Jang YH, An HY, Choi JS, et al. Systemic transplantation of human adipose tissue-derived mesenchymal stem cells for the regeneration of irradiation-induced salivary gland damage. PLoS One. 2013;8(8):e71167.
101. Glucksmann A. The effect of radiation on the production of mucin in salivary glands and the female genital tract of rodents. Ann N Y Acad Sci. 1963;106:631–5.
102. Tamarin A. The leukocytic response in ligated rat submandibular glands. J Oral Pathol. 1979;8(5):293–304.
103. Harrison JD, Fouad HM, Garrett JR. The effects of ductal obstruction on the acinar cells of the parotid of cat. Arch Oral Biol. 2000;45(11):945–9.
104. Hoebers FJ, Kartachova M, de Bois J, van den Brekel MW, van Tinteren H, van Herk M, et al. 99mTc Hynic-rh-Annexin V scintigraphy for in vivo imaging of apoptosis in patients with head and neck cancer treated with chemoradiotherapy. Eur J Nucl Med Mol Imaging. 2008;35(3):509–18.
105. Hall EGA. Radiobiology for the radiologist. 7th ed. Philadelphia: Lippincott Williams & Wilkins; 2011.
106. Li Y, Taylor JM, Ten Haken RK, Eisbruch A. The impact of dose on parotid salivary recovery in head and neck cancer patients treated with radiation therapy. Int J Radiat Oncol Biol Phys. 2007;67(3):660–9.
107. Pramanik R, Osailan SM, Challacombe SJ, Urquhart D, Proctor GB. Protein and mucin retention on oral mucosal surfaces in dry mouth patients. Eur J Oral Sci. 2010;118(3):245–53.
108. Ewert P, Aguilera S, Alliende C, Kwon YJ, Albornoz A, Molina C, et al. Disruption of tight junction structure in salivary glands from Sjogren's syndrome patients is linked to proinflammatory cytokine exposure. Arthritis Rheum. 2010;62(5):1280–9.
109. Barrera MJ, Sanchez M, Aguilera S, Alliende C, Bahamondes V, Molina C, et al. Aberrant localization of fusion receptors involved in regulated exocytosis in salivary glands of Sjogren's syndrome patients is linked to ectopic mucin secretion. J Autoimmun. 2012;39(1–2):83–92.
110. Barrera MJ, Aguilera S, Veerman E, Quest AF, Diaz-Jimenez D, Urzua U, et al. Salivary mucins induce a Toll-like receptor 4-mediated pro-inflammatory response in human submandibular salivary cells: are mucins involved in Sjogren's syndrome? Rheumatology (Oxford). 2015;54(8):1518–27.
111. Ogawa M, Tsuji T. Functional salivary gland regeneration as the next generation of organ replacement regenerative therapy. Odontology. 2015;103(3):248–57.
112. Martinez JR, Bylund DB, Cassity N. Progressive secretory dysfunction in the rat submandibular gland after excretory duct ligation. Arch Oral Biol. 1982;27(6):443–50.
113. Tsuda T, Hamamori Y, Fukumoto Y, Kaibuchi K, Takai Y. Epidermal growth factor increases c-myc mRNA without eliciting phosphoinositide turnover, protein kinase C activation, or calcium ion mobilization in Swiss 3 T3 fibroblasts. J Biochem. 1986;100(6):1631–5.
114. Smaje LH. Spontaneous secretion in the rabbit submaxillary gland. In: Thorn NA, Petersen OH, editors. Secretory mechanisms of exocrine glands. Copenhagen: Munksgaard; 1974. p. 608–22.
115. Harrison JD, Garrett JR. The effects of ductal ligation on the parenchyma of salivary glands of cat studied by enzyme histochemical methods. Histochem J. 1976;8(1):35–44.
116. Cummins M, Dardick I, Brown D, Burford-Mason A. Obstructive sialadenitis: a rat model. J Otolaryngol. 1994;23(1):50–6.
117. Garrett JR. The proper role of nerves in salivary secretion: a review. J Dent Res. 1987;66(2):387–97.
118. Ramos-Casals M, Font J. Primary Sjogren's syndrome: current and emergent aetiopathogenic concepts. Rheumatology (Oxford). 2005;44(11):1354–67.
119. Mavragani CP, Moutsopoulos HM. Sjogren syndrome. CMAJ. 2014;186(15):E579–86.
120. Anaya JM, Talal N. Sjogren's syndrome comes of age. Semin Arthritis Rheum. 1999;28(6):355–9.

121. Dawson LJ, Stanbury J, Venn N, Hasdimir B, Rogers SN, Smith PM. Antimuscarinic antibodies in primary Sjogren's syndrome reversibly inhibit the mechanism of fluid secretion by human submandibular salivary acinar cells. Arthritis Rheum. 2006;54(4):1165–73.
122. Fox RI, Kang HI, Ando D, Abrams J, Pisa E. Cytokine mRNA expression in salivary gland biopsies of Sjogren's syndrome. J Immunol. 1994;152(11):5532–9.
123. Konttinen YT, Sorsa T, Hukkanen M, Segerberg M, Kuhlefelt-Sundstrom M, Malmstrom M, et al. Topology of innervation of labial salivary glands by protein gene product 9.5 and synaptophysin immunoreactive nerves in patients with Sjogren's syndrome. J Rheumatol. 1992;19(1):30–7.
124. Fox RI, Tornwall J, Michelson P. Current issues in the diagnosis and treatment of Sjogren's syndrome. Curr Opin Rheumatol. 1999;11(5):364–71.
125. Delporte C, Steinfeld S. Distribution and roles of aquaporins in salivary glands. Biochim Biophys Acta. 2006;1758(8):1061–70.
126. Dawson LJ, Fox PC, Smith PM. Sjogrens syndrome--the non-apoptotic model of glandular hypofunction. Rheumatology (Oxford). 2006;45(7):792–8.
127. Dawson L, Tobin A, Smith P, Gordon T. Antimuscarinic antibodies in Sjogren's syndrome: where are we, and where are we going? Arthritis Rheum. 2005;52(10):2984–95.
128. Goicovich E, Molina C, Perez P, Aguilera S, Fernandez J, Olea N, et al. Enhanced degradation of proteins of the basal lamina and stroma by matrix metalloproteinases from the salivary glands of Sjogren's syndrome patients: correlation with reduced structural integrity of acini and ducts. Arthritis Rheum. 2003;48(9):2573–84.
129. Molina C, Alliende C, Aguilera S, Kwon YJ, Leyton L, Martinez B, et al. Basal lamina disorganisation of the acini and ducts of labial salivary glands from patients with Sjogren's syndrome: association with mononuclear cell infiltration. Ann Rheum Dis. 2006;65(2):178–83.
130. Aumailley M, Smyth N. The role of laminins in basement membrane function. J Anat. 1998;193(Pt 1):1–21.
131. Giancotti FG, Ruoslahti E. Integrin signaling. Science. 1999;285(5430):1028–32.
132. Hoffman MP, Kibbey MC, Letterio JJ, Kleinman HK. Role of laminin-1 and TGF-beta 3 in acinar differentiation of a human submandibular gland cell line (HSG). J Cell Sci. 1996;109(Pt 8):2013–21.
133. da Costa SR, Wu K, Veigh MM, Pidgeon M, Ding C, Schechter JE, et al. Male NOD mouse external lacrimal glands exhibit profound changes in the exocytotic pathway early in postnatal development. Exp Eye Res. 2006;82(1):33–45.
134. Castle D. Cell biology of salivary protein secretion. In: Dobrosielski-Vergona K, editor. Biology of the salivary glands. Boca Raton: CRC Press Inc; 1993. p. 81–104.
135. Turner RJ. Ion transport related to fluid secretions in salivary glands. In: Dobrosielski-Vergona K, editor. Biology of the salivary glands. Boca Raton: CRC Press Inc; 1993. p. 105–27.
136. Saari H, Halinen S, Ganlov K, Sorsa T, Konttinen YT. Salivary mucous glycoprotein MG1 in Sjogren's syndrome. Clin Chim Acta. 1997;259(1–2):83–96.
137. Castro I, Sepulveda D, Cortes J, Quest AF, Barrera MJ, Bahamondes V, et al. Oral dryness in Sjogren's syndrome patients. Not just a question of water. Autoimmun Rev. 2013;12(5):567–74.
138. Yamada H, Nakagawa Y, Wakamatsu E, Sumida T, Yamachika S, Nomura Y, et al. Efficacy prediction of cevimeline in patients with Sjogren's syndrome. Clin Rheumatol. 2007;26(8):1320–7.
139. Castro I, Aguilera S, Brockhausen I, Alliende C, Quest AF, Molina C, et al. Decreased salivary sulphotransferase activity correlated with inflammation and autoimmunity parameters in Sjogren's syndrome patients. Rheumatology (Oxford). 2012;51(3):482–90.
140. Brockhausen I. Sulphotransferases acting on mucin-type oligosaccharides. Biochem Soc Trans. 2003;31(2):318–25.
141. Yang X, Lehotay M, Anastassiades T, Harrison M, Brockhausen I. The effect of TNF-alpha on glycosylation pathways in bovine synoviocytes. Biochem Cell Biol. 2004;82(5):559–68.
142. Corfield AP, Myerscough N, Bradfield N, Corfield Cdo A, Gough M, Clamp JR, et al. Colonic mucins in ulcerative colitis: evidence for loss of sulfation. Glycoconj J. 1996;13(5):809–22.
143. Yamori T, Kimura H, Stewart K, Ota DM, Cleary KR, Irimura T. Differential production of high molecular weight sulfated glycoproteins in normal colonic mucosa, primary colon carcinoma, and metastases. Cancer Res. 1987;47(10):2741–7.
144. Yamori T, Ota DM, Cleary KR, Hoff S, Hager LG, Irimura T. Monoclonal antibody against human colonic sulfomucin: immunochemical detection of its binding sites in colonic mucosa, colorectal primary carcinoma, and metastases. Cancer Res. 1989;49(4):887–94.
145. Salvini R, Bardoni A, Valli M, Trinchera M. beta1,3-Galactosyltransferase beta 3Gal-T5 acts on the GlcNAcbeta 1-->3Galbeta 1-->4GlcNAcbeta 1-->R sugar chains of carcinoembryonic antigen and other N-linked glycoproteins and is down-regulated in colon adenocarcinomas. J Biol Chem. 2001;276(5): 3564–73.
146. Seko A, Ohkura T, Kitamura H, Yonezawa S, Sato E, Yamashita K. Quantitative differences in GlcNAc:beta1-->3 and GlcNAc:beta1-->4 galactosyltransferase activities between human colonic adenocarcinomas and normal colonic mucosa. Cancer Res. 1996;56(15):3468–73.
147. Seko A, Nagata K, Yonezawa S, Yamashita K. Down-regulation of Gal 3-O-sulfotransferase-2 (Gal3ST-2) expression in human colonic non-mucinous adenocarcinoma. Jpn J Cancer Res. 2002;93(5):507–15.

148. Tzioufas AG, Kapsogeorgou EK, Moutsopoulos HM. Pathogenesis of Sjogren's syndrome: what we know and what we should learn. J Autoimmun. 2012;39(1–2):4–8.
149. Perez P, Aguilera S, Olea N, Alliende C, Molina C, Brito M, et al. Aberrant localization of ezrin correlates with salivary acini disorganization in Sjogren's Syndrome. Rheumatology (Oxford). 2010;49(5):915–23.
150. Barrera MJ, Bahamondes V, Sepulveda D, Quest AF, Castro I, Cortes J, et al. Sjogren's syndrome and the epithelial target: a comprehensive review. J Autoimmun. 2013;42:7–18.
151. Frazao JB, Errante PR, Condino-Neto A. Toll-like receptors' pathway disturbances are associated with increased susceptibility to infections in humans. Arch Immunol Ther Exp (Warsz). 2013;61(6):427–43.
152. Tarang S, Kumar S, Batra SK. Mucins and toll-like receptors: kith and kin in infection and cancer. Cancer Lett. 2012;321(2):110–9.
153. Gallo PM, Gallucci S. The dendritic cell response to classic, emerging, and homeostatic danger signals. Implications for autoimmunity. Front Immunol. 2013;4:138.
154. Brennan TV, Lin L, Huang X, Cardona DM, Li Z, Dredge K, et al. Heparan sulfate, an endogenous TLR4 agonist, promotes acute GVHD after allogeneic stem cell transplantation. Blood. 2012;120(14):2899–908.
155. Termeer C, Benedix F, Sleeman J, Fieber C, Voith U, Ahrens T, et al. Oligosaccharides of hyaluronan activate dendritic cells via toll-like receptor 4. J Exp Med. 2002;195(1):99–111.
156. Muroi M, Tanamoto K. The polysaccharide portion plays an indispensable role in salmonella lipopolysaccharide-induced activation of NF-kappaB through human toll-like receptor 4. Infect Immun. 2002;70(11):6043–7.
157. Lu YC, Yeh WC, Ohashi PS. LPS/TLR4 signal transduction pathway. Cytokine. 2008;42(2):145–51.
158. Jenkins KA, Mansell A. TIR-containing adaptors in Toll-like receptor signalling. Cytokine. 2010;49(3):237–44.
159. Kawakami A, Nakashima K, Tamai M, Nakamura H, Iwanaga N, Fujikawa K, et al. Toll-like receptor in salivary glands from patients with Sjogren's syndrome: functional analysis by human salivary gland cell line. J Rheumatol. 2007;34(5):1019–26.
160. Spachidou MP, Bourazopoulou E, Maratheftis CI, Kapsogeorgou EK, Moutsopoulos HM, Tzioufas AG, et al. Expression of functional Toll-like receptors by salivary gland epithelial cells: increased mRNA expression in cells derived from patients with primary Sjogren's syndrome. Clin Exp Immunol. 2007;147(3):497–503.
161. Horng T, Barton GM, Medzhitov R. TIRAP: an adapter molecule in the Toll signaling pathway. Nat Immunol. 2001;2(9):835–41.
162. Wreschner DH, Hareuveni M, Tsarfaty I, Smorodinsky N, Horev J, Zaretsky J, et al. Human epithelial tumor antigen cDNA sequences. Differential splicing may generate multiple protein forms. Eur J Biochem. 1990;189(3):463–73.
163. Herbert LM, Grosso JF, Dorsey Jr M, Fu T, Keydar I, Cejas MA, et al. A unique mucin immunoenhancing peptide with antitumor properties. Cancer Res. 2004;64(21):8077–84.
164. Ilkovitch D, Handel-Fernandez ME, Herbert LM, Lopez DM. Antitumor effects of Mucin 1/sec involves the modulation of urokinase-type plasminogen activator and signal transducer and activator of transcription 1 expression in tumor cells. Cancer Res. 2008;68(7):2427–35.
165. Grosso JF, Herbert LM, Owen JL, Lopez DM. MUC1/sec-expressing tumors are rejected in vivo by a T cell-dependent mechanism and secrete high levels of CCL2. J Immunol. 2004;173(3):1721–30.
166. Baruch A, Hartmann M, Yoeli M, Adereth Y, Greenstein S, Stadler Y, et al. The breast cancer-associated MUC1 gene generates both a receptor and its cognate binding protein. Cancer Res. 1999;59(7):1552–61.
167. Hartman M, Baruch A, Ron I, Aderet Y, Yoeli M, Sagi-Assif O, et al. MUC1 isoform specific monoclonal antibody 6E6/2 detects preferential expression of the novel MUC1/Y protein in breast and ovarian cancer. Int J Cancer. 1999;82(2):256–67.
168. Zrihan-Licht S, Vos HL, Baruch A, Elroy-Stein O, Sagiv D, Keydar I, et al. Characterization and molecular cloning of a novel MUC1 protein, devoid of tandem repeats, expressed in human breast cancer tissue. Eur J Biochem. 1994;224(2):787–95.
169. Cascio S, Zhang L, Finn OJ. MUC1 protein expression in tumor cells regulates transcription of proinflammatory cytokines by forming a complex with nuclear factor-kappaB p65 and binding to cytokine promoters: importance of extracellular domain. J Biol Chem. 2011;286(49):42248–56.
170. Yamada N, Kitamoto S, Yokoyama S, Hamada T, Goto M, Tsutsumida H, et al. Epigenetic regulation of mucin genes in human cancers. Clin Epigenetics. 2011;2(2):85–96.
171. Ilkovitch D, Carrio R, Lopez DM. Mechanisms of antitumor and immune-enhancing activities of MUC1/sec, a secreted form of mucin-1. Immunol Res. 2013;57(1–3):70–80.
172. Choi S, Park YS, Koga T, Treloar A, Kim KC. TNF-alpha is a key regulator of MUC1, an anti-inflammatory molecule, during airway Pseudomonas aeruginosa infection. Am J Respir Cell Mol Biol. 2011;44(2):255–60.
173. Sheng YH, Triyana S, Wang R, Das I, Gerloff K, Florin TH, et al. MUC1 and MUC13 differentially regulate epithelial inflammation in response to inflammatory and infectious stimuli. Mucosal Immunol. 2013;6(3):557–68.
174. Yen JH, Xu S, Park YS, Ganea D, Kim KC. Higher susceptibility to experimental autoimmune encephalomyelitis in Muc1-deficient mice is associated with increased Th1/Th17 responses. Brain Behav Immun. 2013;29:70–81.

175. Kufe DW. Mucins in cancer: function, prognosis and therapy. Nat Rev Cancer. 2009;9(12):874–85.
176. Yonezawa S, Goto M, Yamada N, Higashi M, Nomoto M. Expression profiles of MUC1, MUC2, and MUC4 mucins in human neoplasms and their relationship with biological behavior. Proteomics. 2008;8(16):3329–41.
177. Nath S, Mukherjee P. MUC1: a multifaceted oncoprotein with a key role in cancer progression. Trends Mol Med. 2014;20(6):332–42.
178. Soares AB, Demasi AP, Altemani A, de Araujo VC. Increased mucin 1 expression in recurrence and malignant transformation of salivary gland pleomorphic adenoma. Histopathology. 2011;58(3):377–82.
179. Yakirevich E, Sabo E, Klorin G, Alos L, Cardesa A, Ellis GL, et al. Primary mucin-producing tumours of the salivary glands: a clinicopathological and morphometric study. Histopathology. 2010;57(3):395–409.
180. Brieger J, Duesterhoeft A, Brochhausen C, Gosepath J, Kirkpatrick CJ, Mann WJ. Recurrence of pleomorphic adenoma of the parotid gland--predictive value of cadherin-11 and fascin. APMIS. 2008;116(12):1050–7.
181. Ferreira JC, Morais MO, Elias MR, Batista AC, Leles CR, Mendonca EF. Pleomorphic adenoma of oral minor salivary glands: an investigation of its neoplastic potential based on apoptosis, mucosecretory activity and cellular proliferation. Arch Oral Biol. 2014;59(6):578–85.
182. Gum Jr JR, Hicks JW, Toribara NW, Siddiki B, Kim YS. Molecular cloning of human intestinal mucin (MUC2) cDNA. Identification of the amino terminus and overall sequence similarity to prepro-von Willebrand factor. J Biol Chem. 1994;269(4):2440–6.
183. Yamada K, Tanaka T, Mori M, Tsubura A, Morii S, Tsubone M, et al. Immunohistochemical expression of MAM-3 and MAM-6 antigens in salivary gland tumours. Virchows Arch A Pathol Anat Histopathol. 1989;415(6):509–21.
184. Siyi L, Shengwen L, Min R, Wenjun Y, Lizheng W, Chenping Z. Increased expression of MUC-1 has close relation with patient survivor in high-grade salivary gland mucoepidermoid carcinoma. J Oral Pathol Med. 2014;43(8):579–84.
185. Handra-Luca A, Lamas G, Bertrand JC, Fouret P. MUC1, MUC2, MUC4, and MUC5AC expression in salivary gland mucoepidermoid carcinoma: diagnostic and prognostic implications. Am J Surg Pathol. 2005;29(7):881–9.
186. Rakha EA, Boyce RW, Abd El-Rehim D, Kurien T, Green AR, Paish EC, et al. Expression of mucins (MUC1, MUC2, MUC3, MUC4, MUC5AC and MUC6) and their prognostic significance in human breast cancer. Mod Pathol. 2005;18(10):1295–304.
187. Velcich A, Yang W, Heyer J, Fragale A, Nicholas C, Viani S, et al. Colorectal cancer in mice genetically deficient in the mucin Muc2. Science. 2002;295(5560):1726–9.

Part II

Glandular Damage and Cell Replacement Therapy

4 Histologic Changes in the Salivary Glands Following Radiation Therapy

Robert S. Redman

Abstract

Therapeutic radiation for cancer of the head and neck damages salivary glands that are situated between the radiation source and the target tumor and its metastases. With moderate to high radiation exposure, salivary glands are devastated and regeneration is limited. The resulting severe reduction in saliva has detrimental effects on the teeth and oral mucosa. The purpose of this review is to describe some of the salient histologic features of salivary gland structures and cells, how these are functionally related to salivary production, and thus how radiation-induced loss and functional impairment of each type of structure may contribute to reduced quantity and quality of saliva.

4.1 Introduction

Ionizing radiation, often augmented with administration of chemotherapeutic agents, is a staple of management of cancers of the head and neck region that are not amenable to successful treatment by surgery alone. The jawbones and salivary glands often suffer moderate to severe damage, the latter resulting in dramatically decreased salivary function [23, 73, 83, 100, 101]. Without diligent professional and home care, rapidly progressive dental caries can ensue, leading to a cascade of infection and extractions with significant risk of osteoradionecrosis. In this chapter, I describe the histology of radiation-induced damage to the salivary glands in the context of how this affects salivary function. I begin with a review of the *functional morphology* of normal, mature salivary glands, i.e., their anatomical, light microscopic, and ultrastructural features, as these relate to the principal functions of the several parenchymal (epithelial) cell types. This is intended to facilitate understanding how radiation-induced physical loss or functional impairment of each cell type affects the quantity and quality of saliva the gland can produce. The review utilizes human and rodent glands as models.

R.S. Redman, DDS
Chief, Oral Pathology Research Laboratory, Department of Veterans Affairs Medical Center, Washington, DC 20422; Adjunct Professor, University of Maryland School of Dentistry, Baltimore, MD; Adjunct Scientist, National Institute of Dental and Craniofacial Research, National Institutes of Health, Bethesda, MD, USA
e-mail: Robert.Redman@med.va.gov

S. Cha (ed.), *Salivary Gland Development and Regeneration*, DOI 10.1007/978-3-319-43513-8_4

4.2 Salivary Gland Nomenclature

Salivary glands are classified by size (major and minor) and secretory product (serous, watery and slightly slippery; mucous, thick, viscous, and very slippery; and mixed, having both serous and mucous secretory units). They are named by location. The major glands are the parotid (*in front of the ear*), sublingual (*under the tongue*), and submandibular (*under the mandible*). In humans, the sublingual and submandibular glands are separate; in rodents, they are enclosed in the same capsule. Human minor salivary glands are located in groups in the buccal and labial mucosae, the soft and posterior hard palate including the tonsillar pillars, the posterior dorsal and anterior ventral surfaces of the tongue, and the anterior floor of the mouth (minor sublingual glands). A few also are located subjacent to the incisive papilla. All are either mucous or mixed, mostly mucous glands except for the serous lingual glands of von Ebner, which secrete into the foliate furrows and the troughs around the vallate papillae. The distribution of minor salivary glands in rats and mice is similar to that of humans except that there is none in the labial mucosa, hard palate, or anterior ventral surface of the tongue. Drawings showing the location and shapes of human major glands can be found in standard textbooks of human anatomy, e.g., Gray's Anatomy [103]. I have published artists' drawings showing the distribution of human palatal glands [70] and all of the minor salivary glands of the rat [69, 74].

4.3 Histology and Ultrastructure

The terminology used here is guided by that of Young and van Lennep [106]. The three-dimensional structure of salivary glands is like that of a bunch of grapes [47], with the grapes being the secretory endpieces (acini) and the "stems" (the pedicels, rachis, branches, and peduncle) being the ducts in order of increasing size (which roughly correlates with distance from the acini), terminating in the main duct.

4.3.1 Illustrations of Normal and Irradiated Glands

Light microscopic descriptions and illustrations are of formaldchyde-fixed, paraffin-embedded, mature rat (Figs. 4.1 and 4.5) and human (Figs. 4.2, 4.6 and 4.7) salivary glands. For transmission electron microscopy (*TEM*, Figs. 4.3 and 4.4), rat salivary glands were fixed in glutaraldehyde, post-fixed in 0S04, embedded in epoxy resin, sectioned at ca. 70 nm, and stained with uranyl acetate and lead citrate. Figure 4.1e is a photomicrograph of a tissue fixed and embedded for TEM but sectioned at 1 μm, mounted on a glass slide, and stained with methylene blue and azure II. All illustrations are of material from previous studies that had been reviewed by the appropriate institutional review bodies.

4.3.2 Acini

The acini provide most of the proteins and fluid of saliva which moisten and initiate digestion (enzymes), lubricate the oral mucosa, teeth, and food (mucins), maintain the minerals of the teeth (statherin), and modulate the oral flora (peroxidase, histatins) [reviewed by Izutsu [33], Tabak [86], and Redman [73]]. Secretory proteins are glycosylated, and the amount and type of sugars attached to the core protein determine the nature of the secretory product. In serous acini, the glycoproteins are lightly glycosylated, carry a predominantly neutral charge, and have a molecular weight of less than 100,000 KD. Mucous acini produce mucins, high molecular weight (1–10 million KD) glycoproteins that are heavily glycosylated with both neutral and acidic sugars. The seromucous acini of rat submandibular glands produce a mucin of 100,000 KD. The greater amount and negative charge of the sugars of mucins attract more water, which is a major reason why mucous secretions are much more viscous and slippery than serous secretions.

By *light microscopy* of sections stained with hematoxylin and eosin (H&E), serous acini take both dyes moderately, resulting in violet to blue-violet staining (Fig. 4.2a, b). Mucous acini

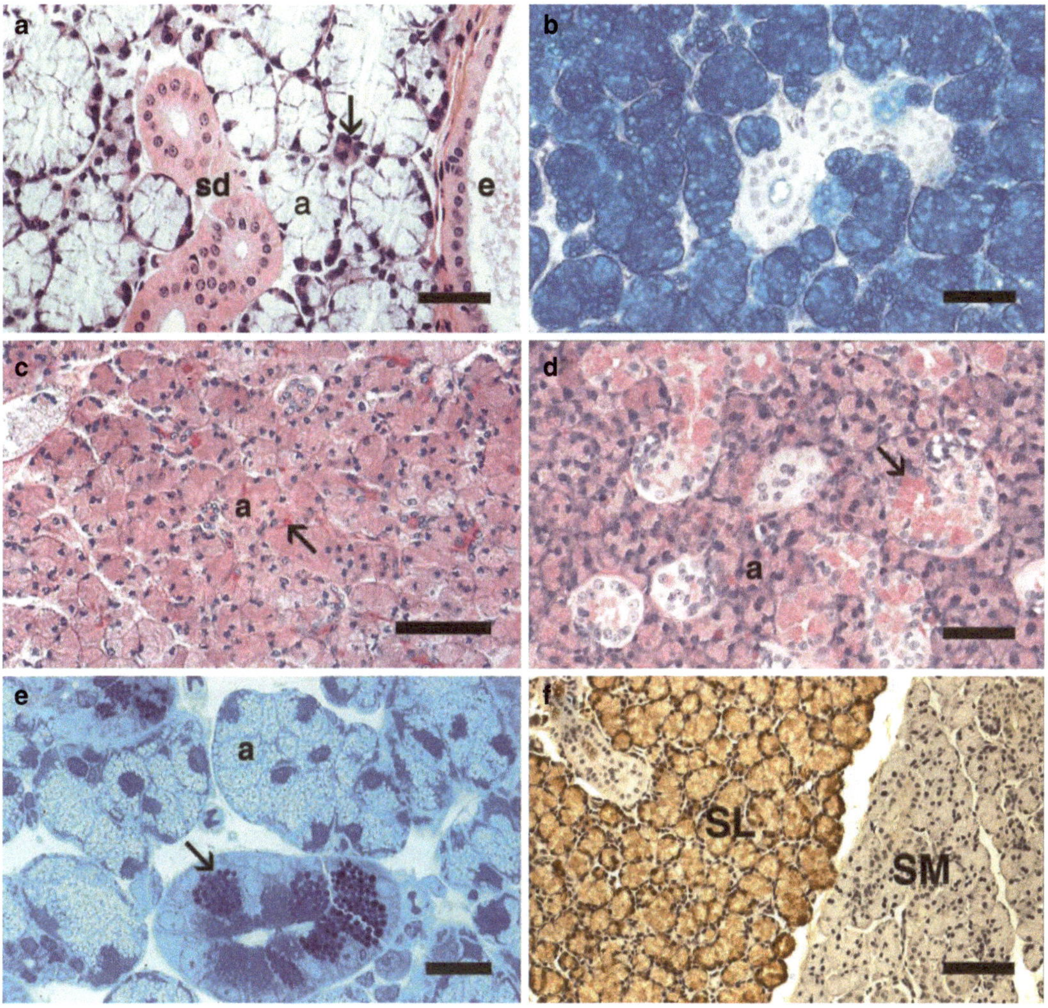

Fig. 4.1 Photomicrographs of mature rat salivary glands. (**a**) SL, H&E. The mucous acini are capped with small serous demilunes (*arrow*). The pink (mainly eosin) tall columnar cells of the striated ducts have prominent radial striations and a single row of centrally placed nuclei. In the first order excretory duct, the cytoplasm of the short columnar cells also is pink, and there are occasional basal cells. (**b**) SL, AB-H. The mucous acini are heavily stained, and the striated ducts and serous demilunes are unstained by AB. (**c**) P, PAS-H. The secretory granules of the serous acini and first-order intercalated ducts (*arrow*) are PAS + (light and moderate magenta, respectively). (**d**) SM, PAS-H. The secretory granules of the seromucous acini (*a*) and serous granular convoluted tubules (*arrow*) are moderate and light magenta, respectively. (**e**) SM, semi-thin (1 μm) epoxy resin section, methylene blue-azure II stain. The contents of the secretory granules of the seromucous acini are empty looking, similar to mucous acini. Those of the granular convoluted tubules (*arrow*) stain blue with varying density, typical of serous granules. (**f**) SL and SM, immunohistochemical localization of Muc 19, hematoxylin counterstain. Only the mucous secretory granules of the sublingual gland reacted with the antibody (brown precipitate). Magnification bars: **a**, **b**, **d** = 50 μm; **c**, 200 μm; **e**, 30 μm; **f**, 100 μm. *Abbreviations for all Figures. P*, *SL*, *SM*, parotid, sublingual, and submandibular glands; *a* acini; *id*, *sd*, *ed*, intercalated, striated, and excretory ducts; *m*, mitochondria; *n*, nerve; *v*, blood vessel. Stains: *AB*, alcian blue; *E*, eosin; *H*, hematoxylin; *PAS*, periodic acid-Schiff.

stain very lightly, giving a pale, cloudy appearance (Fig. 4.1a). Stains that are commonly employed to distinguish between the glycoproteins of serous and mucous acini include alcian blue (AB, Figs. 4.1b and 4.2d) [57] and mucicarmine [43] (Fig. 4.6), for acidic glycoproteins, and periodic acid-Schiff (PAS) for mainly neutral glycoproteins (Figs. 4.1c, d and 4.2d). Serous acini stain a light

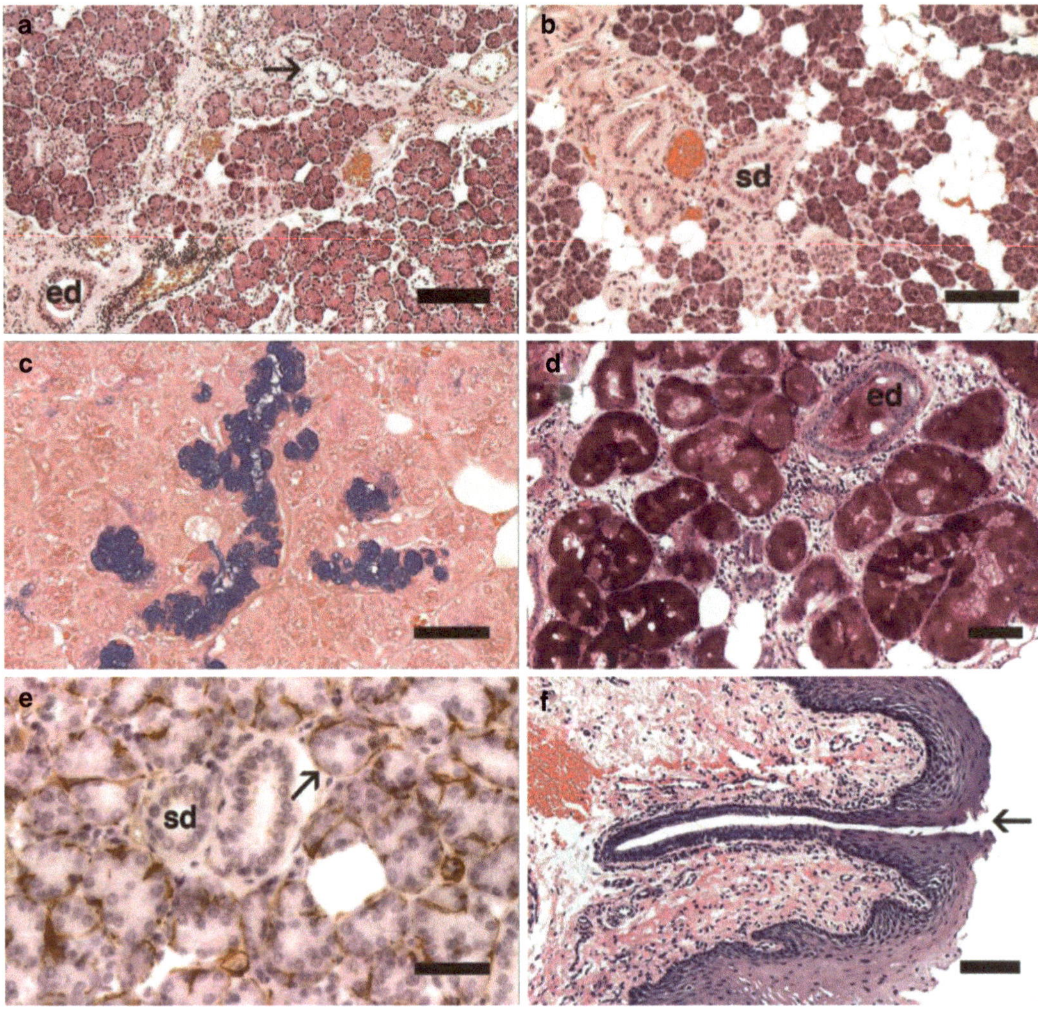

Fig. 4.2 Photomicrographs of human salivary glands. (**a**) SM, (**b**) P, both H&E. Serous acini and demilunes are violet, and mucous acini (*arrow*) are very lightly stained. Empty-looking spaces in (**b**) are adipose cells. (**c**) SM, AB–E. Mucous acini stain strongly with AB; serous acini, and demilunes not at all with AB. (**d**) Palatal (*minor*) gland, PAS-H. The mucous acini and secretory product in the lumen of an excretory duct stained a rich magenta color with PAS. (**e**) P, immunohistochemical localization of smooth muscle actin (SMA). Myofibrils of myoepithelial cells (brown precipitate, *arrow*) invest the acini but not the striated ducts. Small circles center right and bottom mark myofibrils of smooth muscle cells around the small arteries, a built-in positive control. (**f**) Buccal (*minor*) gland, H&E. This shows the transition from pseudostratified columnar to stratified squamous epithelium as the main duct approaches the orifice (*arrow*) to the oral mucosa. Magnification bars: **a**, **f**, 150 μm; **b**, 100 μm; **c**–**e**, 50 μm.

magenta hue with PAS and not at all with AB and mucicarmine. Mucous acini stain an intense magenta with PAS, blue with AB, and pink with mucicarmine. Secretory enzymes, e.g., α-amylase [104] (Fig. 4.6), carbonic anhydrase [39], and salivary peroxidase [77], histatins [3], and specific mucins, e.g. [15] (Fig. 4.1f), are a few of the many secretory products of acini that can be identified by immunohistochemistry (IHC).

Acini have a number of enzymes and structures that work together to move fluid from the interstitial tissue into the lumen. These include Na+/K+−adenosine triphosphatase, Na+/H+ exchangers 1, 2, and 3, and anion exchanger 2, and Na+/K+/2Cl- cotransporter [31], aquaporins [38], and numerous mitochondria (Fig. 4.3b) commensurate with the energy required.

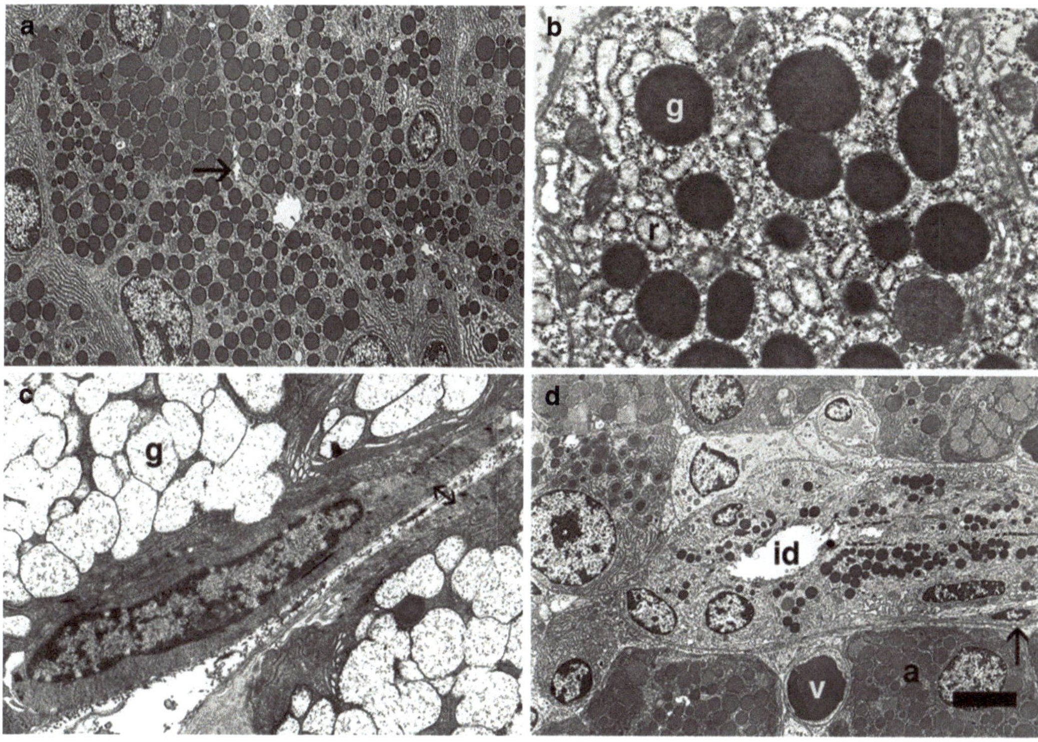

Fig. 4.3 Transmission electron micrographs (TEM) of rat salivary glands. (**a**, **b**) P acini. In (**a**), electron-dense secretory granules occupy the apical three fourths and extensive rough endoplasmic reticulum the basal fourth of the acinar cells. Arrow marks an intercellular extension of the lumen, or secretory canaliculus. In (**b**), rough endoplasmic reticulum (*r*) is rimmed with ribosomes (*dots* on membrane walls), and the lumina are engorged with nascent secretory proteins. (**c**) SL (Reproduced from Redman and Ball [75]). The electron-lucent secretory granules of the mucous acini have a finely particulate substructure. Myoepithelial cells with processes (*double arrow*) filled with bands of myofilaments bunched in dense bodies (*dark spots*) are attached to the acini. One has a typically fusiform (*cigar-shaped*) nucleus. (**d**) First-order (*juxta-acinar*) intercalated duct (*id*). The secretory granules are also electron dense but smaller and more homogeneous than those of the acini (*a*). A myoepithelial cell (*arrow*) lies parallel to the long axis of the duct. The lumen of a small blood vessel (*v*) harbors an erythrocyte. Reference scale bar: **a**, **d** = 4.0 μm, **b** = 0.7 μm, **c** = 1.2 μm

By transmission electron microscopy (TEM), acinar cells are factories for the production, storage, and secretion of secretory proteins. To this end, they have long segments of rough endoplasmic reticulum for protein synthesis, a prominent Golgi apparatus for glycosylation of the nascent proteins, and secretory granules (membrane-bounded sacs) for storage of the secretory glycoproteins between meals (Fig. 4.3). Serous granules may be uniformly electron dense or have patterns of varying density. Mucous granules are electron-lucent with a flocculent substructure of specks or filaments that may be uniformly spaced (monophasic, Fig. 4.3c) or have discrete areas of more or less closely packed substructure (biphasic), e.g. [74]. A feature of acini that is not seen in ducts is intercellular extensions of the lumen called secretory canaliculi [106] (Fig. 4.3a). Acini generally have a dual sympathetic innervation, with cholinergic nerves stimulating mostly water secretion and adrenergic nerves stimulating mostly proteins [24, 29, 107].

4.3.3 Intercalated Ducts

Intercalated ducts (Fig. 4.3d) connect acini with larger ducts in rat and human parotid and submandibular glands.

By *light microscopy*, they are composed of cuboidal cells with fusiform (cigar-shaped) nuclei and, in cross sections, have a smaller diameter than acini and other ducts. The primary (juxta-acinar) segment has small serous secretory granules that stain moderately to strongly with PAS (Fig. 4.1c) but not with any stain for mucins in the major human and rat salivary glands. By *TEM*, the granules are uniformly electron dense in rat parotid glands (Fig. 4.3d). Alpha-amylase, deoxyribonuclease I, and salivary peroxidase have been immunohistochemically localized to the acinar, but not intercalated duct, secretory granules of rat parotid glands [76, 104]. In the rat submandibular gland, by IHC and several histologic stains, the granules are identical to the Types I and III granules in the transitional acini of the neonatal rat [55]. The secondary segment has no secretory granules and a paucity of organelles, thus appearing to be relatively undifferentiated. For this and other reasons, the intercalated duct (and the basal cells of the main excretory duct) long have been regarded as sources of progenitor cells in mature salivary glands [18]. There is favorable evidence for this in mouse [41] and rat [48] submandibular glands, but not in rat parotid gland [72].

Intercalated ducts also may contribute fluid to saliva, though micropuncture studies did not separate the acinar and intercalated duct sources [51]. Some support for a fluid transport rolc lics in the histochemical localization of aquaporin 5 in both acini and intercalated ducts in mouse submandibular gland [4, 38].

4.3.4 Striated Ducts

By *light microscopy* of H&E-stained sections, the smaller striated ducts are composed of a single layer of tall columnar cells with centrally located, round nuclei and pink cytoplasm (Fig. 4.1c). By *TEM*, there are cells with less and more electron-dense cytoplasm called light and dark cells. The cells have deeply infolded and interdigitated basolateral cell membranes accompanied by numerous large, long mitochondria [89, 92] (Fig. 4.4a, b). By light microscopy these appear as radial lines or *striations*, for which these ducts are named. A principal function of the striated ducts is ion exchange with the luminal fluid, Na+ and Cl- being resorbed and HCO3- being luminally transported [33]. These functions are facilitated by the large cell surface areas provided by the basolateral infoldings (Fig. 4.4a, b) and, during active secretion, apical blebbing [54] (Fig. 4.4c) and activity of pinocytotic vesicles. Most of the arteries in salivary glands run parallel with the ducts countercurrent to salivary flow (distally) to the acini, and the veins retrace this route [106]. The presence of a rich vasculature in the surrounding stroma facilitates the clearance of resorbed ions from interstitial tissue along the base of the striated ducts (Fig. 4.4a, b).

Striated duct cells have small, apically located secretory vesicles in rat parotid [30] and human [92] submandibular salivary glands. In rodents such as hamster, mouse, and rat, a secretory duct is interposed between the intercalated and striated ducts (Fig. 4.1e). The secretory granules of these *granular convoluted tubules* store and secrete proteases and bioactive peptides such as epidermal, fibroblast, insulin-like, and nerve growth factors (EGF, FGF, I-LGF, NGF) [26, 62]. These tubules show a sexual dimorphism, being androgen-dependent and thus proportionately much more extensive in males.

In human parotid and submandibular glands, the secretory granules of acini and vesicles in striated ducts also are immunoreactive to EGF [14, 37]. The vesicles are located in both the apical and basal cytoplasm of the striated ducts, suggesting that they secrete not only into the lumen but also into the circulation. Most human salivary EGF is of parotid gland origin and does not differ by gender [12, 97].

In both humans and rodents, EGF has been shown to be secreted in response to oral, esophageal, and gastroduodenal ulceration or injury and to aid in healing by stimulating cellular proliferation [35, 36, 44, 49, 62, 63, 65, 95].

4.3.5 Excretory Ducts

Excretory ducts follow striated ducts and connect the glandular lobules and lobes to the oral cavity via the main excretory duct.

By *light microscopy*, excretory ducts have pink cytoplasm in H&E-stained sections (Fig. 4.1a), and the main excretory duct transitions

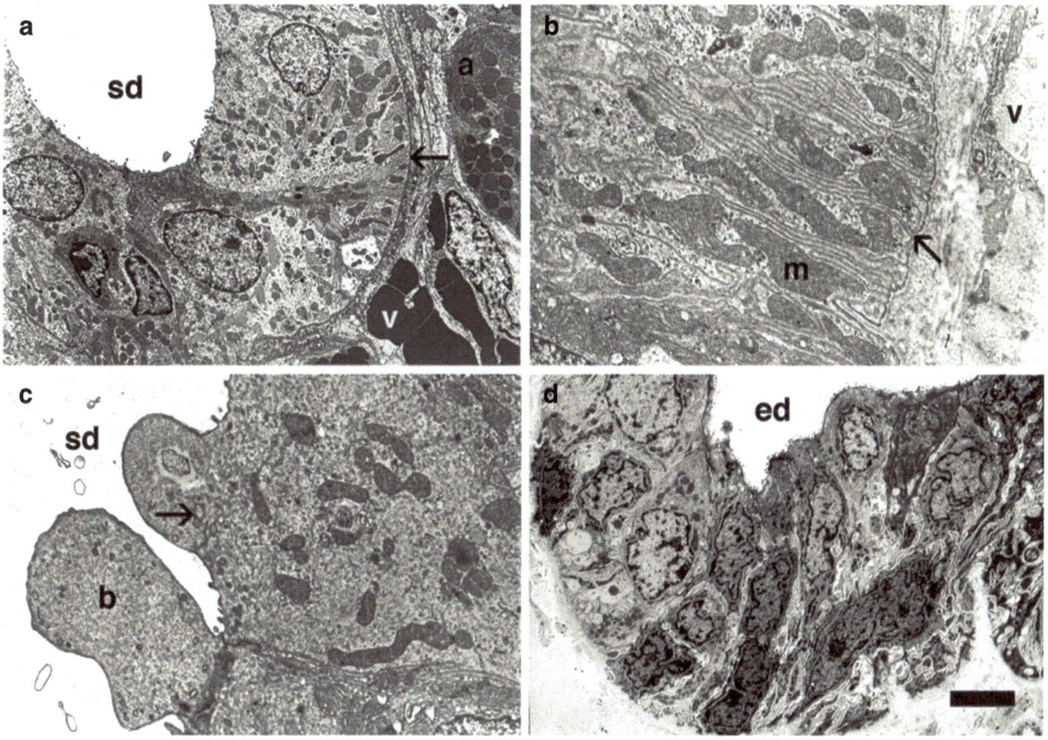

Fig. 4.4 TEM of large ducts in rat parotid glands. (**a**) Striated ducts (*sd*). Large, oblong mitochondria run parallel to the deeply infolded and interdigitated basal cell membranes (*arrows*). Together these make up the radial striations seen by light microscopy. Large, thin-walled veins (*v*) are seen in the supporting stroma. (**c**) This striated duct exhibits apical blebbing (*b*), or transient, focal expansions of the cytoplasm into the lumen, and pinocytotic vesicles (*arrow*). (**d**) Large excretory duct (*ed*). Light, dark, and basal cells are illustrated. Reference scale bar: **a**, **d** = 5.9 μm; **b**, **c** = 1.6 μm

from pseudostratified columnar to stratified squamous epithelium as it merges with the oral epithelium (Fig. 4.2f). Some of the excretory ducts have short columnar cells and occasional basal cells (Fig. 4.1a). Tandler [92] regards these as part of the striated duct population, as some of the columnar cells have infolded basal cell membranes and large mitochondria.

By *TEM*, the main and other large excretory ducts of rodents and human major salivary glands are composed of light, dark, and tuft columnar cells, basal cells (Fig. 4.4d), and scattered mucous ("goblet") and ciliated cells [82, 92, 93, 96, 106]. The tuft cells (not illustrated) have groups of long microvilli projecting into the lumen and secretory granules and vesicles, suggesting reception and secretory functions [79]. Though the columnar cells have many mitochondria, basal membrane infoldings are not prominent. Micropuncture studies have shown that the main excretory duct of rat submandibular, but not parotid, gland resorbs Na+ and K+ from the luminal fluid [51, 105]. The basal cells are dark (relatively electron dense) because of many cytokeratin filaments attached to multiple hemidesmosomes in the basal plasmalemma [79, 93].

The concentrations of the ions in saliva are dependent on the flow rate [81]. In resting (unstimulated) saliva, the concentrations of Na+, Cl-, and HCO3- are low and increase with increasing flow rate, as the resorption of Na+ and Cl- by striated and excretory ducts is unable to keep pace with the increased volume, and HCO3- is increasingly transported into the lumen in exchange for the other ions. On the other hand, the concentration of salivary Ca++ drops with increasing flow rate. A mechanism for perception and resorption of Ca++ to maintain Ca++ within physiological limits in the lumina of the larger ducts of salivary glands has recently been described [6].

4.3.6 Myoepithelium

In human and rat major salivary glands, there are cells shaped like starfish which are situated between the basal lamina and the acini and intercalated but not striated or excretory ducts [59, 71, 88, 91]. In mouse and rat parotid glands, these cells invest only the intercalated ducts. Their processes form a basket-like net around the acini (Fig. 4.2e) and are arranged spirally along the long axis of the intercalated ducts (Fig. 4.3d). The nuclei of myoepithelial cells investing acini are disk shaped [59] but often are sectioned such that they appear as fusiform, or cigar-shaped (Fig. 4.3c), and in myoepithelial cells investing ducts, all are fusiform (Fig. 4.3d). Myoepithelial cells are attached to acini and ducts by desmosomes. The cell processes have bundles of filaments resembling the myofibrils of smooth muscle, as they are interspersed with dense bodies and anchored to the basement membrane with attachment plaques [75]. Though these myofibrils can be labeled via IHC for smooth muscle actin (Fig. 4.2e), the intermediate filaments are made of cytokeratins [56]. Thus, they have features of both epithelial and smooth muscle cells, which is why they are named *myoepithelial cells*. When the gland is stimulated to secrete, the myoepithelial cell processes contract rhythmically in tandem with the terminal web to help expel the secreted fluid out of the acini and through the ducts [71, 90].

4.3.7 Examples of Other Salivary Gland Functional Proteins

There are many other structural and secretory proteins in salivary glands that can be identified by IHC. Though a thorough list is beyond the scope of this review, a few examples follow. Acini, myoepithelium, ducts, and cell types within ducts have characteristic cytoskeletal and membrane proteins by which they can be distinguished via IHC. For example, only myoepithelium (and smooth muscle in blood vessels) labels with antibodies to smooth muscle actin (SMA, [27], Fig. 4.2e); only the basal cells of large ducts, with CK-13 and CK-16 [16]; and both myoepithelium and duct basal cells, with CK-14 [16]. All duct cells, but not acini or myoepithelium, label with CK-5 [9] and CK-19 [25] in human salivary glands and CK-5 in rat salivary glands [41]. Antiepithelial membrane antigen (EMA) labels the luminal membranes of acini and intercalated and striated ducts [94].

4.4 Salivary Gland Damage from Ionizing Radiation

4.4.1 Principles

Ionizing radiation (IR) gives up energy when it is slowed, redirected, or stopped by molecules in tissues. The released energy gives rise to free radicals and, if the initial energy is high enough, secondary radiation. The energy of the radiation determines the depth in tissues where most of it interacts. Therapeutic radiation for cancers of the head and neck is mainly electromagnetic beams such as X-ray and gamma ray. To achieve interaction with the target tissues and to minimize damage to normal tissues, the external radiation is directed through portals from several different directions. However, the distribution of useful portals is limited by the density of minerals such as calcium in the jawbones, which impede delivery to the target tissues and induce secondary radiation in the process. The result is that the major salivary glands often receive moderate to heavy radiation. Computerized radiation dose programs guided by computed tomography can more precisely fit the radiation doses to the site(s) of the cancer and its metastases while minimizing the IR affecting the salivary glands, oral mucosa, and other structures not likely to harbor tumor cells [32, 46, 58]. Another approach is the use of pharmaceuticals such as amifostine (WR2721, Ethyol®, [102]) which spares normal but not malignant tissues from IR damage.

In general, beta particles such as β- (electrons) or β+ (positrons) penetrate much shorter

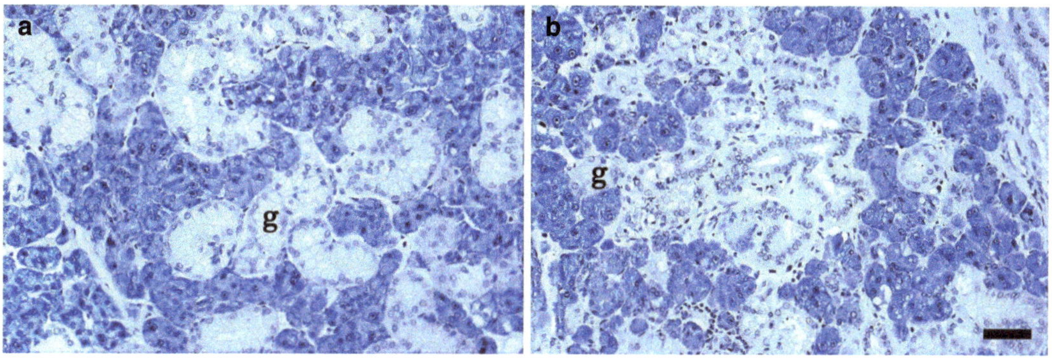

Fig. 4.5 Photomicrographs of submandibular glands of male rats at age 10 months. (**a**) Sham-irradiated gland. Acini (*blue secretory granules*) are tightly spaced in large bunches, granular convoluted tubules (*g*) are numerous and large, and the stroma is scanty except in septa around the larger ducts. (**b**) Gland 8 months after 10 Gy irradiation. Proportionately more stroma and fewer acini and granular convoluted tubules are present, and the granular convoluted tubules are smaller, than in (**a**). Alcian blue and hematoxylin (AB-H). Scale bar = 60 μm (Reproduced from photomicrographs [73] of slides used in a previous study [64])

distances into tissues than electromagnetic radiation of equal energy because they have both mass and charge. To achieve therapeutic radiation, beta particle-emitting radionuclides may be delivered via the circulation to tissues where they are preferentially absorbed and concentrated. The radionuclide used most in the head and neck is ^{131}I, which is concentrated more than 1000-fold in the thyroid gland and therefore is used to destroy papillary and follicular thyroid cancers. Unfortunately the salivary glands, especially the striated ducts, also concentrate ^{131}I much more than do other tissues [50]. The radiation is continuous until the radioactive iodine has been removed from the gland by secretion and circulation. An inflammatory response quickly ensues, and the swelling around the ducts and formation of a ductal plug result in salivary retention, prolonging the radiation exposure for several days. In many patients, the full extent of damage is not reached until a year or more later. Substantial long-term damage to both acini and ducts occurs, as evidenced by reduced salivary secretion and elevated salivary Na+ and Cl− [45].

In this review, all radiation doses are in gray (Gy), which is the international standard unit. It is equivalent to 100 rads, i.e., if the dose in a referenced publication was 4000 rads, it will be listed as 40 Gy here.

4.4.2 Radiation Damage Is Uneven Among and Within Salivary Glands

In both human and rodent salivary glands, serous acini are the structures that are the most vulnerable to ionizing radiation, many undergoing cell death and functional impairment after even moderate exposure [2, 10, 19, 60, 68, 84]. Regeneration of acini is proportional to the dose, being good after the minor damage caused by mild (6–20 Gy) doses, but after more than 50 Gy, the serous acini of the parotid and submandibular glands almost disappear, and little or no regeneration takes place. Less extensive acinar loss and atrophy occur among the mucous acini of the sublingual and submandibular glands, and some regeneration may occur.

The histological effects of mild to moderate (10 Gy) IR on a rat submandibular gland are shown in Fig. 4.5. The stromal area is proportionately increased and the acinar and granular convoluted tubules decreased compared with sham-irradiated glands.

The effects of heavy (70.63 Gy) IR on a human submandibular gland are illustrated in Figs. 4.6 and 4.7. Compared with a normal gland, the stromal area is greatly increased at the expense of the serous acini and demilunes, which have all but disappeared, and the mucous acini are reduced in size. Much of

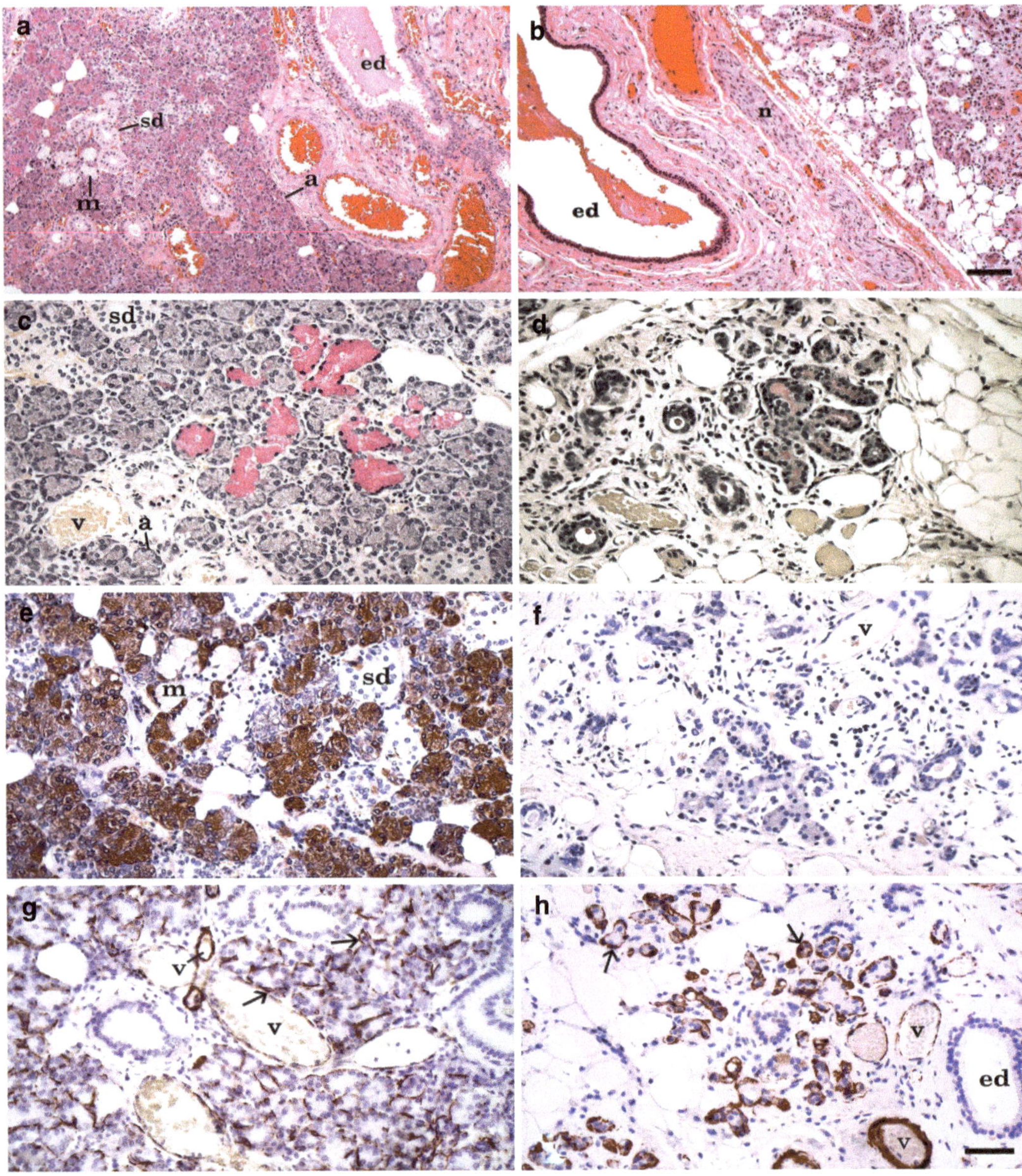

Fig. 4.6 (**a–d**) Normal gland from a 52-year-old male. (**e–h**) Gland removed from a 62-year-old male 4 weeks after being subjected to 70.34 Gy radiation in divided doses concurrent with 8 weeks of chemotherapy as part of a treatment regimen for cancer of the oropharynx. In the normal gland, serous acini contain ample secretory granules that stain darkly with hematoxylin, gray with mucicarmine, blank with anti-smooth muscle actin (SMA), and richly (*brown*) with anti-amylase. Secretion product in mucous acini stains pink with mucicarmine, and myoepithelial cells (*arrows*) and smooth muscle in blood vessels stain brown with anti-SMA. In the irradiated gland, myoepithelial cells are plentiful, but serous acini are not identifiable as such, mucous acini are diminished in number, size, and staining intensity with mucicarmine, the columnar cells of many of the striated and excretory ducts are reduced almost to cuboidal dimensions, and the adipose cells (large, empty-looking cells) and stroma occupy proportionately more space. Large nerves appear to be intact (**e**). *Labels*: *a* serous acini, *ed* excretory ducts, *m* mucous acini, *n* large nerves, *sd* striated ducts, *v* blood vessels. H&E (**a**, **e**), mucicarmine (**b**, **f**), and immunohistochemical localization of α-amylase (**c**, **g**) and SMA (**d**, **h**) counterstained with hematoxylin. Scale bar for **a**, **e** = 100 μm; for **b–d** and **f–h** = 60 μm (Reproduced from Redman [73])

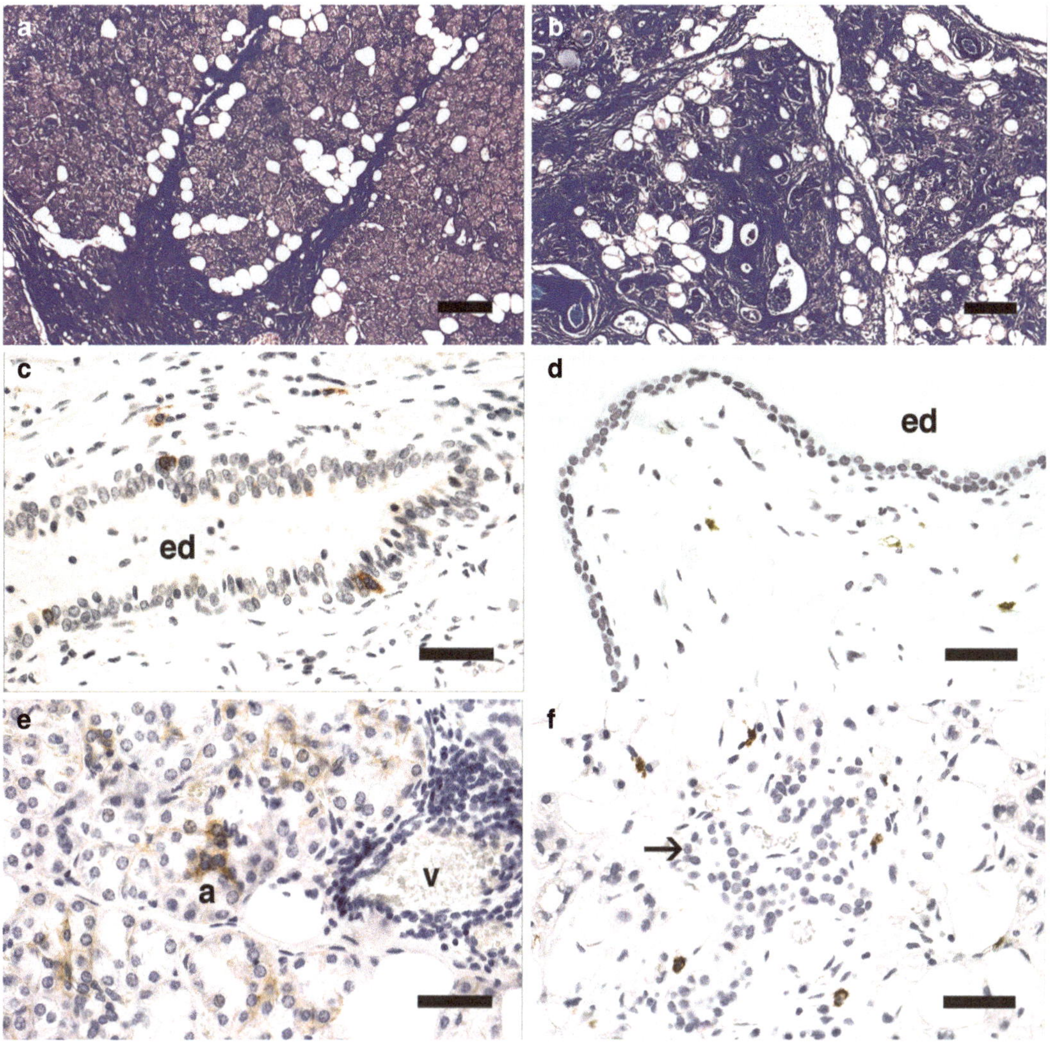

Fig. 4.7 Same glands as in Fig. 4.6. Masson's trichrome stain. (**a**) Normal gland. Mature collagen fibers (*dark blue*) are thin and delicate in the stroma around the acini and thick and dense in the interlobular septa of (**a**). (**b**) Irradiated gland. The collagen fibers are thick and dense in the stroma not only in the septa but also among the septa and in the lobules once occupied by acini, indicating extensive fibrosis has occurred. (**c–f**) Immunohistochemical localization of c-Kit, a marker for stem cells. (**c, e**) Normal gland. Labeled cells (brown cytoplasm) are scattered in the excretory ducts (*ed*), acini (*a*), and intercalated ducts and widely scattered in the stroma. Lymphocytes around the blood vessel (*v*) are unreactive. (**d, f**) Irradiated gland. Positive cells are not found in the large ducts (*ed*) and duct-like structures (remnants of acini and intercalated ducts, *arrow*) but are common in the stroma

the stroma is occupied by fibrosis (Fig. 4.7b) and adipose cells (Fig. 4.6f). In contrast to the acini, many of the myoepithelial cells have survived and invest small duct-like structures (Fig. 4.6h), an arrangement very similar to that observed in duct-ligated rabbit ([52] and rat [87] salivary glands. IHC for c-Kit, a marker for stem cells [41], labeled scattered cells in the stroma, larger ducts, and a few acini in the normal gland (Fig. 4.7c, e). No labeled parenchymal cells were seen in the irradiated gland (Fig. 4.7d), but labeled cells were common in the stroma (Fig. 4.7f). All of the tumor cells in the positive control (an adenocarcinoma of the colon) were strongly labeled (not illustrated).

The mechanism of IR damage to salivary glands has been thoughtfully considered by Nagler [61], Vissink et al. [101], and Dirix et al. [17]. The greater damage to serous acini from IR has been theoretically attributed to generation of free radicals via copper, iron, manganese, and zinc contained in their secretory proteins. However, when the secretory granules of serous acini and granular convoluted tubules were discharged by administration of secretagogues to rats prior to irradiation with 15 Gy, the damage to and loss of acinar cells were not noticeably different from the glands of rats that were not pretreated [13, 67]. Better regeneration of acini and recovery of salivary flow occurred in the pretreated rats, but this was attributed to the stimulation of proliferation by the sialogogues. Moderate to strong secretory stimulation has been shown to induce a significant but transient increase in acinar cell mitosis [80]. Note that in these rat experiments, the mitoses would have been initiated several hours after the administration of *a single dose* of IR.

The dramatic drop in salivary output that occurs during the first few days after a seemingly mild single dose or the first few fractionated doses of IR apparently is not due to immediate death but to widespread dysfunction of the acinar cells. Plausible explanations for this phenomenon include transient damage to the plasmalemma and receptor-signaling apparatuses [13, 17]. The transient increase of salivary amylase in the blood following an initial dose of radiation [8] may be related to leakage from the damaged cells. In any event, if DNA and other cellular damages are not severe enough to cause immediate death of the cell, inaccurately repaired DNA damage may be sufficient to interfere with future proliferative activity or cause delayed death. Thus, as the surviving acinar cells and their precursors in the intercalated and other ducts die during the ensuing weeks, the extent to which the acini can regenerate may be severely compromised.

There may be additional reasons for radiation-induced loss of salivary gland function, such as damage to innervation, blood vessels, and stroma. Chomette et al. [11], using enzyme histochemistry and transmission electron microscopy, reported persistent damage including edema and loss of vesicles in secretory nerve endings of rat submandibular gland through 70 days after administration of single doses of 20, 25, or 30 Gy. This damage was interpreted as sufficient to cause loss of stimulation, thus contributing to the cycle of regeneration, engorgement with secretory granules, and death of acinar cells observed during the 70 days after IR. On the other hand, much of the secretory innervation reportedly survives radiation [1, 13, 22, 34], and surviving acinar cells from rat salivary glands subjected to mild (10 Gy) radiation still respond normally to secretagogues [64]. Furthermore, although neuropeptides such as substance P and bombesin were increased transiently in rat submandibular gland ganglia at 10 days post-radiation of 30–40 Gy IR (given in daily fractionated doses), values had returned to normal by 180 days. Changes in taste after IR also offer a clue to the importance of IR effects on nerves in salivary glands. After IR for cancer of the head and neck, the taste threshold increased dramatically at 1 month but had recovered to normal baseline values by 6 months [78]. These results indicate that the neuroepithelial (taste) cells, nerve endings, or both were destroyed or functionally disabled by the IR. Results with experimental animals have documented damage and destruction of lingual taste buds by IR [20]. Regeneration of taste cells via differentiation from surface epithelial cells has been shown to be dependent on appropriate innervation [20]. From the foregoing, it seems likely that any nerve damage caused by IR for head and neck cancer would have been limited to the proximal portions of the nerve processes, allowing new nerve endings to migrate from the surviving portions of the nerves. Interestingly, taste recovered even when there was marked xerostomia, suggesting that saliva may be less important to taste perception than supposed on theoretical grounds [53]. There was no functional assessment of damage to the lingual glands of von Ebner, however, which serve the majority of taste buds in the troughs of the vallate papillae and foliate folds [28]. In a case pertinent to this point, Fajardo [19] illustrated "a heavily irradiated tongue" in which "all that remained of a lingual salivary gland was a dilated and ulcerated duct forming a mucocele."

Following moderate- to high-dose IR, the stroma of human parotid and submandibular glands has been described as undergoing adiposis and fibrosis, respectively [23] (Fig. 4.7). Some stromal fibrosis also has been observed in rat salivary glands but only after high-dose IR [10, 68, 84]. The endothelium of blood vessels is susceptible to radiation damage, and compromised blood supply has been observed after ^{131}I treatment for thyroid cancer [50]. Thickening of extracellular matrix components in response to high doses of IR also has been reported [7]. These stromal changes may restrict diffusion of nutrients, essential minerals, and oxygen to parenchymal cells and thus may adversely affect late attempts at regeneration and function by surviving acinar cells.

Because of the importance of acini as the origin of the water and most of the organic secretory products in saliva, the more dramatic effects of IR on acinar cells overshadow the considerable effects that moderate to high doses of IR have on excretory and striated ducts [2, 50] (Fig. 3). The diminished function of striated and other large ducts is evident in higher salivary sodium and chloride and lower bicarbonate concentrations and consequently lower pH, in stimulated saliva [100]. In this regard, stimulated saliva from irradiated glands is more like unstimulated saliva from normal glands. The reduced flow and buffering capacity and lower pH of saliva from irradiated glands are major factors in the rapid development of dental caries in the absence of rigorous control measures. In addition to their importance in ion exchange with the luminal fluid as noted above, they have been shown to have stem cells with the capacity to regenerate acinar cells [41].

A plausible explanation for the poor recovery of acinar cells after moderate to high doses of IR is that the lifetime proliferative capacity of the acinar cells and their progenitors is partially depleted by repeated mitoses in attempts to replace cells lost to radiation and is diminished further by DNA/chromosomal damage in still viable cells [13, 61].

Studies in rats have shown that IR induces an increase in cellular proliferation among the parenchymal cells in proportion to their loss. After a single dose of 15 Gy to the parotid and submandibular glands of mature rats, proliferative activity as determined by uptake of bromodeoxyuridine (BrdU) slowed for 24 h and then resumed in the intercalated ducts after 3 days and in the acini, striated ducts, and granular convoluted tubules after 6–10 days [66]. By IHC localization of proliferating cell nuclear antigen (PCNA), proliferative activity increased by factors of 12.6, 3.4, and 2.2 in the acinar, intercalated duct, and striated duct cells, respectively, in rat submandibular glands 7 days after exposure to a single dose of 30 Gy [5]. These observations suggest that the divided doses used in therapeutic IR for cancer of the head and neck in humans compound the damage to salivary glands. Each successive spike in proliferating cells, which are more vulnerable to radiation damage compared to nondividing cells, takes place during the next dose of radiation. C-Kit, a marker of stem cells, has been localized to scattered cells in the excretory and intercalated ducts of mature human salivary glands [21, 42]. Both cell cycle and c-Kit proteins reportedly were reduced in rats 12 weeks post-irradiation [85].

The scattered c-Kit-labeled parenchymal cells in the normal gland and lack of these in the IR gland in Fig. 4.7 are consistent with the finding of Stiubea-Cohen et al. [85], indicating that the stem cell population of salivary glands is seriously depleted by IR. It is noteworthy, however, that in the normal gland, not only relatively undifferentiated basal cells in large ducts but also well-differentiated columnar cells in large ducts and acinar cells were labeled by c-Kit (Fig. 4.7c, e). This suggests that well-differentiated salivary gland cells also may be stem cells. In addition, the enrichment of labeled cells in the stroma is reminiscent of experiments in which mobilized hematopoietic stem cells entered the stroma but not the parenchyma of irradiated mouse submandibular glands [40].

4.5 Perspective

Partial restoration of IR-damaged mouse submandibular glands has been accomplished by injecting a mixture of donor cells enriched in progenitor or stem cells via a needle inserted through the capsule [41]. However, these glands were subjected to only 15 Gy of radiation, much less than the 50–70 Gy that causes serious damage to human salivary glands. Hematopoietic stem cells are attracted to and become part of the normal minor salivary gland and oral epithelia [98, 99], perhaps because labial and buccal mucosae are frequently traumatized. Why, then, do they not home to salivary gland epithelia badly damaged by IR? As noted above, fair to good preservation of salivary glands is being affected by innovations such as computerized IR delivery programs and sparing agents such as amifostine. Perhaps even partial preservation of salivary gland structure, especially the vulnerable serous acini, from therapeutic IR damage may allow techniques such as intraglandular injection of salivary or other stem cells to restore salivary function to nearly normal.

Acknowledgments I thank Su-Wan Chen, Helen Cormier, Edward Flores, Lyvouch Filkoski, Patricia Lewis, Rodney McNutt, and Shirley McLaughlin for preparing the tissues for morphologic evaluation. The research enabling the illustrations was supported in part by grants DE-102, DE-03002, 1-K4-DE-40-019, and DE-14995 from the National Institute of Dental and Craniofacial Research, the National Institutes of Health, Bethesda, MD, the Universities of Colorado, Minnesota, and Washington, and by Research Career and Merit Review Awards from the United States Department of Veterans Affairs. Figures 4.5 and 4.6, the legends to these figures, and some of the text are reproduced from Redman [73], with prior permission.

References

1. Aalto Y, Forsgren S, Kjorell U, Franzen L, Gustafsson H, Henriksson R. Time- and dose-related changes in the expression of substance P in salivary glands in response to fractionated irradiation. Int J Radiat Oncol Biol Phys. 1995;33:297–305.
2. Abok K, Brunk U, Jung B, Ericsson J. Morphologic and histochemical studies on the differing radiosensitivity of ductular and acinar cells of the rat submandibular gland. Virchows Arch Cell Pathol. 1984;45:443–60.
3. Ahmad M, Piludu M, Oppenheim FG, Helmerhorst EJ, Hand AR. Immunocytochemical localization of histatins in human salivary glands. J Histochem Cytochem. 2004;52:361–70.
4. Akamatsu T, Parvin MN, Murdiastuti K, Kosugi-Tanaka C, Yao CJ, Miki O, Kanamori N, Hosoi K. Expression and localization of aquaporins, members of the water channel family, during development of the rat submandibular gland. Pflügers Arch-Eur J Physiol. 2003;446:641–5.
5. Ballagh RH, Kudryk KG, Lampe HB, Moriarity B, Mackay A, Burford-Mason AP, Dardick I. The pathobiology of salivary gland. III. PCNA-localization of cycling cells induced in rat submandibular gland by low-dose x-irradiation. Oral Surg Oral Med Oral Pathol. 1994;77:27–35.
6. Bandyopadhyay BC, Swaim WD, Sarkar A, Liu X, Ambudkar IS. Extracellular Ca (2+) sensing in rat salivary duct cells. J Biol Chem. 2012;287:30305–16.
7. Bartel-Friedrich S, Friedrich RE, Lautenschlager C, Holzhausen HJ. Dose-response relationships and the effect of age and latency period on the expression of laminin in irradiated rat mandibular glands. Anticancer Res. 2000;20:221–5228.
8. Becciolini A, Giannardi CL, Porciani S, Fallai C, Pirtoli L. Plasma amylase activity as a biochemical indicator of radiation injury to salivary glands. Acta Radiol Oncol. 1984;23:353–9.
9. Caselitz J, Osborn M, Wustrow J, Seifert G, Weber K. Immunohistochemical investigations on the epimyoepithelial islands in lymphoepithelial lesions. Use of monoclonal keratin antibodies. Lab Invest. 1986;44:427–32.
10. Cherry CP, Gluckman A. Injury and repair following irradiation of salivary glands in male rats. Br J Radiol. 1959;32:596–608.
11. Chomette G, Auriol M, Vaillant JM, Bertrand JC, Chenal C. Effects of irradiation on the submandibular gland of the rat. Virchows Arch (Pathol Anat). 1981;391:291–9.
12. Christensen ME, Hansesn HS, Poulsen SS, Bretlau P, Nexo E. Immunochemical and quantitative changes in salivary EGF, amylase and haptocorrin following radiotherapy for oral cancer. Acta Otolaryngol. 1996;116:137–43.
13. Coppes RP, Zeilstra LJW, Kampinga HH, Konings AWT. Early to late sparing of radiation damage to the parotid gland by adrenergic and muscarinic receptor agonists. Br J Cancer. 2001;85:1055–63.
14. Cossu M, Perra MT, Piludi M, Lantini MS. Subcellular localization of epidermal growth factor in human submandibular gland. Histochem J. 2000;32:291–4.
15. Culp DJ, Latchney LR, Fallon MA, Denny PA, Denny PC, Couwenhoven RI, Chuang S. The gene encoding mouse Muc19: cDNA, genomic organization and relationship to Smgc. Physiol Genomics. 2004;19:303–18.
16. Dardick I, Parks WR, Little J, Brown DL. Characterization of cytoskeletal proteins in basal cells of human parotid salivary gland ducts. Virchows Arch A Pathol Anat Histopathol. 1988;412:525–32.
17. Dirix P, Nuyts S, Van den Bogaert W. Radiation-induced xerostomia in patients with head and neck cancer. A literature review. Cancer. 2006;107:2525–34.

18. Eversole LR. Histogenetic classification of salivary gland tumors. Arch Pathol. 1971;92:433–43.
19. Fajardo LF. Salivary glands and pancreas. In: Fajardo LF, editor. Pathology of radiation injury. New York: Masson Publishing USA; 1982. p. 77–87.
20. Farbman AI. The taste bud: a model system for developmental studies. In: Slavkin HC, Bavetta LA, editors. Developmental aspects of oral biology. New York: Academic Press; 1972. p. 109–23.
21. Feng J, van der Zwaag M, Stokman MA, van Os R, Coppes RP. Isolation and characterization of human salivary gland stem cells for stem cell transplantation to reduce radiation-induced hyposalivation. Radiother Oncol. 2009;92:466–71.
22. Forsgren S, Franzen L, Liang Y, Gustafsson H, Henriksson R. Effects of irradiation on neuropeptide expression in rat salivary gland and spinal cord. Histochem J. 1994;26:630–40.
23. Frank RM, Herdly J, Phillipe E. Acquired dental defects and salivary gland lesions after irradiation for carcinoma. J Am Dent Assoc. 1965;70:868–83.
24. Garrett JR. The innervation of normal human submandibular and parotid salivary glands. Arch Oral Biol. 1967;12:1417–36.
25. Geiger S, Geiger B, Leitner O, Marshak G. Cytokeratin polypeptides expression in different epithelial elements of human salivary glands. Virchows Arch A Pathol Anat Histopathol. 1987;410:403–14.
26. Gresik EW. The granular convoluted tubule (GCT) of rodent submandibular glands. Microsc Res Tech. 1994;27:1–24.
27. Gugliotta P, Sapino A, Macri L, Skalli O, Gabbiani G, Busolati G. Specific demonstration of myoepithelial cells by anti-alpha smooth muscle actin antibody. J Histochem Cytochem. 1988;36:659–63.
28. Hand AR. The fine structure of von Ebner's gland in the rat. J Cell Biol. 1970;44:340–52.
29. Hand AR. Adrenergic and cholinergic nerve terminals in the rat parotid gland. Electron microscopic observations on permanganate -fixed glands. Anat Rec. 1972;173:131–40.
30. Hand AR. Synthesis of secretory and plasma membrane glycoproteins by striated duct cells of rat salivary glands as visualized by radioautography after ^{3}H-fucose injection. Anat Rec. 1979;195:317–39.
31. He X, Tse C-M, Donowitz M, Alper SL, Gabriel SE, Baum BJ. Polarized distribution of key membrane transport membranes in the rat submandibular gland. Pflügers Arch-Eur J Physiol. 1997;433:260–8.
32. Henson BS, Inglehart MR, Eisbruch A, Ship JA. Preserved salivary output and xerostomia-related quality of life in head and neck cancer patients receiving parotid -sparing radiotherapy. Oral Oncol. 2001;37:84–93.
33. Izutsu KT. Salivary electrolytes and fluid protection in heath and disease. In: Sreebny LM, editor. The salivary system. Boca Raton: CRC Press; 1987. p. 95–122.
34. Kohn WG, Grossman E, Fox PC, Armando I, Goldstein DS, Baum BJ. Effect of ionizing radiation on sympathetic nerve function in rat parotid glands. J Oral Pathol Med. 1992;21:134–7.
35. Konturek SJ, Dembinski A, Warzecha Z, Brzozowski T, Gregory H. Role of epidermal growth factor in healing of chronic gastroduodenal ulcers in rats. Gastroenterology. 1988;84:1300–7.
36. Konturek PG, Ernst EH, Konturek J, Bobrzynski T, Kwiecien N, Faller G, Gedliczka O, Hahn EG. Salivary and gastric luminal release of epidermal growth factor under basal conditions and after pentagastrin stimulation in healthy and in duodenal ulcer patients before and after eradication of Helicobacter pylori. J Physiol Pharmacol. 1997;47:187–94.
37. Lantini MS, Piludi M, Cossu M. Subcellular localization of epidermal growth factor in human parotid gland. Histochem J. 2001;33:427–31.
38. Larsen HS, Aure MH, Peters SB, Larsen M, Messelt EB, Galtung HK. Localization of AQP5 during development of the mouse submandibular salivary gland. J Mol Histol. 2011; doi:10.1007/s10735-010-9308-0.
39. Leinonen J, Parkkila S, Kaunisto K, Koivunen P, Rajaniemi H. Secretion of carbonic anhydrase isoenzyme VI (CA VI) from human and rat lingual serous von Ebner's glands. J Histochem Cytochem. 2001;49:657–62.
40. Lombaert IM, Wierenga PK, Kok T, Kampinga HH, deHaan G, Coppes RP. Mobilization of bone marrow stem cells by granulocyte colony-stimulating factor ameliorates radiation-induced damage to salivary glands. Clin Cancer Res. 2006;12:1804–12.
41. Lombaert IMA, Brunsting JF, Wierenga PK, Faber H, Stokman MA, Kok T, Visser WH, Kampinga HH, de Haan G, Coppes RP. Rescue of salivary gland function after stem cell transplantation in irradiated glands. PLoS One. 2008;3:e2063. doi:10.1371/journal.pone.ooo2603: 1-16.
42. Lombaert IMA, Abrams SR, Li L, Eswarakumar VP, Sethi AJ, Witt RL, Hoffman MP. Combined KIT and FGFR2b signaling regulates epithelial progenitor expansion during organogenesis. Stem Cell Reports. 2013;1:604–19.
43. Luna LG 1968. Manual of histologic staining methods of the Armed Forces Institute of Pathology. 3rd edn. New York: McGraw-Hill. p. 94, 127–128.
44. Ma L, Wang WP, Chow JY, Yuen ST, Cho CH. Reduction of EGF is associated with the delay of ulcer healing by cigarette smoking. Am J Physiol Gastrointest Liver Physiol. 2000;278:G10–7.
45. Maier H, Bihl H. Effect of radioactive iodine therapy on salivary glands. Acta Otolaryngol. 1987;103:318–24.
46. Malouf JG, Aragon C, Henson BS, Eisbruch A, Ship JA. Influence of parotid-sparing radiotherapy on xerostomia in head and neck cancer. Cancer Detect Prev. 2003;27:305–10.
47. Malpighi M. Epistolae anatomicae. De viscerum structura exercitatio anatomica. In: Opera omnia. London: Robert Littlebury; 1687. p. 51–144.

48. Man Y-G, Ball WD, Marchetti L, Hand AR. Contributions of intercalated duct cells to the normal parenchyma of submandibular glands of adult rats. Anat Rec. 2001;263:201–14.
49. Mancini DM, Veit BC. Salivary growth factor in patients with and without acid peptic ulcer disease. Am J Gastroenterol. 1990;85:1102–4.
50. Mandel SJ, Mandel L. Radioactive iodine and the salivary glands. Thyroid. 2003;13:265–71.
51. Mangos JA, Braun G, Hamann KF. Micropuncture study of sodium and potassium excretion in the rat parotid gland. Pflügers Arch ges Physiol. 1966; 291:88–106.
52. Maria OM, Maria SM, Redman RS, Maria AM, El-Din TAS, Soussa EF, Tran SD. Effects of double ligation of Stensen's duct on the rabbit parotid gland. Biotech Histochem. 2014;82:181–98.
53. Matsuo R. Role of saliva in the maintenance of taste sensitivity. Crit Rev Oral Biol Med. 2000;11:216–29.
54. Messelt EB, Berg T. Effect of autonomic nerve stimulation on bleb formation in striated duct cells of the rat submandibular gland. Acta Odontol Scand. 1987;45:275–81.
55. Moreira JE, Hand AR, Ball WD. Localization of neonatal secretory proteins in different cell types of the developing rat submandibular gland from embryogenesis to adulthood. Dev Biol. 1990;139:370–82.
56. Mori M. Histochemistry of the salivary glands. Boca Raton: CRC Press; 1991. p. 1–85.
57. Mowry RW. The special value of methods that color both acidic and vicinal hydroxyl groups in the histochemical study of mucins. With revised directions for the colloidal iron stain, the use of alcian blue 8GX and their combinations with the periodic acid-Schiff reaction. Ann N Y Acad Sci. 1963;106:402–23.
58. Murdoch-Kinch CA, Russo N, Griffith S, Braun T, Eisbruch A, D'Silva NJ. Recovery of salivary epidermal growth factor in parotid saliva following sparing radiation therapy: a proof of principle study. Oral Surg Oral Med Oral Pathol Oral Radiol Endod. 2011;111:64–70.
59. Nagato T, Yoshida H, Yoshida A, Uehara Y. A scanning electron microscope study of myoepithelial cells in exocrine glands. Cell Tissue Res. 1980;209:1–10.
60. Nagler RM, Baum BJ, Miller G, Fox PC. Long-term salivary effects of single-dose head and neck irradiation in the rat. Arch Oral Biol. 1998;43:297–303.
61. Nagler RM. The enigmatic mechanism of irradiation-induced damage to the major salivary glands. Oral Dis. 2002;8:141–6.
62. Nagy A, Nagashima H, Cha S, Oxford GE, Zelles T, Peck AB, Humphreys-Beher MG. Reduced wound healing in the NOD mouse model for Type I autoimmune diabetes and its reversal by epidermal growth factor supplementation. Diabetes. 2001;50:2100–4.
63. Noguchi S, Ohba Y, Oka T. Effect of salivary growth factor on wound healing of tongue in mice. Am J Physiol. 1991;260:E620–5.
64. O'Connell AC, Redman RS, Evans RL, Ambudkar IS. Irradiation-induced progressive decrease in fluid secretion in rat salivary glands is related to decreased acinar volume and not impaired signaling. Radiat Res. 1999;151:150–8.
65. Oxford GE, Jonsson R, Olofsson J, Zelles T, Humphreys-Beher MG. Elevated levels of human salivary epidermal growth factor after oral and juxta-oral surgery. J Oral Maxillofac Surg. 1999;57:154–9.
66. Peter B, Van Waarde MAWH, Vissink A, 's-Gravenmade EJ, Konings AWT (1994) Radiation-induced cell proliferation in the parotid and submandibular glands of the rat. Radiat Res 140: 257–265.
67. Peter B, Van Waarde MAWH, Vissink A, 's-Gravenmade EJ, Konings AWT (1995) The role of secretory granules in radiation-induced dysfunction of rat salivary glands. Radiat Res 141: 176–182.
68. Phillipe RM. X-ray-induced changes in function and structure of the rat parotid gland. J Oral Surg. 1970;28:431–7.
69. Redman RS. The anterior buccal gland of the rat: a mucous salivary gland which develops as a branch of stensen's duct. Anat Rec. 1972;172:167–78.
70. Redman RS. The importance of the minor salivary glands. Northwest Dent. 1974;53:19–23. 1974)
71. Redman RS. Myoepithelium of the salivary glands. Microsc Res Tech. 1994;27:25–45.
72. Redman RS. Proliferative activity by cell type in the developing rat parotid gland. Anat Rec. 1995; 241:529–40.
73. Redman RS. On approaches to the functional restoration of salivary glands damaged by therapeutic irradiation for head and neck cancer, with a review of related aspects of salivary gland morphology and development. Biotech Histochem. 2008;83:103–30.
74. Redman RS. Morphologic diversity of the minor salivary glands of the rat: fertile ground for studies in gene function and proteomics. Biotech Histochem. 2012;87:273–87.
75. Redman RS, Ball WD. Differentiation of myoepithelial cells in the developing rat sublingual gland. Am J Anat. 1979;156:543–66.
76. Redman RS, Field RB. Chronology of salivary peroxidase activity in the developing rat parotid gland. Anat Rec. 1993;235:611–21.
77. Riva A, Puxeddu P, Del Fiacco M, Testa-Riva F. Ultrastructural localization of endogenous peroxidase activity in human parotid and submandibular glands. J Anat. 1978;127:181–91.
78. Sandow PI, Hejrat-Yazdi M, Heft MW. Taste loss and recovery following radiation therapy. J Dent Res. 2006;85:608–11.
79. Sato A, Miyoshi S. Fine structure of tuft cells of the main excretory duct in the rat submandibular gland. Anat Rec. 1997;248:325–31.
80. Schneyer CA. Mitosis induced in adult rat parotid following normal activity of the gland. Proc Soc Exp Biol Med. 1970;134:603–7.
81. Schneyer LH, Schneyer CA. Inorganic composition of saliva. In: Code CF, editor. Secretion. Handbook of physiology, section 6, vol. 2. Washington, DC: American Physiological Society; 1967. p. 497–529.

82. Shackleford JM, Schneyer LH. Ultrastructural aspects of the main excretory duct of rat submandibular gland. Anat Rec. 1971;169:679–96.
83. Ship JA, Hu K. Radiotherapy-induced salivary dysfunction. Semin Oncol. 2004;31(Suppl. 18):29–36.
84. Sholley MM, Sodicoff M, Pratt NE. Early radiation injury in the rat parotid gland. Reaction of acinar cells and vascular endothelium. Lab Invest. 1974;31:340–54.
85. Stiubea-Cohen R, David R, Neumann Y, Krief G, Deutsch O, Zacks B, Aframian DJ, Palmon A. Effect of irradiation on cell transcriptome and proteome of rat submandibular salivary glands. PLoS One. 2012;7(7):e40636. doi:10.1371/journal pone.0040636.
86. Tabak LA. In defense of the oral cavity. Pediatr Dent. 2006;28:110–7.
87. Takahashi S, Nakamura S, Suzuki R, Islam N, Domon T, Yamamoto T, Wakita M. Changing myoepithelial cell distribution during regeneration of rat parotid glands. Int J Exp Pathol. 1999;80:283–90.
88. Tamarin A. Myoepithelium of the rat submaxillary gland. J Ultrastruct Res. 1966;16:320–38.
89. Tamarin A, Sreebny LM. The rat submaxillary gland. A correlative study by light and electron microscopy. J Morphol. 1965;117:295–352.
90. Tamarin A, Walker J. A longitudinal study of parotid secretory kinematics by cinematography and computer analysis. Pflügers Arch. 1976;366:101–6.
91. Tandler B. Ultrastucture of the human submaxillary gland. III. Myoepithelium. Z Zellforsch. 1965; 68:852–63.
92. Tandler B. Structure of the human parotid and submandibular glands. In: Sreebny LM, editor. The salivary system. Boca Raton: CRC Press; 1987. p. 21–41.
93. Tandler B. Structure of the duct system in mammalian major salivary glands. Microsc Res Tech. 1993;26:57–74.
94. Tatemoto Y, Kumasa S, Watanabe Y, Mori M. Epithelial membrane antigen as a marker of human salivary gland acinar and ductal cell function. Acta Histochem. 1987;82:219–26.
95. Tepperman BL, Kiernan JA, Soper BD. The effect of sialoadenectomy on gastric mucosal integrity in the rat: roles of epidermal growth factor and prostaglandin E2. Can J Physiol Pharmacol. 1989;67:1512–9.
96. Teste-Riva F, Puxeddu P, Riva A, Diaz G. The epithelium of the excretory duct of the human submandibular gland: a transmission and scanning electron microscopic study. Am J Anat. 1981;160:361–93.
97. Thesleff I, Vinikka L, Saxén S, Lehtonen E, Perheentupa J. The parotid gland is the main source of human salivary epidermal growth factor. Life Sci. 1988;43:13–8.
98. Tran SD, Pillemer SR, Dutra A, Barrett AJ, Brownstein MJ, Keys S, Pak E, Leakan RA, Kingman A, Yamada KM, Baum BJ, Mezey E. Differentiation of human bone marrow-derived cells into buccal epithelial cells in vivo: a molecular analytical study. Lancet. 2003;361:1084–8.
99. Tran SD, Redman RS, Barrett AJ, Pavletic SZ, Key S, Liu Y, Carpenter A, Nguyen HM, Sumita Y, Baum BJ, Pillemer SR, Mezey É. Microchimerism in salivary glands after blood- and marrow-derived stem cell transplantation. Biol Blood Marrow Transplant. 2011;17:429–33.
100. Valdez IH, Atkinson JC, Ship JA, Fox PC. Major salivary gland function in patients with radiation-induced xerostomia: flow rates and sialochemistry. Int J Radiat Oncol Biol Phys. 1992;25:41–7.
101. Vissink A, Jansma J, Spijkervet F, Burlage FR, Coppes RP. Oral sequelae of head and neck radiotherapy. Crit Rev Oral Biol Med. 2003a;14:199–212.
102. Vissink A, Burlage FR, Spijkervet F, Jansma J, Coppes RP. Prevention and treatment of the consequences of head and neck radiotherapy. Crit Rev Oral Biol Med. 2003b;14:213–25.
103. Williams PL, editor. Gray's anatomy. New York: Charles Livingston; 1996.
104. Yamashina S. Immunohistochemical study of amylase and deoxyribonuclease in rat parotid gland. Acta Histochem Cytochem. 1981;14:236–60.
105. Young JA, Schlögel E. Micropuncture investigation of sodium and potassium excretion in the rat submaxillary gland. Pflügers Arch ges Physiol. 1966;291:85–98.
106. Young JA, van Lennep EW. The morphology of salivary glands. New York: Academic Press; 1978.
107. Yopilko A, Caillou B. Fine structural localization of acetylcholinesterase activity in the rat submandibular gland. J Histochem Cytochem. 1989;33:439–45.

Adult Stem Cell Therapy for Salivary Glands, with a Special Emphasis on Mesenchymal Stem Cells

5

Simon D Tran, Yoshinori Sumita, Dongdong Fang, and Shen Hu

Abstract

Mesenchymal stromal/stem cell (MSC) therapy with the goal of restoring salivary function following irradiation injury or in Sjögren's syndrome (SS) has made significant advances within the past 5 years. The majority of studies used MSCs obtained from the bone marrow or adipose tissue, but MSCs isolated from the salivary gland, dental pulp, and umbilical cord also demonstrated a therapeutic efficacy in reestablishing salivary function. Based on the amount of stimulated saliva secretion as a functional quantitative measure, irradiated mice/rats that received MSC therapy restored their salivary flow rate (SFR) to 60–90 % of normal age-matched animals, while SFR of irradiated animals without treatment remained at 35–50 % of secretory function. Thus, there was 25–40 % therapeutic improvement in animals receiving MSC therapy versus those that did not. This would be clinically significant because patients with severe salivary hypofunction (dry mouth) due to head and neck irradiation have no improvement in SFR, if left untreated. In the SS-like disease mouse model, MSC therapy restored SFR 80–100 % when treatment was given at an initial phase of SS-like disease, while its effectiveness decreased to 50–60 % when given at an advanced stage of disease. In SS patients, MSC therapy improved SFR by 40–50 %. When tested in the rodent model, MSC therapy was successful in restoring/maintaining the gland normal weights and histology (acinar cells, blood vessels) and upregulated the expression of genes favorable for salivary gland development and regeneration while downregulating inflammation and cell apoptosis, promising positive effects of MSC therapy.

S.D. Tran, DMD, PhD (✉) • D. Fang, BDS, MSc
McGill Craniofacial Tissue Engineering and Stem Cells Laboratory, Faculty of Dentistry, McGill University, 3640 University Street, Montreal H3A 0C7, Canada
e-mail: simon.tran@mcgill.ca

Y. Sumita, DDS, PhD
Department of Regenerative Oral Surgery, Nagasaki University, Nagasaki, Japan

S. Hu, PhD, MBA
Division of Oral Biology and Medicine, School of Dentistry, University of California, Los Angeles, CA, USA

S. Cha (ed.), *Salivary Gland Development and Regeneration*, DOI 10.1007/978-3-319-43513-8_5

5.1 Experimental Approaches for the Functional Restoration of Salivary Glands

Clinically, there are two severe conditions that greatly reduce the secretion/output of saliva. First, 40,000 new cases of head and neck cancer are diagnosed each year in the United States. Worldwide, this amounts to 640,000 new cases yearly [12]. Irradiation (IR, radiation therapy, radiotherapy) is a key component of therapy for these cancer patients. Salivary glands (SGs), particularly the acinar cells, in the ionizing radiation field suffer severe damage [49, 50]. These cells are the principal site of fluid secretion in SGs, and such patients cannot produce adequate levels of saliva, leading to considerable morbidity and extreme discomfort [31]. Second, patients with Sjögren's syndrome (SS), an autoimmune disorder, suffer similar irreversible damage to their salivary glands. In the United States, there are an estimated 4 million SS patients, mostly perimenopausal women.

Both IR and SS lead to the destruction of the salivary glandular parenchyma, and saliva production is drastically reduced as a result. Both groups of patients experience severe SG hypofunction causing symptoms such as xerostomia (dry mouth), dysphagia (impaired chewing and swallowing), dental caries, altered taste, oropharyngeal infections (candidiasis), mucositis, pain, and discomfort [5, 10]. These patients suffer considerable morbidity as xerostomia leads to reduced nutritional intake and weight loss, significantly affects general health, and severely reduces their quality of life [5]. For many SS and IR patients, in particular those whose salivary epithelial cells have been replaced by fibrotic tissue, there is no available adequate treatment. Current pharmacological approaches (such as saliva-stimulating drugs) require the presence of some surviving epithelial tissue [3]. Experimental approaches tested to date for functional restoration of SGs are the use of electrostimulation, acupuncture, gene therapy, tissue engineering, and cell therapy [2, 3, 15, 36, 39, 48]. Electrostimulation and acupuncture require the presence of some surviving acinar cells. Gene therapy can target remaining ductal cells. Tissue engineering and cell therapy-based methods can theoretically be utilized in the absence of surviving acinar cells and can be viewed as regenerative methods.

The first regenerative approach is building an artificial SG using tissue engineering [1, 4]. We and others have reported feasibility in culturing SG epithelial cells for a proposed artificial SG prototype [4, 47]. This strategy generates mainly one portion of the SG parenchymal tissue (ductal cells), and it is still challenging to regenerate a fully fluid secretory tissue (both ductal and acinar cells) [28, 30]. Thus, the second regenerative approach, using (stem) cell-based therapy, has been enthusiastically pursued by several research groups, including ours [13]. Currently, there are two types of stem/progenitor cells that have shown promises as experimental treatments for SG hypofunction: a) salivary epithelial stem cells and b) multipotent mesenchymal stromal cells (MSCs) from either the bone marrow (BM) or from the adipose tissue. Within the past 5 years, there have been an increasing number of studies reporting the successful use of adult stem/progenitor cell-based therapies in treating SG hypofunction for both IR- and SS-injured SGs. The main emphasis of this chapter will be on MSC cell-based therapies. The topic of salivary epithelial stem cell therapies against IR-induced SG damage will only be mentioned briefly below, as there is already an excellent review covering this topic [41].

5.2 Salivary Epithelial Stem Cell Therapy for Irradiation-Injured Salivary Glands

For the past several years, Robert Coppes' group has steadily made significant advances in the study and use of salivary epithelial stem cells in the treatment of IR-injured SGs. This group of researchers initially developed a culture system to enrich and characterize salivary stem cells (termed as "salispheres") [26]. The salispheres rescued SG functions after transplantation into IR mice.

The same group isolated salispheres (c-Kit+ cells) from the human SG excretory ducts, which when placed in 3D culture developed into acinar-like cells. Subsequent studies identified the salispheres with additional stem cell markers such as CD24(high), CD29(high), and/or CD49f [7, 9, 33, 34]. Recently, the same researchers reported successful isolation of adult human SG stem/progenitor cells at the single-cell level with evidence of in vitro self-renewal and differentiation into multilineage organoids. These human SG stem/progenitor cells repaired IR-injured SG in mice [40]. This novel and promising approach is currently being tested in a clinical study of patients undergoing IR for their head and neck cancers (personal communication from RP Coppes). This strategy was initially thought to be limited for clinical use due to an insufficient number of salispheres obtained from patients' SG biopsies. However, Coppes' group recently demonstrated that FACS-sorted $CD24^{high}$ and $CD29^{high}$ SG and salisphere-derived cells from young and old mice exhibited similar in vitro expansion and in vivo regeneration potential. Although older mice (22–26 months old) had a reduced number of SG stem cells, they were indistinguishable in vitro from SG stem cells from younger mice (8–12 weeks old) [27]. Quynh Thu Le's group also investigated salispheres by isolating murine Lin^- $CD24^+$ c-Kit^+ $Sca1^+$ salivary stem cells [51].

Gene expression analysis revealed that "glial cell line-derived neurotrophic factor" (GDNF has a role in neuron survival, growth, differentiation, and migration) was highly expressed in Lin^- $CD24^+$ c-Kit^+ $Sca1^+$ salivary stem cells. Administration of GDNF improved saliva production and increased the number of acinar cells in submandibular glands of IR mice and enhanced salisphere formation in vitro. These promising data indicated that the GDNF pathway may have potential therapeutic benefits for IR-induced xerostomia. Another group of researchers led by Aaron Palmon and Doron Aframian have also isolated salivary epithelial progenitor cells using the integrin alpha6 beta1 cell marker [37]. They demonstrated that SG integrin alpha6 beta1-expressing cells contained a subpopulation of SG-specific progenitor cells [35]. This group is clinically oriented and has succeeded in incorporating a cell-separation method based on magnetic-affinity cell sorting microbeads, as well as demonstrating that cultured single alpha6 beta1 cell-originated clones could be cryopreserved for 3 years without exhibiting genetic or functional instability when compared to non-cryopreserved salivary cells.

5.3 Multipotent Mesenchymal Stromal Cell (MSCs) Therapy for Salivary Glands

A second source of adult stem cells that has shown promises in reestablishing salivary function is the bone marrow (BM) mesenchymal stromal/stem cells (MSCs). More recently, stromal cells from the adipose tissue have demonstrated a comparable success for repairing SGs. Transplantation methods for bone marrow- and adipose tissue-derived stromal cells are successfully being used in the clinic because these cells can now be harvested easily and safely from patients. Considering that methods for harvesting MSCs from the adipose tissue are less invasive than those used for the BM, the adipose-derived stromal cells may gain more popularity among clinicians.

In this chapter, the following definitions, taken from two positional papers of the International Society for Cellular Therapy (ISCT), will be used [6, 11]. A "stromal cell" is a connective tissue cell of any organ. A "stem cell" has the ability to self-renew and is multipotent. The "multipotent mesenchymal stromal cells" (MSCs) are fibroblast-like plastic-adherent cells, regardless of the tissue from which they are isolated. The term "mesenchymal stem cells" (also MSCs) is used only for cells that meet specified stem cell criteria. The widely recognized acronym, MSC, may be used for both cell populations, but investigators must clearly define the scientifically correct designation in their publications [11]. Adipose tissue, either resected as an intact tissue or aspirated using tumescent liposuction, is minced and digested by enzymes. The released cells are defined as the adipose tissue-derived "stromal vascular fraction" (SVF). The SVF consists of a heterogeneous population of cells that includes adipose stromal, hematopoietic stem/progenitor cells, endothelial cells, erythrocytes, fibroblasts, lymphocytes, monocyte/macrophages, and pericytes [6]. When SVF cells are cultured, a subset of fibroblast-like plastic-adherent cells appears (as observed with bone marrow MSCs). These cells are defined as "adipose tissue-derived stromal cells" (ASCs). Preclinical and clinical studies on the use of these adult stromal cell populations (ASCs, SVF cells, bone marrow MSCs) in regenerative medicine have increased in recent years. This trend has been followed as well in SG research. Within the past 5 years (2011–2015), there were 14 studies using stromal cells as a therapy for SS-like diseased (n = 5 studies) or IR-injured SGs (n = 9 studies) [14, 16–18, 20, 22–25, 32, 43, 52–54]. The mesenchymal stromal cells were either unselected from whole bone marrow [16, 18, 25, 32, 43], cultured from BM [17, 24, 53], cultured from adipose tissue [20, 22, 23, 52], cultured from the SG [14], cultured from the dental pulp [54], or isolated from the umbilical cord [53]. This book chapter will initially review recent advances in the treatment of SG hypofunction using ASCs, then BM MSCs, and finally the authors' experience with the soluble contents/factors from these cells. For the sake of brevity, this chapter will list studies published from year 2010 and forward because our group has previously reviewed cell-based therapies in the treatment of xerostomia with studies from 2000 to 2010 [48].

5.4 Adipose Tissue-Derived Stromal Cell (ASC) Therapy for Irradiation-Injured Salivary Glands

Four studies presented transplanted adipose tissue-derived stromal cells (ASCs) for the regeneration of IR-damaged SG [20, 22, 23, 52]. All four studies indicated that ASCs were effective in improving salivary function. ASC-treated mice reestablished their salivary flow rate (SFR) to 70–75 % of non-IR controls with follow-up times between 5 and 24 weeks post-transplantation. Irradiated but non-ASC-treated mice remained at 35–50 % when compared to non-IR mice (i.e., healthy mice). All studies also reported that ASC-treated mice increased SG weights and preserved their histological morphologies with more acini, a higher cell proliferation activity, reduced apoptosis, higher amylase/mucin levels, and higher microvessel densities than untreated irradiated SGs. The Lim and the Xiong studies were of particular interest because they used human ASCs [23, 52].

Other treatment variables/factors that could be observed from these four ASCs studies were that: (a) One-time injection of ASCs directly into the submandibular glands [20] [52] was as successful as with multiple i.v. injections (1x a week for 3 weeks [23], 2x a week for 6 weeks [22]). (b) Injections of ASCs immediately following the IR injury restored SG function (18 Gy IR) in the studies of [22, 52] and 15 Gy in the Lim study [23], or as far as 10 weeks post-IR were as equally successful, but the IR dose of the latter study was much less (10 Gy) [20]. (c) It is important to note that both mouse and human ASCs could be injected into an immune-competent rodent (mouse and rat) model. For instance, the Lim study [23] and the Xiong study [52] demonstrated that human ASCs could repair and engraft into mouse and rat salivary glands, respectively. In addition, both of these studies were carried out until 12–24 weeks posttransplantation of human ASCs, which differ from other studies that typically followed up transplanted animals for 6–8 weeks.

5.5 Multipotent Mesenchymal Stromal Cell (MSC) Therapy to Restore Irradiation-Induced Salivary Hypofunction

Within the past 5 years (2011–2015), there were five studies using mesenchymal stromal cells as a therapy for IR-injured SGs. The stromal cells were either unselected from whole bone marrow

[25, 43], cultured from BM [24], cultured from SG [14], or cultured from the dental pulp [54]. Both the Lin [25] and the Sumita studies [43] used bone marrow cells (which contained a small fraction of stromal cells) and reported successful functional restoration of SG function to 80–90 % of non-IR controls. Lim and colleagues built their study [24] upon the Lin and the Sumita studies by isolating a homogeneous group of MSCs from BM through clonal cell culture. These cells were named "bone marrow-derived clonal mesenchymal stem cells," BM-cMSCs, and were positive for CD44, Sca-1, while negative for CD34, CD45. Injected BM-cMSCs ameliorated IR-induced SG in mice. At 12 weeks posttransplantation SFR was reestablished to ~80 % of nonirradiated controls (analogous to findings from the Sumita study). Thus, all three studies (the Lin, the Sumita, and then the Lim studies) reported that administration of BM cells (either whole BM, co-cultured with salivary epithelial cells, or clonally cultured as MSC) provided successful restoration of SG function to 80–90 % of non-IR controls and improved histological features such as preserved acinar cells, less apoptotic cells, and increased microvessel density.

Two studies reported the use of MSCs as a cell therapy for IR-damaged SG, but these cells were not isolated from the BM or adipose tissue. Rather MSCs were isolated from human SGs [14] or from the mouse dental pulp [54]. Jeong and colleagues cultured cells digested from human parotid and submandibular glands [14]. These cultured adherent fibroblastic-like cells expressed CD44, CD49f, and CD90 but not CD105, CD34, and CD45. These human salivary MSCs underwent adipogenic, osteogenic, and chondrogenic differentiation as well as differentiation, in vitro, into amylase-expressing cells. Intraglandular injections of human salivary MSCs into rat SGs 1 day following IR were found to reestablish their salivary flow rate to ~ 65 % at the 8 weeks follow-up, while irradiated non-MSC-treated mice remained at ~40 % SFR. Yamamura and colleagues tested if cultured mouse dental pulp cells when differentiated in vitro into dental pulp endothelial cells (DPECs) could be used as a cell source (for their capacity to participate in neovascularization) in the treatment of SG hypofunction following head and neck IR [54]. DPECs were injected into the submandibular glands of 15 Gy irradiated mice at 4 and 14 days postirradiation. At 8 weeks post-IR, mice injected with DPECs reestablished their SFR to 60 % SFR when compared to control (non-IR) mice, while irradiated PBS-injected mice remained at ~40 % SFR. In summary, all the five abovementioned studies suggested that BM or its mesenchymal stromal cells could be used for restoration of IR-induced salivary hypofunction.

5.6 Multipotent Mesenchymal Stromal Cell (MSC) Therapy for Sjögren's Syndrome (SS)

Within the past 5 years (2010–2015), there were four studies using either whole BM, spleen cells, MSCs from BM, or umbilical cord as a therapy to repair SG in SS [16–18, 53]. Our group has been testing the regenerative capacity of BM cells in SGs of NOD mice [16–18, 32, 44], the most frequently used animal model to study SS-like disease. NOD mice display infiltrates of lymphocytes and a gradual loss of salivary function, and the reduced saliva output mimics in part the condition seen in patients with SS [48]. We initially tested if a therapy that reverses end-stage diabetes in NOD mice would affect their SS-like disease [19, 44]. This therapy had two components. The first component was an injection of complete Freund's adjuvant (CFA) to induce endogenous TNF-α to kill disease-causing activated T cells. The second component of the therapy was the transplantation (reintroduction) of major histocompatibility complex (MHC) class I-matched normal splenocytes. We initially chose splenocytes as these were comparable to BM cells in mice. NOD mice that received CFA + splenocytes therapy could reverse both diabetes and xerostomia in SS [44]. Untreated NOD mice showed a continuous decline in salivary flow, followed by hyperglycemia and death. In a subsequent study [18], we tested the long-term 52 weeks post-therapy effects of CFA and MHC class I-matched normal BM cells to 7-week-old

NOD mice (i.e., these mice had not yet developed SS-like disease). At week 52 posttreatment, CFA+BM cell-treated mice were normoglycemic compared to 10 % in the control group. BM cell-treated mice had their SG function (SFR) restored both quantitatively and qualitatively, compared to control NOD mice which continued to have their saliva secretion deteriorate over time.

In humans, SS is usually not diagnosed until an advanced stage when clinical symptoms (such as a decrease in saliva secretion) are manifested. It was unknown if cell-based therapies would be effective in restoring salivary function when given at an advanced stage of sialadenitis and loss of salivary secretory function [16]. Thus, we compared the efficacy of two cell-based therapies (BM versus spleen cells) in halting/reversing salivary hypofunction at two critical time points of SS-like, which were (a) at the "initial phase of SS-like" using 7–8-week-old NOD mice which had normal saliva output and (b) at the "advanced clinical disease phase" using 20-week-old NOD mice with minimal saliva output. Either BM or spleen cell therapies were effective during the initial phase of SS-like disease as SFRs were maintained between 80 and 100 % of presymptomatic levels (baseline SFR). When cell therapies were given at an advanced phase of SS-like disease (20 weeks and older), SFRs improved but were at best 50 % of presymptomatic levels.

The low immunogenicity and (high) immunoregulatory potential of MSCs offer new treatment possibilities for autoimmune diseases. Songlin Wang's group reported that the immunoregulatory activities of BM MSCs were impaired in SS-like disease in NOD mice and SS patients [53]. These authors elegantly demonstrated that injections of mouse BM MSCs or of human umbilical MSCs suppressed the autoimmunity and restored SG secretory function in both the NOD mouse model (SFR was maintained at pre-disease levels, 100 % in 6-week-old NOD and 50–65 % SFR improvement in 16-week-old treated NOD mice) and in SS patients (~40–50 % improvement in unstimulated and stimulated SFR during the 12 months follow-up time). MSC treatment alleviated disease symptoms and directed T cells toward Treg and Th2 while suppressing Th17 and Tfh responses. Intravenously infused MSCs migrated toward the inflammatory regions in a stromal cell-derived factor-1 (SDF-1)-dependent manner, while neutralization of the SDF-1 ligand CXCR4 abolished the effectiveness of BM MSC treatment in NOD mice. Thus, Songlin Wang's group discovered a critical role of the SDF-1/CXCR4 axis (C-X-C chemokine receptor 4 is a receptor for SDF-1) in directing MSC trafficking toward inflammation sites, to exert suppressive activities and improve SG function.

5.7 Possible Mechanisms for the Observed Therapeutic Efficacy of MSC Therapy in Restoring Salivary Function

Four in vivo studies have shown that transplanted BM cells, MSCs, or ASCs differentiated into salivary epithelial, acinar, ductal, or endothelial cells, and thus have proposed the differentiation of transplanted cells as a possible mechanism of action for the measured therapeutic effect in IR-injured SGs [24, 25, 43, 52]. However, few studies quantified this observed phenomenon, and thus the mechanism of action for the therapeutic efficacy of BM and MSC cell therapies remained to be investigated. Two human (observational) studies reported that the frequency of cell differentiation/chimerism (without an IR injury) in SGs was much lower (~1 %) in patients who received a BM or hematopoietic stem cell transplant [42, 46]. This cell transdifferentiation phenomenon was further characterized in vitro by using cell culture models of human MSCs co-cultured with human salivary cells [29], human ASCs co-cultured with mouse salivary cells [23], mouse MSCs co-cultured with mouse salivary cells [38], and mouse ASCs co-cultured with mouse salivary cells [21]. These in vitro studies indicated that stromal cells adopted an epithelial phenotype when co-cultured with salivary epithelial cells or with the conditioned media of salivary cells [21].

There has been progress in the putative molecular cues responsible for the restoration of salivary function observed following salivary epithelial stem cells, MSCs, or ASCs transplantation. Seunghee Cha's group investigated the regulatory factors by proteomics that differentiate MSCs into salivary cells in vitro [38]. These authors reported that, of the 58 proteins detected, three transcription factors involved in SG embryogenesis were selected as potential regulatory molecules in driving the transdifferentiation of multipotent MSCs into salivary epithelial cells: ankyrin repeat domain-containing protein 56, high mobility group protein 20B, and transcription factor E2a. In mouse IR-injured SGs, Quynh Thu Le's group reported that glial cell line-derived neurotrophic factor (GDNF) was highly expressed in Lin^- $CD24^+$ $c\text{-}Kit^+$ $Sca1^+$ salispheres and the GDNF pathway may have potential therapeutic benefits for IR-induced xerostomia [51]. In the SS-like NOD model, Songlin Wang's group discovered the SDF-1/CXCR4 axis directed MSC trafficking toward inflammation sites to exert suppressive activities and to improve SG function [53].

Our group has pursued a relatively more general approach in searching for these molecular cues. We tested if a paracrine mechanism (i.e., transplanted cells secrete cytokines and growth factors to repair salivary tissue) could be the main effect behind the reported improvement in salivary function following cell transplantation [45]. Whole BM cells were lysed, and their soluble intracellular contents, which we coined the term as "bone marrow soup" *(BM Soup)*, were injected into mice with IR-injured SGs. We use the term BM Soup to represent all the yet-to-be-identified soluble components of the cell lysate from whole bone marrow cells. Eight weeks post-IR, BM Soup restored salivary flow rates to normal levels, protected salivary acinar, ductal, myoepithelial, and progenitor cells, increased cell proliferation and blood vessels, and upregulated expression of several tissue remodeling/repair/regenerative genes (MMP2, CyclinD1, BMP7, EGF, NGF). BM Soup was as an efficient therapeutic agent as injections of live whole BM cells (which so far were thought to be essential). Because the BM Soup is an extract from a cell lysate, it is theoretically less immunogenic and tumorigenic. However, the components which were responsible for these promising therapeutic actions remained unknown. In a recent study, to demonstrate that proteins were the active ingredients, we devised a method using proteinase K followed by heating to deactivate proteins and for safe injections into mice. BM Soup and its "deactivated BM Soup" form were injected into mice that had their SGs injured by IR [8]. Results at week 8 post-IR showed the "deactivated BM Soup" was no better than injections of saline, while injections of native BM Soup restored saliva flow and protected salivary cells and blood vessels from IR damage. We demonstrated that the protein components but not the nucleic acids, lipids, or carbohydrates in the BM Soup were the active/therapeutic ingredients for functional salivary restoration following IR. Protein arrays were used to preliminarily identify important cytokines and growth factors in the BM Soup. The protein arrays detected several angiogenesis-related factors (CD26, FGF, HGF, MMP-8, MMP-9, OPN, PF4, SDF-1) and cytokines (IL-1ra, IL-16) in BM Soup.

Our group has also tested BM Soup in the SS-like disease NOD mouse model [32]. This study investigated if injections of BM Soup versus whole BM cells would provide comparable improvements to diseased SGs, and the molecular alterations associated with these treatments [32]. BM Soup was found to restore SFR to normal levels and significantly reduced the focus scores in SGs of NOD mice. More than 1800 proteins in SG cells were quantified by a proteomic approach. Many salivary proteins involved in inflammation and apoptosis were found to be downregulated, whereas those involved in SG biology and development/regeneration were upregulated in the BM Soup-treated mice.

Conclusions

In summary, MSC therapy with the goal of restoring SG function following IR injury or in SS is an efficient and promising approach. When tested in an animal model and using stimulated saliva secretion as a functional quantitative measure, mice/rats with IR-injured SGs which received the MSC therapy restored their SFR to 60–90 % of that measured in normal age-matched animals, while the SFR of IR animals without treatment remained at 35–50 %. Thus, there was a 25–40 % therapeutic improvement between animals receiving the MSC therapy or not. This would be a clinically significant outcome because patients with severe salivary hypofunction due to head and neck IR have 0 % improvement in SFR, if left untreated. In the NOD mouse model, MSC therapy restored SFR 80–100 % if the treatment was given at an initial phase of SS-like disease, while its effectiveness decreased to 50–60 % when given at an advanced phase of SS-like disease. In SS patients, MSC therapy improved SFR by 40–50 %. When tested in the rodent model, MSC therapy was successful in restoring/maintaining the gland normal weights and histology (acinar cells, blood vessels) and upregulated the expression of genes favorable to SG development and regeneration while downregulating inflammation and cell apoptosis. The majority of studies used MSCs obtained from the bone marrow and adipose tissue, but MSCs isolated from the SG and dental pulp also demonstrated a therapeutic efficacy in reestablishing SG function. From that perspective, MSCs from tissues that can be obtained with a less invasive procedure (such as adipose tissue versus bone marrow) or from tissues already removed during the surgical procedure (such as SG or teeth) would be more advantageous to both the patient and the clinician. We were excited that two studies reported the successful transplantation of human adipose tissue-derived MSCs in restoring salivary function in immune-competent mice and rats and that one study (with a follow-up of 12 months) reported a therapeutic effect of umbilical cord MSCs infused to SS patients. We perceived that there were major advances within the past 5 years because all studies using MSC therapies were moderately to highly successful in reestablishing function to IR- and SS-injured SGs. Despite these encouraging results, the cellular and molecular mechanisms that support the use of MSC in repairing injured SGs will need to be deciphered in greater detail.

Acknowledgments We thank Professor Bruce J Baum for his invaluable advice, input, and encouragements during these past several years on the research pertaining to bone marrow cells and its cell soup in the repair of salivary glands. These studies were supported in part by the Canadian Institutes of Health Research (CIHR).

References

1. Aframian DJ, Palmon A. Current status of the development of an artificial salivary gland. Tissue Eng Part B Rev. 2008;14:187–98.
2. Alajbeg I, Falcao DP, Tran SD, Martin-Granizo R, Lafaurie GI, Matranga D, Pejda S, Vuletic L, Mantilla R, Leal SC, Bezerra AC, Menard HA, Kimoto S, Pan S, Maniegas L, Krushinski CA, Melilli D, Campisi G, Paderni C, Mendoza GR, Yepes JF, Lindh L, Koray M, Mumcu G, Elad S, Zeevi I, Barrios BC, Lopez Sanchez RM, Lassauzay C, Fromentin O, Beiski BZ, Strietzel FP, Konttinen YT, Wolff A, Zunt SL. Intraoral electrostimulator for xerostomia relief: a long-term, multicenter, open-label, uncontrolled, clinical trial. Oral Surg Oral Med Oral Pathol Oral Radiol. 2012;113:773–81.
3. Baum BJ, Kok M, Tran SD, Yamano S. The impact of gene therapy on dentistry: a revisiting after six years. J Am Dent Assoc. 2002;133:35–44.
4. Baum BJ, Tran SD. Synergy between genetic and tissue engineering: creating an artificial salivary gland. Periodontol 2000. 2006;2000(41):218–23.
5. Bhide SA, Miah AB, Harrington KJ, Newbold KL, Nutting CM. Radiation-induced xerostomia: pathophysiology, prevention and treatment. Clin Oncol (R Coll Radiol). 2009;21:737–44.
6. Bourin P, Bunnell BA, Casteilla L, Dominici M, Katz AJ, March KL, Redl H, Rubin JP, Yoshimura K, Gimble JM. Stromal cells from the adipose tissue-derived stromal vascular fraction and culture expanded adipose tissue-derived stromal/stem cells: a joint statement of the International Federation for Adipose Therapeutics and Science (IFATS) and the International Society for Cellular Therapy (ISCT). Cytotherapy. 2013;15:641–8.
7. Coppes R, Stokman M. Stem cells and the repair of radiation-induced salivary gland damage. Oral Dis. 2011;17:143–53.
8. Fang D, Hu S, Liu Y, Quan VH, Seuntjens J, Tran SD. Identification of the active components in bone marrow soup: a mitigator against irradiation-injury to salivary glands. Sci Rep. 2015;5:16017.
9. Feng J, van der Zwaag M, Stokman MA, van Os R, Coppes RP. Isolation and characterization of human

salivary gland cells for stem cell transplantation to reduce radiation-induced hyposalivation. Radiother Oncol. 2009;92:466–71.
10. Fox PC. Acquired salivary dysfunction. drugs and radiation. Ann N Y Acad Sci. 1998;842:132–7.
11. Horwitz EM, Le Blanc K, Dominici M, Mueller I, Slaper-Cortenbach I, Marini FC, Deans RJ, Krause DS, Keating A. Clarification of the nomenclature for MSC: the international society for cellular therapy position statement. Cytotherapy. 2005;7:393–5.
12. Jemal A, Siegel R, Xu J, Ward E. Cancer statistics, 2010. CA Cancer J Clin. 2010;60:277–300.
13. Jensen DH, Oliveri RS, Trojahn Kolle SF, Fischer-Nielsen A, Specht L, Bardow A, Buchwald C. Mesenchymal stem cell therapy for salivary gland dysfunction and xerostomia: a systematic review of preclinical studies. Oral Surg Oral Med Oral Pathol Oral Radiol. 2014;117:335–42. e1
14. Jeong J, Baek H, Kim YJ, Choi Y, Lee H, Lee E, Kim ES, Hah JH, Kwon TK, Choi IJ, Kwon H. Human salivary gland stem cells ameliorate hyposalivation of radiation-damaged rat salivary glands. Exp Mol Med. 2013;45:e58.
15. Kagami H. The potential use of cell-based therapies in the treatment of oral diseases. Oral Dis. 2015;21: 545–9.
16. Khalili S, Faustman DL, Liu Y, Sumita Y, Blank D, Peterson A, Kodama S, Tran SD. Treatment for salivary gland hypofunction at both initial and advanced stages of Sjogren-like disease: a comparative study of bone marrow therapy versus spleen cell therapy with a 1-year monitoring period. Cytotherapy. 2014;16:412–23.
17. Khalili S, Liu Y, Kornete M, Roescher N, Kodama S, Peterson A, Piccirillo CA, Tran SD. Mesenchymal stromal cells improve salivary function and reduce lymphocytic infiltrates in mice with Sjogren's-like disease. PLoS One. 2012;7:e38615.
18. Khalili S, Liu Y, Sumita Y, Maria OM, Blank D, Key S, Mezey E, Tran SD. Bone marrow cells are a source of undifferentiated cells to prevent Sjogren's syndrome and to preserve salivary glands function in the non-obese diabetic mice. Int J Biochem Cell Biol. 2010;42:1893–9.
19. Kodama S, Kuhtreiber W, Fujimura S, Dale EA, Faustman DL. Islet regeneration during the reversal of autoimmune diabetes in NOD mice. Science. 2003;302:1223–7.
20. Kojima T, Kanemaru S, Hirano S, Tateya I, Ohno S, Nakamura T, Ito J. Regeneration of radiation damaged salivary glands with adipose-derived stromal cells. Laryngoscope. 2011;121:1864–9.
21. Lee J, Park S, Roh S. Transdifferentiation of mouse adipose-derived stromal cells into acinar cells of the submandibular gland using a co-culture system. Exp Cell Res. 2015;334:160–72.
22. Li Z, Wang Y, Xing H, Wang Z, Hu H, An R, Xu H, Liu Y, Liu B. Protective efficacy of intravenous transplantation of adipose-derived stem cells for the prevention of radiation-induced salivary gland damage. Arch Oral Biol. 2015;60:1488–96.
23. Lim JY, Ra JC, Shin IS, Jang YH, An HY, Choi JS, Kim WC, Kim YM. Systemic transplantation of human adipose tissue-derived mesenchymal stem cells for the regeneration of irradiation-induced salivary gland damage. PLoS One. 2013a;8:e71167.
24. Lim JY, Yi T, Choi JS, Jang YH, Lee S, Kim HJ, Song SU, Kim YM. Intraglandular transplantation of bone marrow-derived clonal mesenchymal stem cells for amelioration of post-irradiation salivary gland damage. Oral Oncol. 2013b;49:136–43.
25. Lin CY, Chang FH, Chen CY, Huang CY, Hu FC, Huang WK, Ju SS, Chen MH. Cell therapy for salivary gland regeneration. J Dent Res. 2011;90:341–6.
26. Lombaert IM, Brunsting JF, Wierenga PK, Faber H, Stokman MA, Kok T, Visser WH, Kampinga HH, de Haan G, Coppes RP. Rescue of salivary gland function after stem cell transplantation in irradiated glands. PLoS One. 2008;3:e2063.
27. Maimets M, Bron R, de Haan G, van Os R, Coppes RP. Similar ex vivo expansion and post-irradiation regenerative potential of juvenile and aged salivary gland stem cells. Radiother Oncol. 2015;16:443–8.
28. Maria OM, Maria O, Liu Y, Komarova SV, Tran SD. Matrigel improves functional properties of human submandibular salivary gland cell line. Int J Biochem Cell Biol. 2011a;43:622–31.
29. Maria OM, Tran SD. Human mesenchymal stem cells cultured with salivary gland biopsies adopt an epithelial phenotype. Stem Cells Dev. 2011;20:959–67.
30. Maria OM, Zeitouni A, Gologan O, Tran SD. Matrigel improves functional properties of primary human salivary gland cells. Tissue Eng Part A. 2011b;17: 1229–38.
31. Melvin JE, Yule D, Shuttleworth T, Begenisich T. Regulation of fluid and electrolyte secretion in salivary gland acinar cells. Annu Rev Physiol. 2005;67: 445–69.
32. Misuno K, Tran SD, Khalili S, Huang J, Liu Y, Hu S. Quantitative analysis of protein and gene expression in salivary glands of Sjogren's-like disease NOD mice treated by bone marrow soup. PLoS One. 2014;9:e87158.
33. Nanduri LS, Baanstra M, Faber H, Rocchi C, Zwart E, de Haan G, van Os R, Coppes RP. Purification and ex vivo expansion of fully functional salivary gland stem cells. Stem Cell Reports. 2014;3:957–64.
34. Nanduri LS, Lombaert IM, van der Zwaag M, Faber H, Brunsting JF, van Os RP, Coppes RP. Salisphere derived c-Kit+ cell transplantation restores tissue homeostasis in irradiated salivary gland. Radiother Oncol. 2013;108:458–63.
35. Neumann Y, David R, Stiubea-Cohen R, Orbach Y, Aframian DJ, Palmon A. Long-term cryopreservation model of rat salivary gland stem cells for future therapy in irradiated head and neck cancer patients. Tissue Eng Part C Methods. 2012;18:710–8.
36. Nguyen TT, Mui B, Mehrabzadeh M, Chea Y, Chaudhry Z, Chaudhry K, Tran SD. Regeneration of tissues of the oral complex: current clinical trends and research advances. J Can Dent Assoc. 2013;79:d1.

37. Palmon A, David R, Neumann Y, Stiubea-Cohen R, Krief G, Aframian DJ. High-efficiency immunomagnetic isolation of solid tissue-originated integrin-expressing adult stem cells. Methods. 2012;56:305–9.
38. Park YJ, Koh J, Gauna AE, Chen S, Cha S. Identification of regulatory factors for mesenchymal stem cell-derived salivary epithelial cells in a co-culture system. PLoS One. 2014;9:e112158.
39. Patel VN, Hoffman MP. Salivary gland development: a template for regeneration. Semin Cell Dev Biol. 2014;25-26:52–60.
40. Pringle S, Maimets M, van der Zwaag M, Stokman MA, van Gosliga D, Zwart E, Witjes MJ, de Haan G, van Os R, Coppes RP. Human salivary gland stem cells functionally restore radiation damaged salivary glands. Stem Cells. 2016;34:640–52.
41. Pringle S, van Os R, Coppes RP. Concise review: adult salivary gland stem cells and a potential therapy for xerostomia. Stem Cells. 2013;31:613–9.
42. Souza LN, Faria DR, Dutra WO, Gomes CC, Gomez RS. Microchimerism in labial salivary glands of hematopoietic stem cell transplanted patients. Oral Dis. 2011;17:484–8.
43. Sumita Y, Liu Y, Khalili S, Maria OM, Xia D, Key S, Cotrim AP, Mezey E, Tran SD. Bone marrow-derived cells rescue salivary gland function in mice with head and neck irradiation. Int J Biochem Cell Biol. 2011;43:80–7.
44. Tran SD, Kodama S, Lodde BM, Szalayova I, Key S, Khalili S, Faustman DL, Mezey E. Reversal of Sjogren's-like syndrome in non-obese diabetic mice. Ann Rheum Dis. 2007;66:812–4.
45. Tran SD, Liu Y, Xia D, Maria OM, Khalili S, Wang RW, Quan VH, Hu S, Seuntjens J. Paracrine effects of bone marrow soup restore organ function, regeneration, and repair in salivary glands damaged by irradiation. PLoS One. 2013;8:e61632.
46. Tran SD, Redman RS, Barrett AJ, Pavletic SZ, Key S, Liu Y, Carpenter A, Nguyen HM, Sumita Y, Baum BJ, Pillemer SR, Mezey E. Microchimerism in salivary glands after blood- and marrow-derived stem cell transplantation. Biol Blood Marrow Transplant. 2011a;17:429–33.
47. Tran SD, Sugito T, Dipasquale G, Cotrim AP, Bandyopadhyay BC, Riddle K, Mooney D, Kok MR, Chiorini JA, Baum BJ. Re-engineering primary epithelial cells from rhesus monkey parotid glands for use in developing an artificial salivary gland. Tissue Eng. 2006;12:2939–48.
48. Tran SD, Sumita Y, Khalili S. Bone marrow-derived cells: a potential approach for the treatment of xerostomia. Int J Biochem Cell Biol. 2011b;43:5–9.
49. Vissink A, Burlage FR, Spijkervet FK, Jansma J, Coppes RP. Prevention and treatment of the consequences of head and neck radiotherapy. Crit Rev Oral Biol Med. 2003a;14:213–25.
50. Vissink A, Jansma J, Spijkervet FK, Burlage FR, Coppes RP. Oral sequelae of head and neck radiotherapy. Crit Rev Oral Biol Med. 2003b;14:199–212.
51. Xiao N, Lin Y, Cao H, Sirjani D, Giaccia AJ, Koong AC, Kong CS, Diehn M, LE QT. Neurotrophic factor GDNF promotes survival of salivary stem cells. J Clin Invest. 2014;124:3364–77.
52. Xiong X, Shi X, Chen F. Human adipose tissue derived stem cells alleviate radiationinduced xerostomia. Int J Mol Med. 2014;34:749–55.
53. Xu J, Wang D, Liu D, Fan Z, Zhang H, Liu O, Ding G, Gao R, Zhang C, Ding Y, Bromberg JS, Chen W, Sun L, Wang S. Allogeneic mesenchymal stem cell treatment alleviates experimental and clinical Sjogren syndrome. Blood. 2012;120:3142–51.
54. Yamamura Y, Yamada H, Sakurai T, Ide F, Inoue H, Muramatsu T, Mishima K, Hamada Y, Saito I. Treatment of salivary gland hypofunction by transplantation with dental pulp cells. Arch Oral Biol. 2013;58:935–42.

6 Directed Cell Differentiation by Inductive Signals in Salivary Gland Regeneration: Lessons Learned from Pancreas and Liver Regeneration

Yun-Jong Park and Seunghee Cha

Abstract

Xerostomia (dry mouth) is a deleterious condition that patients with radiation therapy for head and neck cancer or autoimmune Sjögren's syndrome suffer from. Current remedies for this condition provide no substantial relief of xerostomia. As a result, new alternatives to these palliative remedies, such as artificial salivary glands, gene therapy, or cell-based interventions, are on the horizon. An urgent demand for acquisition of knowledge on stem cell regulation, which is critical for salivary gland regeneration, has allowed systematic and mechanistic research endeavor focusing on the identification of key regulators for cell lineage determination. This book chapter summarizes the key inductive signals, which include extrinsic factors secreted from the microenvironment and cell intrinsic factors that drive differentiation of the stem cells into the cells of the pancreas, liver, and salivary glands as they share the endodermal origin during development. The plethora of information available in pancreas and liver regeneration studies provides insight into key signals that govern vital processes during orchestrated stem cell differentiation for salivary epithelial cells. Some examples of transdifferentiation between differentiated cells of different organs and in vivo applications of inductive factors offer perspectives on future clinical applications with improved safety and efficacy.

6.1 Introduction

Saliva contains various components that play critical roles in oral health. The production of saliva from the salivary glands (SGs) can be significantly altered by pathological conditions, such as autoimmune Sjögren's syndrome (SjS) or radiation therapy for head and neck cancer [1]. SjS is characterized by severe dry eyes (keratoconjunctivitis sicca) and dry mouth (xerostomia) with various

S. Cha, DDS, PhD (✉)
Department of Oral and Maxillofacial Diagnostic Sciences, University of Florida College of Dentistry, Gainesville, FL 32610, USA
e-mail: scha@dental.ufl.edu

Y.-J. Park
Department of Pharmacology, The University of North Carolina at Chapel Hill, Chapel Hill, NC 27599-3280, USA

S. Cha (ed.), *Salivary Gland Development and Regeneration*, DOI 10.1007/978-3-319-43513-8_6

complications in multiple organs. Although SjS can develop at any age, it is more prevalent in elderly females than in males. Radiation therapy for patients with head and neck cancer results in acinar cell death in the SGs causing pathological xerostomia. Roughly 50,000 cases of head and neck cancer are reported in the United States every year ranking as the fifth most common malignancy worldwide [2]. Upon diagnosis, the standard of care is dictated by tumor stage, and, for locally advanced tumors, surgical resection followed by radiotherapy is recommended. Due to the positioning of many oral tumors, non-affected tissues, such as the SGs, are often exposed to therapeutic radiation. Radiation damage to the oral tissues results in significant morbidity and diminished quality of life [3–5].

Recently, studies have reported that SGs are amenable to a stem cell-based approach for regeneration following injury [6, 7]. The emerging concepts from recent studies suggest that different types of stem cells, such as embryonic stem (ES) cells, adult stem cells from the bone marrow (BM) and adipocyte, and SG stem cells/progenitor cells (SSPCs), and induced pluripotent stem cells (iPS) cells, have great potential for SG regeneration [7, 8]. Of those stem cells, SSPCs have provided insight into future therapeutics as they have demonstrated a notable capacity to differentiate into salivary epithelial-like cells [9], and BM-derived mesenchymal stem cells (MSCs) have added their superb value to current research endeavor to restore glandular function [10–12].

In this book chapter, we describe molecules that are known to trigger or promote stem cell differentiation into SG progenitors (SPCs) or SG epithelial cells (SECs) by specifically focusing on factors available in the SG microenvironment (extrinsic factors) and transcription factors (TFs) (intrinsic factors) expressed in SSPCs or BM-MSCs. In addition, we detail TFs that are known to determine cell fate in the field of pancreas and liver regeneration to exploit the knowledge available. The rationale for our focus on these molecules is straightforward. In real clinical cases, patients with irradiation or SjS rarely recover their secretory function. In other words, once harmful radiation is applied or autoreactive cells infiltrate the glands, the damage is most likely permanent. This yields two possible scenarios. In the first, SSPCs are completely depleted. This will require supplement of in vitro grown SSPCs, which can be procured prior to irradiation or SPCC-derived and differentiated SECs into the glands. In the second, SSPCs are spared from damage. However, there is a lack of critical inductive signals for differentiation from SSPCs to SECs in the injured SGs. Interventions for the second scenario will require providing necessary inductive signals to the SSPCs, which include extrinsic factors, intrinsic factors, and molecules that relay the signals generated by extrinsic and/or intrinsic factors. In most advanced clinical cases, considering the extent of radiation/inflammatory damage and fibrotic/atropic changes that follow, we can indisputably assume that neither SSPCs nor inductive signals may exist in the severely damaged SGs.

Therefore, orchestration of these signals in a timely and a hierarchical manner in conjunction with replenishing SSPCs can be an attractive option for reprograming of stem cells to achieve functional SG regeneration. Currently, information on the inductive signals for stem cell differentiation in the field of SG research is extremely scarce, especially TFs in SG regeneration, in contrast to the plethora of information available in the pancreas or liver regeneration. In this book chapter, we will present extrinsic factors and TFs with an emphasis on pancreatic/liver TFs for their importance in lineage determination. The latter information provides insights into the overview of stem cell research in general and can be utilized to confidently guide future applications of SG stem cells. In addition, we add some of our recent findings on TFs identified in mouse BM-MSC-derived SPCs in vitro. Some examples of in vivo applications of inductive signals to mouse models were described as well. We conclude this chapter with important lessons that we can learn from currently available studies.

6.2 Inductive Signals for Stem or Nonstem Cell Lineage Determination in Endodermal Organs

6.2.1 Extrinsic Factors

Understanding a SG microenvironment that involves complex processes of cell–cell communication, cell–matrix interaction, and cell signal transduction within a three-dimensional structure is a critical step toward stem cell reprogramming and SG reconstruction [13–15]. Extrinsic factors available in or from the SG microenvironment contribute to SG development or SSPC differentiation by generating appropriate scaffolds or key inductive signals [16]. Specifically, the extracellular matrix (ECM) is an important element of the cellular niche and supplies critical biochemical and physical signals to initiate cellular functions in the tissue [17]. Therefore, researchers in the field of tissue engineering focus on various biomaterial scaffolds for tissue regeneration or repair that mimic some of the critical properties of the ECM-like cellular microenvironment. For instance, the scaffold material Matrigel®, which contains basement membrane proteins secreted by mouse sarcoma cells, has been used to reconstitute tissue-specific ECM to control stem cell fate [18]. Although varying levels of success have been achieved with this product, there are some limitations compared to natural scaffold materials such as silk, which has a wide range of elasticity and pore size (allowing tissue-specific scaffold formation and nutrition access), the ability to biodegrade, and low toxicity and immunogenicity [19, 20].

Other studies have suggested the use of polyethylene glycol hydrogel (PEG-hydrogel) for SG regeneration [21]. It is inert due to its highly hydrated and uncharged structure for introduction of growth factors, ECM proteins, or mimetics to control cell behavior. Furthermore, the stiffness, degradability, and mesh size of PEG-hydrogel can be controlled by the composition and relative amounts of PEG macromers [21–26]. PEG-hydrogel also has been utilized successfully to culture and control the behavior of various cell types, including MSC [22, 27], pancreatic β-cells [24, 25] and neurons [26]. Recent research has demonstrated that PEG-hydrogels are bioinert and enhance cell–matrix and cell–cell interactions that are commonly utilized to maintain survivability of sensitive cell types [22, 24, 28, 29]. As cell–cell interactions, in particular, play a vital role in SG cell functions in vitro and during gland development [30–33], utilization of PEG-hydrogel for SG cell aggregation into microspheres increases long-term viability of hydrogel-encapsulated SG cells [34]. The detailed information on these scaffold materials can be found in Chap. 8 of this book.

Endodermal-originated organs, such as major SG (submandibular and sublingual), pancreas, and liver, are regulated by complex processes of development, which involves the balancing of interplay among common signaling pathways. Specific growth factors and differentiation factors are released from the microenvironment, which controls the expression of a set of TFs resulting in the generation of targeted endodermal cell types. The initial interactions usually confer competence to respond to additional inductive signals that establish organ determination and specification at particular time points referred to as competence windows [35]. The importance of the competence windows is exemplified during liver and pancreatic development. For instance, when bone morphogenetic protein (BMP) activity generated by the adjacent mesoderm-derived structures operates on the ventral endoderm, hepatic induction ensues during early embryonic development. With active BMP signaling, ventral pancreas progenitors are located away from the midline endoderm, whereas BMP inhibition in the dorsal endoderm plays a crucial function in the acquisition of a pancreatic fate [36, 37]. This interesting phenomenon illustrates that despite our limited knowledge of extrinsic factors controlling patterning, proliferation, and differentiation in the pancreas development, there is strong evidence observed from all vertebrate model organisms that complex spatiotemporal combinations of common signaling pathways are important in the specification of endodermal cell fate [38–40]. A list of such important pathways includes the sonic hedgehog, Wnt, retinoid, and notch pathways, activin/BMP, fibroblast growth factor (FGF), vascular endothelial growth factor (VEGF), epidermal growth factor (EGF), and hepatocyte growth factor (HGF) signaling pathways [41]. FGFs, such as FGF7, FGF8, and FGF10, are known to generate MSC-derived SPCs, proving the capacity of cell differentiation into c-Kit$^+$ ductal cells after transplantation, with the involvement of epithelial sonic hedgehog signaling pathways [8, 42]. Some of these extrinsic and intrinsic factors are summarized in Table 6.1.

In a rather unconventional approach as a proof of concept, BM soup, which consists of the soluble contents of lysed BM cells, was delivered via

Table 6.1 Examples of inducing/promoting factors for target cell differentiation in vitro

Extrinsic	Expression Pancreas	Liver	Salivary glands	Functions	Ref.
FGFs	Dalvi et al. (2009) Hardikar et al. (2003)	Si-Tayeb et al. (2010)	Lombaert et al. (2011) Patel et al. (2006)	β-cell differentiation and cluster formation Haploinsufficiency of FGF or its receptor showing severe defects in the survival and function of SPCs Induced hepatic specification when pluripotent stem cells were used in monolayer culture	[8, 42–45]
EGF	Cras-Meneur et al. (2001) Dalvi et al. (2009)	Kitade et al. (2013)	Kashimata et al. (2000)	High concentration being inhibitory to β-cell differentiation Acted as a regulating factor for proliferation and/or differentiation of liver progenitor cells (LPCs) Showed important role in SG organogenesis, especially in branching morphogenesis	[43, 46–48]
Betacellulin	Demeterco et al. (2000)			Formation of islet-like clusters Induction of β-cell differentiation	[49]
Activin A	Demeterco et al. (2000)			Increased insulin content	[49]
Nicotinamide	Chen et al. (2004), Segev et al. (2004), Tang et al. (2004), Sun et al. (2007), Sun et al. (2007), Wu (2007), Chao et al. (2008), Gabr et al. (2008), Gao et al. (2008), Otonkoski et al. (1993), Gao et al. (2008)			Differentiation of stem cells of various origins into iPS cells Increased insulin content, DNA content, expression of insulin, glucagon, and somatostatin genes	[50–60]
Exendin-4	Tang et al. (2004) Timper et al. (2006) Wu et al. (2007) Gabr et al. (2008) Aquayo-Mazzucato (2011)			Differentiation of murine BM stem cells into iPS cells Formation of insulin expression cells generated from adipose tissue-derived MSCs Differentiation of BM-MSCs into iPS cells Increased insulin release by iPS cells generated from mouse ES cells	[52, 55, 57, 61, 62]
HGF	Mashima et al. (1996) Otonkoski et al. (1996)	Onitsuka et al. (2010) Si-Tayeb et al. (2010) Ishikawa et al. (2012)		Differentiation of pancreatic acinar cells into iPS cells Increased number of iPS cells in cultured human islets Induced differentiation into LPCs	[63, 64] [45, 65, 66]
Gastrin	Wang et al. (1993, 1997)			Stimulation of islet differentiation and islet growth	[67, 68]
Glucose	Halban et al. (1987)			Low concentration increased insulin content High concentrations increased β-cell replication	[69]

Oncostatin M		Kamiya et al. (1999, 2002) Matsui et al. (2002)		Promoted hepatocyte maturation in vitro by inducing the formation of adherent junctions	[70–72]
Dexamethasone		Banas et al. (2007) Lee et al. (2004)		BM-MSC or $CD105^+$ adipose-derived MSC differentiation into hepatocyte-like cells	[73, 74]
Intrinsic	**Expression**				
	Pancreas	**Liver**	**Salivary gland**	**Functions**	**Ref.**
Ascl3			Arany et al. (2011)	Characterized as a SG progenitor marker Involved in the regeneration of SG cells	[75]
C/EBPα/β		Courtois et al. (1987) Cereghini et al. (1988) Tadashi et al. (2003)		Promoted hepatoblast differentiation into mature hepatocyte	[76–78]
Foxa2	Gao et al. (2007) Gao et al. (2008)	Xu et al. (2012)		Played a key role in hepatogenesis in early development	[79–81]
Gata4	Carrasco et al. (2012)	Carrasco et al. (2012)		Highly expressed in early pancreatic budding stage and remained in mature acinar cells Played a key role in liver development during embryogenesis and regulated Hnf4 expression	[82]
Hmg20b			Park et al. (2015)	Increased expression in cocultured MSCs into SPCs	[12]
Hnf1b (Tcf2)	Haumaitre et al. (2005)			Expressed during pancreatic development Pancreatic agenesis in E13.5 in Tcf2 deficient mice	[83]
Hnf4		Watt et al. (2003)		Regulated key regulatory genes involved in hepatocyte maturation	[84]
Hnf6	Pierreux et al. (2004)	Pierreux et al. (2004)		Broad expression throughout development and regulate α and acinar cells in the liver and pancreas	[85]
Isl1	Du et al. (2009)			Stimulate essential growth factor for pancreatic development Initiate expression in mouse pancreatic mesenchyme at E9	[86]
MafA	Aguayo-Mazzucato et al. (2011)	Aguayo-Mazzucato et al. (2011)		Detected in late developmental stage and involved in stimulation of mature β-cell differentiation	[61]

(continued)

Table 6.1 (continued)

Mist1	Pin et al. (2000)		Park et al. (2015)	Expressed in a wider array of tissues including the acinar cells of SGs and the serous-secreting cells found in the stomach and prostate Defect in mice resulting a loss of correct cellular organization in pancreas and SGs during embryogenesis	[12, 87]
NeuroD1	Imai et al. (2007) Gu et al. (2010)			Found in all endocrine cell of mature islets	[88, 89]
Ngn3	Magenheim et al. (2011)		Magenheim et al. (2011)	Required for endocrine cell specification in mice and also initiates endocrine commitment in pancreatic and SGECs	[90]
Nkx2.2	Papizan et al. (2011)			Detected in pancreatic precursor cells and stimulate β-cell differentiation	[91]
Nkx6.1	Taylor et al. (2013)			Expressed in early multipotent pancreatic progenitors and have key role in β-cell differentiation	[92]
Mnx1	Flanagan et al. (2014) Bonnefond et al. (2013)			Identified their expression in early stage of the developing pancreas Expressed only in adult β-cells at final developmental stage	[93, 94]
Pax6	Hart et al. (2013)			Activated in early pancreas and maintained throughout the adult islet cells	[95]
Pdx1	Jennings et al. (2013) Gu et al. (2002) Brissova et al. (2005)			Plays a role throughout all phases in development of pancreas and highly expressed in adult β-cell Decreased expression in both human and rodent model type 2 diabetes islets	[96–98]
Ptf1a	Yoshitomi et al. (2004) Weedon et al. (2014)		Park et al. (2015)	Broad expression in dorsal and ventral pancreatic bud in mice and their mutation leads to pancreatic agenesis Highly increased in differentiating MSCs during coculture	[12, 99, 100]
Sox2			Lombaert et al. (2011)	Highly abundant and involved in duct lineage differentiation in the SGs	[8]
Sox9	Kawaguchi et al. (2013)	Mead et al. (2013) Kawaguchi et al. (2013)	Mead et al. (2013)	Essential for the maintaining multipoint progenitor population in mice and have effective role in endodermal cells such as the pancreas, lung, liver, SG, and gut	[101, 102]
Sox17	Kanai-Azuma et al. (2002)			Detected for a short time period during development Necessary in endoderm formation and pancreatic fate	[103]
Tcf3			Park et al. (2015)	Protein and gene expression highly increased in cocultured MSCs converting to SPCs	[12]

intraglandular delivery or systemic delivery, resulting in restored saliva flow in mice with irradiated SG [104]. The paracrine effects of MSCs were reconfirmed in various studies where BM-MSCs exerted the capacity for organ repair by secretion of cytokines, chemokines, and growth factors [105]. In a subsequent study, the same group reported that deactivated soup treated with proteinase K digestion had no effect on the recovery of secretory function in irradiated mice at 8 weeks postirradiation [106]. Protein arrays revealed that native proteins, especially angiogenesis-related factors such as CD26, FGF, HGF, MMP-8, MMP-9, OPN, PF4, SDF-1, and cytokines IL-1ra and IL-16, in the BM soup were the therapeutic ingredients [106]. The studies once more emphasize the importance of extrinsic factors that promote tissue regeneration and affirm MSCs as trophic mediators.

Two rare genetic syndromes, aplasia of lacrimal and salivary glands (ALSG) and lacrimo-auriculo-dento-digital syndrome (Ladd syndrome), in humans shed light on a growth factor and its receptor pathway that are essential for progenitor cells to initiate and form SGs. These diseases, caused by mutations resulting in haplo insufficiency of Fgf10 or its receptor Fgfr2B, are characterized by severe defects in the survival and function of SPCs [107, 108]. Embryos do not develop SGs in mice that have both copies of Fgf10 or Fgfr2b deleted ($Fgf10^{-/-}$ and $Fgfr2b^{-/-}$). Therefore, SPCs require Fgf10/Fgfr2b signaling to survive and initiate organogenesis of SGs. It has also been demonstrated in vivo that Fgf7, another Fgfr2b ligand, has an effect on SPCs [109]. Fgf7 injections before and after gland irradiation enhanced the number of progenitor cells. As a consequence, a higher number of progenitor cells remained after radiation, forming more saliva-producing cells that prevented radiation-induced hyposalivation [8]. Addressing sequential events in development and differentiation can be challenging due to the fact that signaling events are tightly associated with each other and they function redundantly or intermittently at several developmental stages. It is also important to note that the developmental effects elicited from these common signals can differ greatly according to the developmental stage in which the signal occurs [8].

6.2.2 Intrinsic Factors (Transcription Factors)

Cellular programs regarding proliferation, potency, and cell fate determination can be mediated by signal transduction events that modulate TFs expression and/or activation. As we previously mentioned, the specification of cell types under normal development is controlled by extrinsic factors that impose regional characteristics on specific progenitor cells at early developmental stages. Signaling molecules play key roles by controlling the establishment of distinct progenitor domains that can be defined by unique expression profiles of TFs. Some of these TFs function as more broad region-specifying determinants, while others are expressed exclusively in individual progenitor domains in which they mediate highly selective functions [110].

Many of these factors, but not all, contain a protein structure motif called basic helix-loop-helix (bHLH). Proteins of the bHLH superfamily have two highly conserved and functionally distinct domains of the basic domain located at the amino-terminal end and of the HLH domain located at the carboxy-terminal end [111]. The basic domain binds the TF to DNA at consensus hexanucleotide sequence known as the E-box, and the HLH domain facilitates interactions with other protein subunits to form homeodimers or heterodimers. The heterogeneity in the E-box sequence that is recognized and the dimers formed by different bHLH proteins determine how TFs of the bHLH family control diverse developmental functions through transcriptional regulation [112].

TFs promote lineage specification by cross-repressive interactions between TFs driving alternative lineage programs in multipotent progenitors and by induction of additional TFs to further execute the differentiation process [113]. Thus, interplay between TFs that act downstream of extracellular stimuli further enhances the complexity of lineage determination. Due to such complexity in signal pathways, this chapter mainly explains critical TFs that are known to be involved in stem cell reprogramming as well as organogenesis toward endodermal organs,

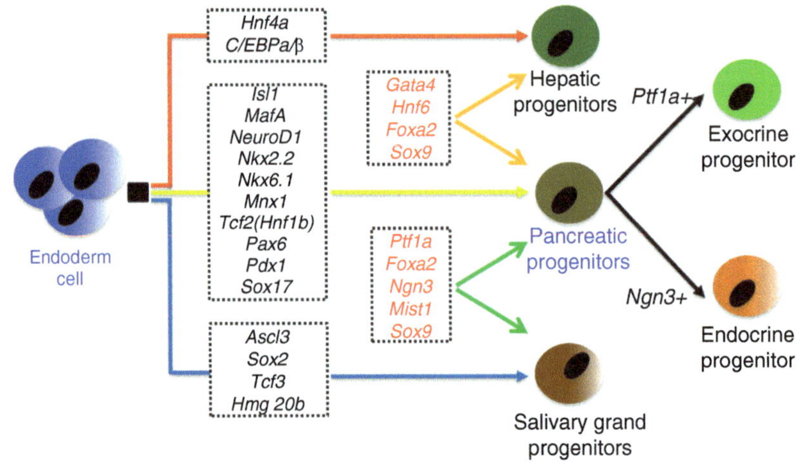

Fig. 6.1 Schematic diagram of endodermal cell lineage differentiation and TFs that are known to be involved in the process. TFs in red are reported to be involved in both organs. Studies identified that the expression of some TFs such as Ptf1a and Pdx1 can drive the cells into pancreatic lineage and promote pancreatic exocrine progenitor or endocrine progenitor differentiation in combination with other TFs

such as the liver, pancreas, and the SGs. Some of exemplary TFs are listed in Fig. 6.1 and Tables 6.1 and 6.2.

6.2.2.1 Transcription Factors Involved in Pancreas Regeneration

Understanding how TFs control early development of pancreatic cells can yield insight into transcriptional regulation of cell fate determination and guide protocols for therapeutic stem cell reprogramming. TFs that are reported to be involved in pancreas cell differentiation and development are listed below and exemplified in Fig. 6.1. Whether these pancreatic TFs exert similar or identical functions in SG cell lineage determination remains unknown.

Isl1 (ISL LIM homeobox 1) appears to be an essential growth factor for pancreatic development in both humans and mice. Isl1 is a pan-endocrine cell marker. When Isl1 is mutated, it exhibits characteristics of diabetic, impaired islet cell maturation, and reduced postnatal islet mass expansion [86]. Isl1 is first expressed in wide range in the pancreatic mesenchyme at E9.0, and expression of Isl1 is maintained in the mature hormone-positive endocrine cells [120, 121].

MafA (V-Maf avian musculoaponeurotic fibrosarcoma oncogene homolog A) in mice is detected in late developmental stages and expressed only in the second wave of insulin positive cells which ultimately become mature β-cells. MafA is known as a maturation marker and is crucial for glucose-responsive β-cells by regulating insulin and glucose transporter 2 [61]. Recent studies have found reduction of MafA levels in mice and in humans under diabetic conditions [122]. Type 2 diabetes mellitus islets may be significantly related to dysfunctional β-cells caused by loss of nuclear MafA [123, 124].

NeuroD1 (neurogenic differentiation factor 1) is found in all endocrine cell types of mature islets. Homozygous NeuroD1 mutations are predicted to cause autosomal recessive neonatal diabetes [88]. Without the NeuroD1, β-cells generated are immature with increased glycolytic gene expression of neuropeptide Y (a hormone that decreases expression after birth) and elevated basal insulin secretion [89].

Nkx2.2 (NK2 Homeobox 2) is a TF involved with β-cell differentiation [125, 126]. Nkx2.2 expression is limited to only pancreatic α- and β-cells along with some portion of pancreatic precursor cells [91].

Nkx6.1 (NK6 Homeobox 1) is broadly expressed in early multipotent pancreatic progenitors [127]. Mice lacking Nkx6.1 have been shown to have a dramatic reduction in β-cells [92]. To maintain the β-cell lineage in conditional

Table 6.2 Examples of inducing/promoting factors for target cell/organ regeneration in animal models

Factors	Manipulations	Target cell /organ	Animal strains	Delivery methods	Outcomes	Ref.
Factors derived from inducers in coculture	mEES-6 (mouse early ES) cocultured with human SG-derived fibroblast	SG (SGECs)	SMG of 8-week-old female SCID/Jcl mice	Direct transplantation	Reconstituted acinar-like or duct like structure	[114]
Ngn3, Pdx1, and MafA	Tissue-specific stem cell-like cells generated by introduction of TFs	Pancreas (β-cell)	$Rag1^{-/-}$, $Rag1^{-/-}$; NOD	Purified adenovirus directly into the splenic lobe of the dorsal pancreas	Ameliorated hyperglycemia by remodeling local vasculature and secreting insulin (over 20 % of infected cells positive for insulin)	[115, 116]
Activin A and betacellulin (BTC)	Treated cells with the factors for 3D culture	Pancreas (β-cell)	Streptozotocin-treated neonatal SD rat	Subcutaneous injection	β-cell mass increased by 69 % at 2 months post-injection	[117]
Expression of Gata4, Hnf1a, Foxa3 & inactivation $p19^{Arf}$	Isolated tail fibroblast and expressed TFs to induce functional hepatocyte-like cells	Liver (Hepatocyte)	$Fah^{-/-}Rag2^{-/-}$ mice (liver failure mice w/ immunodeficiency)	Direct transplantation	iHep cells engrafted into liver sinusoid comprised between 5 % and 80 % of total hepatocytes in iHep-$Fah^{-/-}Rag2^{-/-}$ liver and substantially improved liver functions	[118]
Hnf1b & Foxa3	Induced hepatocytic stem cell (iHepSC) from mouse embryonic fibroblast (mMEF) by TF transfection	Liver (hepatocyte)	$Fah^{-/-}$ mice and 3,5-diethoxycarbonyl-1,4-dihydrocollidine (DDC)-fed mice	Direct transplantation	iHepSCs possessed the capacities of both self-renewal and bipotency of differentiation into both hepatocytes and cholangiocytes	[119]

Nkx6.1 mutants, specifying endocrine precursors are required by marker protein expression, such as Ngn3, Pax4, and Pdx1. In mature pancreas, Nkx6.1 plays a key role in identification of β-cells [92]. Nkx6.1 expression is severely reduced in diabetic and obese mice [128].

Mnx1 (motor neuron and pancreas homeobox 1) is expressed in early stages of the developing pancreas and then decreased to lower levels when entering into gestation. Mouse Mnx1 expression was found in the E9.5 endoderm with detailed expression analysis [128–130]. The expression was gradually restricted to the $Pax6^+$ endocrine population by E15.5. At final stage, it is expressed only in adult β-cells. A recent study reported that homozygous mutation within the DNA-binding homeodomain of Mnx1 caused permanent neonatal diabetes [93]. Similar to the null mouse model, patients with a defective Mnx1 gene did not show obvious exocrine defects, but β-cell numbers were reduced [94].

Tcf2 (transcription factor 2): During early stages of embryogenesis, a high level of Tcf2 (a.k.a. Hnf1b, hepatocyte nuclear factor 1-β) expression persists, and homozygous mutations cause diabetes [83]. Heterozygous loss-of-function Tcf2 mutations result in diabetes in humans, but only homozygous mutations produced diabetes in mice [83]. This could be due to a potentiated single wave of human endocrine differentiation versus the two phases observed in rodents, rendering these human cells more sensitive to Tcf2 dosage. Tcf2-deficient mice exhibit pancreatic agenesis by E13.5, suggesting that the role of Tcf2 in pancreatic development is evolutionarily conserved [83].

Pax6 (paired box 6) is activated in early pancreas and maintained throughout the adult islet cells [131–133]. Pax6 null mice die at birth from brain abnormalities, but embryos have other side effects such as reduced islet cell numbers, impaired hormone synthesis, and defective islet morphogenesis, which indicates a role in allocation and differentiation of endocrine cells [95].

Pdx1 (pancreatic and duodenal homeobox 1), which is also known as insulin promoter factor 1, plays a role throughout all phases in development of pancreas. In early stages, Pdx1 is highly expressed, but expression is restricted in adult β-cells. Pdx1 also regulates the expression of Ins1 and MafA [96]. Besides its main role in β-cells, reports about mouse lineage tracing and Pdx1 expression in mouse acinar tissue demonstrated that $Pdx1^+$ cells mark progenitors of all the mature pancreatic cell types including endocrine, acinar, and ductal cells [97]. A targeted disruption causing homozygous inactivating mutation in the Pdx1/Ifp1 resulted in pancreatic agenesis [98]. An autosomal recessive mutation in the Pdx1 locus is known to cause permanent neonatal diabetes [134]. Pdx1 levels were decreased in both human model and rodent model islets with type 2 diabetes mellitus [135].

Sox17 (SRY(Sex determining region Y)-box 17) is a TF for expression of the high mobility group (HMG) box. It is only observed for a short time period and excluded during mouse pancreatic development [136]. Studies found that Sox17 expression is necessary in early stages of endoderm formation, but it later represses the pancreatic fate [103].

6.2.2.2 Transcription Factors Involved in Liver Regeneration

Combinatorial protein interactions among liver-specific TFs are required to achieve synergistic cellular stimulation of tissue-specific gene expression for liver regeneration. Moreover, recent advances in hepatocyte and β-cell transdifferentiation research have provided valuable insight into how to regenerate and restore normal functions of liver and pancreas under pathological conditions. The definition and the applications of transdifferentiation are discussed in Sect. 6.3. Hepatocyte-specific TFs that are involved in generating functional hepatocytes are listed herein, which would provide an invaluable resource to understand the roles of TFs in cells within the same developmental lineage.

Hnf4α (hepatocyte nuclear factor 4α) is capable of modulating hepatocyte gene expression in differentiating hepatoma cells [84, 137]. Gene knockout studies of Hnf4α, whether in fetuses or in adults, have been found to disrupt expression of a large number of genes involved in most aspects of mature hepatocyte function [137, 138].

These functions include control of energy metabolism, xenobiotic detoxification, bile acid synthesis, and serum protein production [84].

C/EBPα/β (CAAT/enhancer-binding proteins α and β) is a basic leucine zipper TF, which plays an important role in liver function and development. C/EBPα is enriched in the liver, governing expression of liver-specific genes, and downregulated after partial hepatectomy [139, 140]. Hepatocytes of C/EBPα-deficient mice have increased proliferative potential [141, 142]. C/EBPβ is upregulated after partial hepatectomy and also upregulated by TNFα and IL-6 in stellate cells [143–145]. These findings indicate that C/EBPα and C/EBPβ are major players in liver regeneration. Hepatocytes of mice deficient in C/EBPα show a pseudoglandular structure [146]. Hepatocytes lining the structure have potential for differentiation into both hepatocytes and bile duct epithelial cells, suggesting that C/EBPα promotes differentiation of hepatoblasts to mature hepatocytes [147, 148].

6.2.2.3 Transcription Factors Involved in Salivary Gland Regeneration

In contrast to TFs investigated in the pancreas and liver, critical roles of SG TFs for cell differentiation, lineage commitment, and fate determination toward secretory acinar or duct cells are understudied and poorly understood. In this section, we will describe SG TFs, typically identified as SSPC marker proteins during mouth SG development, along with a brief introduction of TFs that our group reported following the proteomic analyses of mouse MSC-derived SPCs in coculture.

Ascl3 (achaete-scute family bHLH transcription factor 3, a.k.a. Sgn1) was reported as a marker of SPCs for both acinar and ductal cells in mouse SGs [149] and as a determinant of ductal cell lineage in mice [150]. The Ascl3 knockout mouse model showed the complexity of SG maintenance and regeneration. Ascl3 is considered to be an attractive molecule for induction of SECs as it has been found to be involved in the regeneration of acinar, ductal, and myoepithelial salivary cells in vitro [75] and known to be expressed in ductal neonatal progenitor cells [151].

Sox2 (SRY(sex-determining Y)-box 2) is essential in ES cell self-renewal [152]. However, in SG development, it is differentially expressed within the $K5^+$ population [153]. Sox2 is expressed by a subpopulation of $K5^+$ SPCs and is highly abundant in the committed $K5^+K19^+$ and $K5^-K19^+$ duct cells in the SG. This suggests that their potential to self-renewal, driven by Sox2, may be present even as $K5^+$ cells differentiate along the duct lineage [8].

Tcf3 (transcription factor 3) is the most abundant Tcf/Lef member in mouse ES cells [154]. It was reported that heterodimers between Tcf3 and tissue-specific bHLH proteins play major roles in determining tissue-specific cell fate during embryogenesis [155]. Tcf3 is also known to be closely involved in Wnt/β-catenin signing to control self-renewal and regulates the lineage differentiation of ES cells [156–158]. Interestingly, Tcf3-β-catenin interaction may indirectly affect submandibular SG during mouse embryogenesis. Furthermore, vascular integrity defects in organs such as the submandibular glands and liver were reported in the Tcf3 knock-in mutation model, which specifically lacks Tcf3-β-catenin interaction [159]. However, the exact functions of Tcf3 in the SGs during development and stem cell differentiation remain largely unknown.

Hmg20b (high mobility group 20B) is known to be expressed in various tissues. Many researchers suggest that breast cancer susceptibility gene 2 and Hmg20b complex may have a role in cell cycle regulation and affect cell fate determination [160]. Previously, our lab identified that Hmg20b and Tcf3 gene and protein expression was upregulated as mBM-MSCs transdifferentiated into SPC in a coculture system [12]. Their exact roles in MSC transdifferentiation are currently under investigation.

6.2.2.4 Transcription Factors Involved in the Pancreas, Liver, and Salivary Gland Regeneration

Ptf1α (pancreas transcription factor 1α, pancreas and SG) is better characterized in mice with broad expression in dorsal and ventral pancreatic buds, as the name indicates [161]. Ptf1α is known to be an exocrine lineage determinant in the pancreas and is predominantly detected in pancreatic acinar cells [99]. Ptf1α locus mutation activates autosomal recessive permanent neonatal diabetes which requires insulin for survival. Similar behavior has been reported in $Ptf1a^{-/-}$ mice, which die postnatally with impaired pancreatic and major SG development. A recent study reported that human Ptf1α enhancer mutation leads to pancreatic agenesis [100]. Mutant phenotypes of human and mouse support an evolutionarily conserved role during early pancreatic formation. Interestingly, our recent study determined that Ptf1α expression was highly increased in transdifferentiating BM-MSC in coculture although the expression of Ptf1 α in the SGs was never reported before [12]. Nomenclature for this molecule may need to be revised once its broader role than originally anticipated in cell fate determination in SG regeneration is explored and confirmed.

Gata4 (GATA binding protein 4, pancreas and liver) is expressed during the early pancreatic budding stage but later dramatically decreased in expression in the pancreatic progenitors and only remained in mature acinar cells [82]. Although Gata4 mutations are known to be associated with congenital heart defects, only the pancreatic phenotype is documented with mouse models [162, 163]. This leads to a belief that there may be another Gata TF in human pancreas [162, 163]. Moreover, as briefly shown in Fig. 6.1, the TF plays a key role in the liver development. Each factor in Fig. 6.1 is important for the expansion of the liver bud during embryogenesis but serves redundant functions in hepatic specification. Based on mouse studies, Gata4 expression is first detected in the ventral foregut endoderm and cardiac mesoderm at E8.0. Gata4 is required for the full expression of select hepatic genes, including albumin and Hnf4 [164].

Hnf6 (hepatocyte nuclear factor 6, pancreas and liver): mRNA analysis of Hnf6 demonstrated that mouse Hnf6 expression at E8.5 has broad expression throughout development and directs endocrine allocation until before the birth when it is restricted by α-cells and acinar cells in the liver and pancreas [85].

Foxa2 (forkhead box A2, pancreas and liver) is a TF that is expressed throughout the definitive endoderm, which persists into adulthood in endodermal derivatives such as the liver, pancreas, lung, and thyroid during mouse development. In pancreas development, Foxa2 is a major upstream regulator of Pdx1 and continues to be active in all mature pancreatic cell types of mice and humans [79]. Foxa2 deficiency in a mouse model resulted in the absence of mature α-cells and a reduction of Pdx1 expression and β-cell differentiation [165, 166]. Given that Foxa1 is a close homolog to Foxa2 and contains an identical DNA-binding domain, it is also presumed to control pancreas development [79]. Moreover, the early activation of the Foxa2 genes in the hepatogenic region of the foregut endoderm, combined with the abundance of liver-specific Foxa2 target genes, has been interpreted as evidence that Foxa2 genes play a key role in regulating hepatogenesis [80, 137, 167]. The early lethal phenotype observed in $Foxa2^{-/-}$ embryos, owing mostly to node and notochord defects, precluded analysis of Foxa2 at later stages of development, thus necessitating the derivation of mice harboring a conditional Foxa2 allele to directly assess its role during liver development [168].

Ngn3 (neurogenin 3, pancreas and SG) is required for endocrine cell specification in mice [90]. It also initiates endocrine commitment in pancreatic and/or SG epithelial cells. Isl1, NeuroD1, MafB, Nkx2.2, and Pax6 are islet-enriched factors that activate downstream of Ngn3 in mice, and these factors are integrated in late endocrine cell differentiation [169]. Ngn3 null mutation is a rare mutation with permanent neonatal diabetes with histologically detectable islets [170]. $Ngn3^{-/-}$ mice completely lack endocrine cells and thus develop diabetes and die only few days after birth [171].

Mist1 (muscle, intestine, and stomach expression-1, a.k.a. Bhlha15 for basic helix-loop-helix family member A15, pancreas and SG) is a known bHLH TF with acinar cell-specific expression [87, 172]. It acts as a regulator of differentiation and morphogenesis of exocrine cells by negative regulation of bHLH-mediated transcription through an NH2-terminus repressor domain [87]. Mist1 gene expression is observed in a wider array of tissues including the acinar cells of SG and the serous-secreting cells found in the stomach, prostate, and seminal vesicles [173]. An essential function that is shared by all Mist1-positive cells is regulated exocytosis, which involves the temporary storage of zymogen granules at the cell's apical surface and the establishment of specific signaling pathways through which external cues induce regulated secretion [87]. Based on knockout studies, the primary outcome of its expression defect reveals a loss of correct cellular organization in pancreas and SGs [174, 175].

Sox9 (SRY (sex-determining region Y)-Box9, pancreas, liver, and SG) is located in $Pdx1^{+}$ cells in early stages of pancreas development [96, 162, 176, 177]. Later, it is excluded from mature endocrine cells. Sox9 is essential for maintaining the multipoint progenitor population in mice [101]. Deletion of Sox9 resulted in islet hypoplasia by unsuccessful maintenance of endocrine progenitors. In Sox knockout mice studies, Sox9 has been demonstrated to have an effective role in endodermal cells such as the pancreas, lung, SG, kidney, gut, and liver [102].

6.3 Transdifferentiation

6.3.1 Definition

The term "transdifferentiation" first surfaced to describe the conversion of cuticle-producing cells into salt-secreting cells in the silkmoth during metamorphosis from the larval to the adult moth [178]. Subsequently, Eguchi and Okada elegantly demonstrated the switch of pigmented epithelial cells to lens fibers in their in vitro clonal cell culture system [179], setting the standard for determining a true transdifferentiation event [180, 181]. A review by Eberhard and Tosh explicitly defines "transdifferentiation" as in "nonstem cell" [182], supporting the well-accepted definition of the conversion of one specialized cell type into another without reversion to pluripotent cells or progenitor cells [179]. However, in recent years this traditional definition has been broadened by the evidence that adult stem cells not only differentiated into anticipated and committed lineages but also differentiated into cell types beyond the expected lineage of the respective stem cells [183]. This plasticity can be seen in many examples of stem cell research, including the ones described in Sect. 6.3.2. We use this broadened definition in this chapter to cover fascinating studies from the organs of interest.

This type of conversion can be induced experimentally, or it can normally occur in tissues that arise from the neighboring regions of the developing embryo [178]. This ability to interconvert suggests that the two tissues are related at the molecular level, sharing some common TFs, but in addition there is one or two that are different. Therefore, adjacent tissues presumably differ in the expression of one or a few key TFs [184]. As for the experimentally induced conversion, delivery of TFs via viral vectors altered cell fate by transdifferentiating cells into cell types of interest [101]. For instance, Sox2 expression by lentivirus in the mouse brain directly converted astrocytes into neuroblast [185]. Human fibroblasts differentiating into cardiac cells by forced expression of cardiac TFs with muscle-specific microRNAs would be another example [102]. Transdifferentiation may occur either directly or through a de-differentiation step before cells re-differentiate to a new mature phenotype [186].

6.3.2 Examples

Tissue injury can lead to the appearance of multipotent stem cells for the liver and the pancreas [187, 188]. Dabeva et al. reported that pancreatic epithelial progenitor cells isolated from the copper-deficient diet rat pancreas can differentiate into hepatocytes when they are transplanted into the liver [189]. Recent studies have shown that there is a tissue stem cell in the pancreas that can differentiate into hepatocytes in addition to pancreatic cell types [190–192]. Zulewski et al. reported that multipotent stem cells from rat and human islets differentiated into hepatic cell types in culture when the cell density increased [193]. In other experiments, a cell line

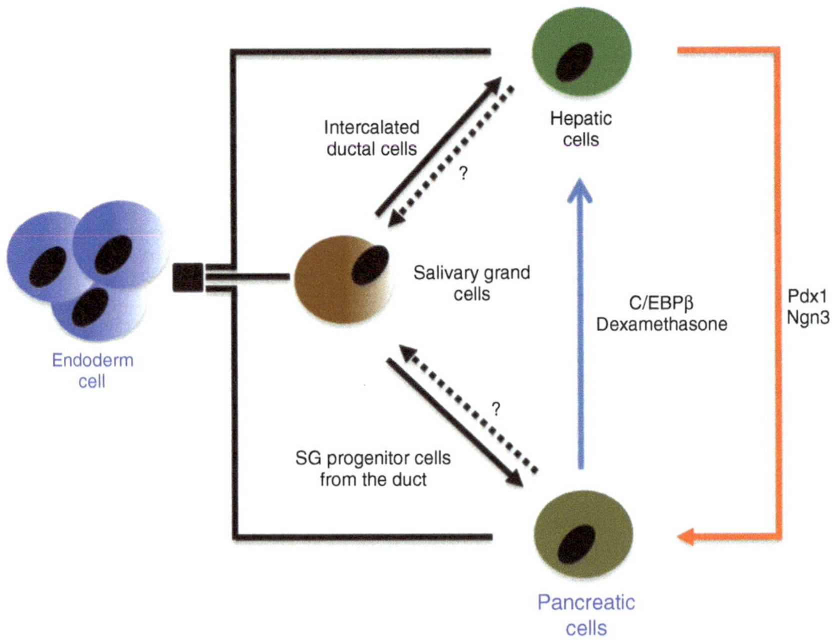

Fig. 6.2 Schematic diagram of transdifferentiation among pancreatic, hepatic, and salivary gland cells and TFs involved

from pancreatic acinar cells differentiated into hepatocytes in the presence of dexamethasone (Fig. 6.2) [194]. Thus, resident stem cells in the pancreas or the liver have demonstrated their capacity to transdifferentiate into a different type of cells in other organs.

Ptf1α, the bHLH TF, was first described in exocrine-specific pancreas development [99] as mentioned earlier. Recent results have expanded the role of Ptf1α, with involvement in both exocrine and endocrine cell lineages in murine and zebrafish models [161]. Kawaguchi et al. demonstrated conversion of pancreatic progenitors into duodenal epithelium through the inactivation of Ptf1α [195], thereby suggesting a close developmental relationship between intestine and pancreas similar to the liver and pancreas. Also, it provides a possible route for creating new pancreatic cells following the overexpression of Ptf1α in other cell types such as duodenal epithelial cells. It will be interesting to examine if the switch between hepatocytes and pancreatic cells is also feasible with the modulation of Ptf1α. There have been reports on cell transdifferentiation of hepatocytes into pancreatic β-cells involving several key pancreatic TFs. Of those, Pdx1 seems to play a key role in the hepatic-pancreatic cell fate conversion [196]. In mice, Pdx1 expression is known to be critical in the conversion of hepatocytes into pancreatic β-cell-like cells that secret insulin [197]. Interestingly, Pdx1 and Ngn3 are known to synergistically induce expression of β-cell factors and insulin biosynthesis in the liver and drastically ameliorated glucose tolerance in mouse [198].

The SGs are known to be derived from the endoderm and the ectoderm (parotid glands) that participate in organogenesis, although the origins of these glands are still controversial [188, 199–201]. SGs consist of many cell types such as three different parenchymal cell types found in mice: (1) acinar, (2) intercalated duct and granular intercalated duct, and (3) granular and striated duct cells. The relative proportions of these three cell types are 43 %, 18 %, and 39 %, respectively [202]. Intercalated ducts are small ducts connecting the terminal acini and striated duct, and they are known to contain the tissue stem cells or progenitor cells [202]. Studies have suggested that the SG might contain stem cells that can differentiate into the cell types of other endodermal organs, such as the liver and pancreas [199].

Furthermore, the features of SG resemble the pancreatic exocrine system. Transplantation of intercalated ductal cells prepared from damaged SG led to identification of epithelium-like cells that settled in the oval cell response area of the rat liver [201]. Moreover, progenitor cells isolated from the duct-ligated SG differentiate into pancreatic cell types characterized by CD44 expression that is specific for the developing pancreas [201], implying a capacity of transdifferentiation that SPCs might possess. We have not encountered any studies yet that demonstrated generation of SECs from hepatocytes, pancreatic β-cells, or their progenitors (Fig. 6.2).

6.4 Salivary Gland Stem Cells/Progenitor Cells

Stem cell studies in SG regeneration in recent years have focused on SSPCs. As explained earlier, the inability to form new acinar cells in the SGs with irradiation or SjS is associated with loss of progenitors and/or unfavorable microenvironment in the glands. Irradiation-surviving progenitors in vivo are no longer stimulated to form acinar cells because the loss of parasympathetic function and innervation prevents proper regeneration [153]. Current treatments for hyposalivation after radiotherapy include administration of saliva substitutes, secretagogues, or palliative approaches. Researchers have proposed an ideal therapeutic approach involving SSPCs for treating irradiated patients, which requires isolating SSPCs prior to radiation therapy, expanding the cells in vitro, and then transplanting them back into the patients [203]. SSPCs are characterized based on their topographical position in the gland, expression of TFs and specific marker proteins that are known to be found on stem cells in other systems, or unique stem cell/progenitor properties. SSPCs are also considered as an important source for constructing an artificial SG as they are known to be involved in the regeneration of acinar, ductal, and myoepithelial salivary cells in vitro [8]. In addition, an emerging concept in the field is that multiple glandular progenitors may exist with overlapping functions [204].

Currently, active investigation is underway to define SSPCs and investigate their roles in SG regeneration. For instance, Ascl3 is essential for the determination of cell fate, development, and differentiation of numerous tissues [75, 149]. Transplanted Ascl3-expressing cells were able to induce differentiation and tissue repair in SGs, indicating that these cells are potent inducers of SG regeneration [153]. Furthermore, Ascl3 was found to be expressed in ductal cells, and they contribute to the maintenance of mature SG tissues [149], pointing to a notion that $Ascl3^+$ cells may be a subpopulation of SSPCs.

Other studies demonstrated that cells located in the striated ducts of the SGs express other stem cell markers (i.e., CD24, CD49f, CD133, and c-Kit^+) [205]. Particularly, c-Kit was definitively established as a stem cell marker and therefore gained the highest priority. However, flow cytometric analysis of cells obtained from submandibular glands indicated that only a small number of salivary cells expressed c-Kit [205]. These results suggest that c-Kit may not be a practical marker for stem cell isolation from SGs, as it would not yield workable levels of stem cells. In addition, structural protein $K5^+$, which is a cytokeratin to form cytoskeleton intermediate filaments, has been established as a marker of progenitor cells [149, 153]. Initially, $K5^+$ has been established as a marker for progenitor cells in tracheal and lung epithelial cells. Additionally, $K5^+$ cells express Sox2, a TF involved in the self-renewal of stem cells [206]. However, $K5^+$ cells make up a very small percentage (5–9 %) of Sox2 cells. These studies also suggest that K5 may not be a practical marker to isolate SPCs either [207, 208]. Recently, a long-term in vitro expansion of SG stem cells driven by Wnt signals was reported in a study [209], whose practicality is forthcoming to be determined.

α6β1 integrin has been found to be a marker for SPCs in rats. Interestingly, this marker was only expressed after duct ligation [201]. These α6β1 integrin-expressing cells were used to establish an immortalized cell line of rat SPCs [210]. This cell line is able to differentiate into both acinar-like and ductal-like structures and has the ability to regulate transdifferentiation when grown on Matrigel®-based 3D scaffolds. However, cells grown under these conditions display uncontrolled growth. Further studies are needed to determine whether acinar-like and ductal-like structures generated from this cell line are capable of responding to salivary secretory agonists. Additionally, the mechanisms of uncontrolled cell growth need to be well understood and regulated before these structures can be used for transplantation in vivo.

6.5 Applications of Stem Cells for In Vivo Applications

Predictable and precise controlling cell proliferation and differentiation requires comprehensive profiling of the molecular and genetic signals that regulate cell division and specification. While recent developments with stem cells suggest specific roles of differentiation factors, techniques must be devised to introduce these factors safely into the cells and induce maximum outcomes without adverse side effects. Stem cells, such as ES cells, iPS cells, MSC, and tissue adult stem cell, are currently being used to screen and develop new drugs in preclinical studies [211]. To screen drugs effectively, the testing conditions must be identical for proper comparisons of drugs, and precise controlling of stem cell fate is required for consistency and reproducibility. We summarize published in vivo studies with inductive signals in the pancreas, liver, and SG regeneration in Table 6.2, although the field is still in its infancy with a diverse range of preliminary approaches. A brief discussion of stem cell use with or without inductive signals is provided herein.

6.5.1 Pancreas Regeneration

One of the major pancreatic diseases, diabetes mellitus, is a metabolic disorder caused by an insufficient number of insulin-producing β-cells. Theoretically, replenishment of β-cells by cell transplantation can restore normal metabolic control. Exogenous insulin administration in diabetes patients is not sufficient to mimic the normal function of β-cells. Consequently, diabetes mellitus often progresses and can lead to major chronic complications and morbidity [212]. New studies indicate that it may be possible to direct the differentiation of stem cells from human in ex vivo cell culture to form insulin-producing cells for transplantation. Strategies involving pluripotent stem cells appear to have the highest translational potential [213], allowing sufficient numbers of pancreatic progenitor cells with a potential to differentiate into β-cells. Differentiated cells within the adult pancreas retain sufficient plasticity to transdifferentiate into β-cells as well. The most promising examples are α-cells and acinar exocrine cells that transdifferentiated into β-cells with the introduction of TFs [115]. However, stem cell-based therapeutic approaches have some limitations when it comes to clinical applications, such as teratoma formation, ethical issues related to ES cells, and safety issues of gene expression associated with iPS. Furthermore, the future success of ES-derived β-cell transplantation or regeneration of endogenous β-cells depends on effective prevention of infiltration of autoreactive immune cells and/or alloimmune rejection [213].

6.5.2 Liver Regeneration

Most liver diseases lead to hepatic dysfunction with organ failure. Liver transplantation is the best curative therapy, but it has some limitations, such as donor shortage, possibility of rejection, and maintenance of immunosuppression [214]. New therapies have been actively searched for over several decades, primarily in the form of artificial liver-support devices and hepatocyte transplantation, but both of these modalities remain experimental. For this reason, scientific interests have switched to the use of stem cells to regenerate the liver. Numerous stem cell types have been used for the differentiation of hepatocytes both in vivo and in vitro [215]. Plasticity in adult MSC repopulated liver cells in a liver injury animal model, which appeared to improve liver function. For instance, multipotent adult progenitor cells (MAPCs) derived from human, mouse, and rat postnatal BM can differentiate into hepatocyte-like cells in vitro [216]. The differentiated cells expressed hepatic markers such as Hnf-3β, Gata4, Afp, Alb, and CK-18, and obtained functional characteristics of hepatocytes as indicated by p450 activity, LDL uptake, ALB secretion, urea secretion, and glycogen storing [216]. Historically, MAPCs had been identified as a subpopulation of MSCs [217]. Human umbilical cord blood-derived MSCs (UCB-MSCs) were also proven to have potential for hepatocyte-like cell differentiation in vitro by the treatment of special cytokine mixtures [218]. Recently, MSCs were isolated from human thigh adipose tissue and induced to differentiate into hepatocyte-like cells using growth factors, cytokines, and hormones [219–221]. These hepatocyte-like cells expressed hepatocyte-specific genes and exhibited hepatocyte functions [222]. Moreover, a study examined the impact of the coculture with adult liver cells or fetal liver cells on hepatic differentiation of rat BM-MSCs [223] and found that liver-specific gene expression was induced when treated with gadolinium chloride [224]. Although all cells differentiated into hepatocyte-like cells, the cocultured ones expressed more hepatic gene markers and exhibited higher metabolic function (P450 activity) [216]. These pioneering research studies led to the application of human BM-derived cells xenografted to damaged liver of Sprague Dawley rats by allylalcohol, resulting in differentiation of hepatocyte-like cells [225].

Other groups reported plasticity of MSCs after human adipose tissue MSCs were transplanted into mice with liver damage for hepatocyte differentiation [222]. Following the study, human MSCs were transplanted into preimmune fetal sheep, providing further evidence for feasibility [226]. As a result, the animals exhibited a widespread distribution of human hepatocyte-like cells throughout the liver parenchyma, indicating that MSCs are a valuable source of cells for liver repair and regeneration [227]. Although the hepatogenic capacity of MSCs seems to be strongly established, the mechanism by which MSCs restore liver functionality is still not clear. In addition, significant challenges still exist before these cells can be used in humans. These challenges include the lack of consensus about the immunophenotype of liver progenitor cells, uncertainty about the physiological roles of reported candidate stem/progenitor cells, the practicality of obtaining sufficient quantity of cells for clinical use, and concerns over ethics, long-term efficacy, and safety [228]. Nonetheless, cells from BM are increasingly recognized as being major therapeutic players for liver fibrosis and regeneration, indicating that the BM positively contributes to liver fibrosis [229].

6.5.3 Salivary Gland Regeneration

Two major approaches for xerostomia include (1) utilization of combinations of cells, biomaterials, and biochemical cues for tissue engineering and (2) utilization of cell-based therapy for SG regeneration. Studies have investigated the role of MSCs as a therapeutic option for treatment of SjS [80, 230]. Investigators used NOD mice with a SjS-like disease to investigate the effect of MSCs in reducing lymphocytic infiltrates in the SG and restoring salivary function. Authors found that intravenous injection of MSCs reduced lymphocytic infiltrate and inflammation in the SG compared to untreated controls, including a tenfold decrease in the inflammatory cytokine TNF-α. MSC injection also preserved the saliva flow rate over the 14-week posttreatment period; moreover, when MSCs were administered in conjunction with complete Freund's adjuvant (CFA), the SG regenerative potential increased. These findings indicate that direct injection of MSC for therapy alone reduced inflammation, but there was additional tissue repair and regeneration when administered in conjunction with CFA [230].

SSPCs are known to be located in intercalated ducts and exocrine ducts [203, 231, 232]. Stem cells isolated from the human SG have revealed a certain capacity for in vivo recovery of SG function in irradiated rat SGs by differentiating into amylase-expressing cells [233]. Sixty days after human SSPC were transplanted into the irradiated glands of rats, the average saliva flow rate of the human SG stem cell-treated group was twice that of the PBS-treated irradiated group but was still lower than the undamaged group [233]. By using a floating sphere culture, further in vitro characterization of submandibular-derived SG stem cells showed that cultured spheres contained acinar and ductal cells characterized by specific cell marker expression [234]. Interestingly, acinar cells mostly disappeared by the third day but reappeared within the existing ductal spheres by the fifth day in culture. By day ten, acinar cells dominated sphere composition and amylase expression. Intraglandular transplantation of 3-day cultured salispheres into irradiated mice resulted in the formation of ductal structures near the injection site. There was increased acinar cell surface area in salisphere-treated mice compared to the untreated group. Ninety days after irradiation, saliva production in salisphere-treated mice was higher than the untreated counterparts and correlated strongly with acinar surface area [234]. A major limitation for SSPC therapy in the treatment of radiation-induced hyposalivation continues to be the difficulty with isolating autologous stem cells from a severely injured gland unless the isolation and purification are carried out prior to radiation. As mentioned earlier, a recent research progress by Coppes et al. reported the expansion of SSPCs ex vivo via the stimulation of Wnt signaling [209].

A coculture system of mouse ES cells and human SG fibroblasts was developed to facilitate differentiation of mouse ES cells to SG stem cells [114]. After 1 week in coculture, a significant change in cell morphology was found, and RT-PCR results showed a sudden appearance of amylase and bFGF. These GFP-expressing SG cells were transplanted into SG of normal mice, and histology was performed after 1 month. H&E and PAS staining of SG stem cell-treated mice showed normal formation of ductal and acinar structures. Even though this method is not limited by the need for autologous stem cells from a radiation-damaged gland, it is limited by the ethical concerns over acquisition of ES cells in clinical settings.

6.6 Conclusion and Future Directions

Xerostomia due to SG hypofunction has a severe impact on the oral health of patients with SjS and radiation therapy. The regeneration of functional SG tissue is an important therapeutic goal for the field of regenerative medicine and will likely involve cell therapies such as stem cells and/or tissue engineering. Despite research endeavors, critical information on signaling pathways, master regulators, and molecular mechanisms that direct stem cell differentiation and tissue engineering is lacking in the SG research field, compared to information available in other organ regeneration. Nevertheless, researchers have isolated SSPCs from rodent and human SGs and demonstrated that SPCs can differentiate into both salivary epithelial-like cells and endodermal progeny, such as pancreatic β-cells and hepatocytes [33]. Whether SG cells can be regenerated with hepatic or pancreatic progenitor cells transplanted into the SGs or whether hepatic or pancreatic TFs can contribute to directed differentiation of SPCs or SECs is completely unknown. Considering that most TFs studied in those organs are critical in determining the lineage of endocrine cells for enzyme secretion, SG may require a set of distinct TFs that define and drive exocrine cells for more organized structural differentiation for proper secretion.

Several reports have suggested that the conversion of pancreatic progenitor cells toward the hepatocyte endocrine fate (reciprocal transdifferentiation) is achievable [213]. Moreover, forced expression of pancreatic TFs elicits insulin expression in the liver and corrects experimental diabetes [235]. Taken together, these findings suggest that the adult organs contain cells with epigenetic memory of their common embryonic origin. Based on these studies, SG regeneration or progenitor cell differentiation might be closely related with pancreas or liver organogenesis and stem cell regeneration. Therefore, identification of intrinsic regulatory factors and external factors for stem differentiation/transdifferentiation and the sequential and timely introduction of those factors via viral or nonviral methods, as similar approaches have shown in pancreas and liver regeneration, may benefit SG stem cell differentiation and regeneration. This will ultimately allow translational applications for some tissues that normally lack regeneration capacity. Similarly, mapping and profiling of those factors and understanding their critical roles in differentiation will enable us to control cell fate to devise therapeutic interventions for xerostomia.

As for clinical translation of stem cell therapy, safety is the major issue, given that a small number of contaminating undifferentiated cells could remain after cell purification and potentially cause teratoma formation. Human iPS cells do not pose ethical concerns unlike ES cells but do present a potential danger associated with DNA transfer of cancer-associated genes [213]. Generation of individualized ES or iPS cell lines to avoid ethical issues will be expensive, given the relatively low efficiency of these procedures. Storing a wide range of human ES and iPS cell lines matched to the most common HLA haplotypes might be a better option than the generation of patient-specific cell lines [236]. Allorejection is another issue to consider, although both issues could theoretically be solved by immunoisolation of the graft [237]. Immunoisolation could be achieved by macroencapsulation with or without a biopolymer in devices that prevent the passage of stem cells into the recipient and immune cells into the graft [238]. Alternatively, MSCs lessen immunoreactivity as they express the human leukocyte antigen (HLA)-G, which is a nonclassical HLA class I molecule that mediates the suppressive effect of MSCs through the induction and proliferation of regulatory T cells, along with other immunosuppressive properties, which adds to the benefits of MSCs for autoimmune SjS [239]. In addition, HLA compatibility between a MSC donor and a recipient is not a major concern due to the lack of HLA-DR surface expression [240].

In conclusion, stem cell approaches hold great promise for future clinical trials for xerostomia despite existing challenges. Increasing knowledge and information that is currently being generated from in vitro, ex vivo, and in vivo applications of stem cells will continue to provide the foundations for future clinical trials. Comprehensive and multidisciplinary research endeavor focusing on the epigenetic and molecular regulation of cell fate determination and transdifferentiation of stem cells and differentiated cells will allow orchestrated and/or directed reprogramming of cells for SG regeneration. Harnessing TF-directed differentiation to classical approaches of stem cell reprogramming will expedite clinical translation of cell replacement therapy.

Acknowledgments This work is supported by the NIH/NIDCR grants DE019644 and DE025726 (S.C.). The authors thank Dr. Kathleen M. Berg for critical proofreading.

References

1. Voulgarelis M, Tzioufas AG. Pathogenetic mechanisms in the initiation and perpetuation of Sjogren's syndrome. Nat Rev Rheumatol. 2010;6:529–37.
2. Seiwert TY, Cohen EE, Haraf DJ, Stenson K, Blair EA, et al. A phase I trial of docetaxel based induction and concomitant chemotherapy in patients with locally advanced head and neck cancer. Cancer Invest. 2007;25:435–44.
3. Malouf JG, Aragon C, Henson BS, Eisbruch A, Ship JA. Influence of parotid-sparing radiotherapy on xerostomia in head and neck cancer patients. Cancer Detect Prev. 2003;27:305–10.
4. Trotti A, Bellm LA, Epstein JB, Frame D, Fuchs HJ, et al. Mucositis incidence, severity and associated outcomes in patients with head and neck cancer receiving radiotherapy with or without chemotherapy: a systematic literature review. Radiother Oncol. 2003;66:253–62.
5. Dirix P, Nuyts S, Van den Bogaert W. Radiation-induced xerostomia in patients with head and neck cancer: a literature review. Cancer. 2006;107:2525–34.
6. Baum BJ. Prospects for re-engineering salivary glands. Adv Dent Res. 2000;14:84–8.
7. Feng J, van der Zwaag M, Stokman MA, van Os R, Coppes RP. Isolation and characterization of human salivary gland cells for stem cell transplantation to reduce radiation-induced hyposalivation. Radiother Oncol. 2009;92:466–71.
8. Lombaert IM, Knox SM, Hoffman MP. Salivary gland progenitor cell biology provides a rationale for therapeutic salivary gland regeneration. Oral Dis. 2011;17:445–9.
9. Pringle S, Van Os R, Coppes RP. Concise review: adult salivary gland stem cells and a potential therapy for xerostomia. Stem Cells. 2013;31:613–9.

10. Maria OM, Tran SD. Human mesenchymal stem cells cultured with salivary gland biopsies adopt an epithelial phenotype. Stem Cells Dev. 2011;20:959–67.
11. Sumita Y, Liu Y, Khalili S, Maria OM, Xia D, et al. Bone marrow-derived cells rescue salivary gland function in mice with head and neck irradiation. Int J Biochem Cell Biol. 2011;43:80–7.
12. Park YJ, Koh J, Gauna AE, Chen S, Cha S. Identification of regulatory factors for mesenchymal stem cell-derived salivary epithelial cells in a co-culture system. PLoS One. 2014;9:e112158.
13. Watt FM, Hogan BL. Out of Eden: stem cells and their niches. Science. 2000;287:1427–30.
14. Xu YX, Chen L, Wang R, Hou WK, Lin P, et al. Mesenchymal stem cell therapy for diabetes through paracrine mechanisms. Med Hypotheses. 2008;71:390–3.
15. Ichim TE, Alexandrescu DT, Solano F, Lara F, Campion Rde N, et al. Mesenchymal stem cells as anti-inflammatories: implications for treatment of Duchenne muscular dystrophy. Cell Immunol. 2010;260:75–82.
16. Sequeira SJ, Larsen M, DeVine T. Extracellular matrix and growth factors in salivary gland development. Front Oral Biol. 2010;14:48–77.
17. Gattazzo F, Urciuolo A, Bonaldo P. Extracellular matrix: a dynamic microenvironment for stem cell niche. Biochim Biophys Acta. 2014;1840:2506–19.
18. Maria OM, Zeitouni A, Gologan O, Tran SD. Matrigel improves functional properties of primary human salivary gland cells. Tissue Eng Part A. 2011;17:1229–38.
19. Leal-Egana A, Scheibel T. Silk-based materials for biomedical applications. Biotechnol Appl Biochem. 2010;55:155–67.
20. Kundu B, Kurland NE, Yadavalli VK, Kundu SC. Isolation and processing of silk proteins for biomedical applications. Int J Biol Macromol. 2014;70:70–7.
21. Anseth KS, Metters AT, Bryant SJ, Martens PJ, Elisseeff JH, et al. In situ forming degradable networks and their application in tissue engineering and drug delivery. J Control Release. 2002;78: 199–209.
22. Nuttelman CR, Tripodi MC, Anseth KS. In vitro osteogenic differentiation of human mesenchymal stem cells photoencapsulated in PEG hydrogels. J Biomed Mater Res A. 2004;68:773–82.
23. Benoit DS, Durney AR, Anseth KS. The effect of heparin-functionalized PEG hydrogels on three-dimensional human mesenchymal stem cell osteogenic differentiation. Biomaterials. 2007;28:66–77.
24. Weber LM, He J, Bradley B, Haskins K, Anseth KS. PEG-based hydrogels as an in vitro encapsulation platform for testing controlled beta-cell microenvironments. Acta Biomater. 2006;2:1–8.
25. Lin CC, Raza A, Shih H. PEG hydrogels formed by thiol-ene photo-click chemistry and their effect on the formation and recovery of insulin-secreting cell spheroids. Biomaterials. 2011;32:9685–95.
26. Mahoney MJ, Anseth KS. Three-dimensional growth and function of neural tissue in degradable polyethylene glycol hydrogels. Biomaterials. 2006;27:2265–74.
27. Benoit DS, Schwartz MP, Durney AR, Anseth KS. Small functional groups for controlled differentiation of hydrogel-encapsulated human mesenchymal stem cells. Nat Mater. 2008;7:816–23.
28. Weber LM, Anseth KS. Hydrogel encapsulation environments functionalized with extracellular matrix interactions increase islet insulin secretion. Matrix Biol. 2008;27:667–73.
29. Lin CC, Anseth KS. Cell-cell communication mimicry with poly(ethylene glycol) hydrogels for enhancing beta-cell function. Proc Natl Acad Sci U S A. 2011;108:6380–5.
30. Pradhan S, Liu C, Zhang C, Jia X, Farach-Carson MC, et al. Lumen formation in three-dimensional cultures of salivary acinar cells. Otolaryngol Head Neck Surg. 2010;142:191–5.
31. Walker JL, Menko AS, Khalil S, Rebustini I, Hoffman MP, et al. Diverse roles of E-cadherin in the morphogenesis of the submandibular gland: insights into the formation of acinar and ductal structures. Dev Dyn. 2008;237:3128–41.
32. Davis MA, Reynolds AB. Blocked acinar development, E-cadherin reduction, and intraepithelial neoplasia upon ablation of p120-catenin in the mouse salivary gland. Dev Cell. 2006;10:21–31.
33. Okumura K, Shinohara M, Endo F. Capability of tissue stem cells to organize into salivary rudiments. Stem Cells Int. 2012;2012:502136.
34. Shubin AD, Felong TJ, Graunke D, Ovitt CE, Benoit DS. Development of poly(ethylene glycol) hydrogels for salivary gland tissue engineering applications. Tissue Eng Part A. 2015;21:1733–51.
35. Sander, M., Seymour, P., Kopp, J., Shih, B., Patel, N. & Lim, C. Molecular cues regulating segregation of pancreatic, hepatic and intestinal lineages. The FASEB Journal 2011;25, 303.3.
36. Zaret KS. Regulatory phases of early liver development: paradigms of organogenesis. Nat Rev Genet. 2002;3:499–512.
37. Wandzioch E, Zaret KS. Dynamic signaling network for the specification of embryonic pancreas and liver progenitors. Science. 2009;324:1707–10.
38. Bonal C, Herrera PL. Genes controlling pancreas ontogeny. Int J Dev Biol. 2008;52:823–35.
39. Gittes GK. Developmental biology of the pancreas: a comprehensive review. Dev Biol. 2009;326:4–35.
40. Kim SK, Hebrok M. Intercellular signals regulating pancreas development and function. Genes Dev. 2001;15:111–27.
41. Mfopou JK, Chen B, Sui LN, Sermon K, Bouwens L. Recent advances and prospects in the differentiation of pancreatic cells from human embryonic stem cells. Diabetes. 2010;59:2094–101.
42. Patel VN, Rebustini IT, Hoffman MP. Salivary gland branching morphogenesis. Differentiation. 2006;74:349–64.

43. Dalvi MP, Umrani MR, Joglekar MV, Hardikar AA. Human pancreatic islet progenitor cells demonstrate phenotypic plasticity in vitro. J Biosci. 2009;34:523–8.
44. Hardikar AA, Marcus-Samuels B, Geras-Raaka E, Raaka BM, Gershengorn MC. Human pancreatic precursor cells secrete FGF2 to stimulate clustering into hormone-expressing islet-like cell aggregates. Proc Natl Acad Sci U S A. 2003;100:7117–22.
45. Si-Tayeb K, Noto FK, Nagaoka M, Li JX, Battle MA, et al. Highly efficient generation of human hepatocyte-like cells from induced pluripotent stem cells (vol 51, pg 297, 2010). Hepatology. 2010;51:1094.
46. Cras-Meneur C, Elghazi L, Czernichow P, Scharfmann R. Epidermal growth factor increases undifferentiated pancreatic embryonic cells in vitro – a balance between proliferation and differentiation. Diabetes. 2001;50:1571–9.
47. Kitade M, Factor VM, Andersen JB, Tomokuni A, Kaji K, et al. Specific fate decisions in adult hepatic progenitor cells driven by MET and EGFR signaling. Genes Dev. 2013;27:1706–17.
48. Kashimata M, Sayeed S, Ka A, Onetti-Muda A, Sakagami H, et al. The ERK-1/2 signaling pathway is involved in the stimulation of branching morphogenesis of fetal mouse submandibular glands by EGF. Dev Biol. 2000;220:183–96.
49. Demeterco C, Beattie GM, Dib SA, Lopez AD, Hayek A. A role for activin A and betacellulin in human fetal pancreatic cell differentiation and growth. J Clin Endocrinol Metab. 2000;85:3892–7.
50. Chen LB, Jiang XB, Yang L. Differentiation of rat marrow mesenchymal stem cells into pancreatic islet beta-cells. World J Gastroenterol. 2004;10:3016–20.
51. Segev H, Fishman B, Ziskind A, Shulman M, Itskovitz-Eldor J. Differentiation of human embryonic stem cells into insulin-producing clusters. Stem Cells. 2004;22:265–74.
52. Tang DQ, Cao LZ, Burkhardt BR, Xia CQ, Litherland SA, et al. In vivo and in vitro characterization of insulin-producing cells obtained from murine bone marrow. Diabetes. 2004;53:1721–32.
53. Sun Y, Chen L, Hou XG, Hou WK, Dong JJ, et al. Differentiation of bone marrow-derived mesenchymal stem cells from diabetic patients into insulin-producing cells in vitro. Chin Med J (Engl). 2007;120:771–6.
54. Sun B, Roh KH, Lee SR, Lee YS, Kang KS. Induction of human umbilical cord blood-derived stem cells with embryonic stem cell phenotypes into insulin producing islet-like structure. Biochem Biophys Res Commun. 2007;354:919–23.
55. Wu XH, Liu CP, Xu KF, Mao XD, Zhu J, et al. Reversal of hyperglycemia in diabetic rats by portal vein transplantation of islet-like cells generated from bone marrow mesenchymal stem cells. World J Gastroenterol. 2007;13:3342–9.
56. Chao KC, Chao KF, Fu YS, Liu SH. Islet-like clusters derived from mesenchymal stem cells in Wharton's Jelly of the human umbilical cord for transplantation to control type 1 diabetes. PLoS One. 2008;3:e1451.
57. Gabr MM, Sobh MM, Zakaria MM, Refaie AF, Ghoneim MA. Transplantation of insulin-producing clusters derived from adult bone marrow stem cells to treat diabetes in rats. Exp Clin Transplant. 2008;6:236–43.
58. Gao F, Wu DQ, Hu YH, Jin GX. Extracellular matrix gel is necessary for in vitro cultivation of insulin producing cells from human umbilical cord blood derived mesenchymal stem cells. Chin Med J (Engl). 2008;121:811–8.
59. Otonkoski T, Beattie GM, Mally MI, Ricordi C, Hayek A. Nicotinamide is a potent inducer of endocrine differentiation in cultured human fetal pancreatic cells. J Clin Invest. 1993;92:1459–66.
60. Gao F, Wu DQ, Hu YH, Jin GX, Li GD, et al. In vitro cultivation of islet-like cell clusters from human umbilical cord blood-derived mesenchymal stem cells. Transl Res. 2008;151:293–302.
61. Aguayo-Mazzucato C, Koh A, El Khattabi I, Li WC, Toschi E, et al. Mafa expression enhances glucose-responsive insulin secretion in neonatal rat beta cells. Diabetologia. 2011;54:583–93.
62. Timper K, Seboek D, Eberhardt M, Linscheid P, Christ-Crain M, et al. Human adipose tissue-derived mesenchymal stem cells differentiate into insulin, somatostatin, and glucagon expressing cells. Biochem Biophys Res Commun. 2006;341:1135–40.
63. Mashima H, Shibata H, Mine T, Kojima I. Formation of insulin-producing cells from pancreatic acinar AR42J cells by hepatocyte growth factor. Endocrinology. 1996;137:3969–76.
64. Otonkoski T, Cirulli V, Beattie GM, Mally MI, Soto G, et al. A role for hepatocyte growth factor/scatter factor in fetal mesenchyme-induced pancreatic beta-cell growth. Endocrinology. 1996;137:3131–9.
65. Onitsuka I, Tanaka M, Miyajima A. Characterization and functional analyses of hepatic mesothelial cells in mouse liver development. Gastroenterology. 2010;138:1525–U1395.
66. Ishikawa T, Factor VM, Marquardt JU, Raggi C, Seo D, et al. Hepatocyte growth factor/c-met signaling is required for stem-cell-mediated liver regeneration in mice. Hepatology. 2012;55:1215–26.
67. Wang TC, Bonnerweir S, Oates PS, Chulak M, Simon B, et al. Pancreatic gastrin stimulates islet differentiation of transforming growth-factor alpha-induced ductular precursor cells. J Clin Investig. 1993;92:1349–56.
68. Wang RN, Rehfeld JF, Nielsen FC, Kloppel G. Expression of gastrin and transforming growth factor-alpha during duct to islet cell differentiation in the pancreas of duct-ligated adult rats. Diabetologia. 1997;40:887–93.
69. Halban PA, Powers SL, George KL, Bonnerweir S. Spontaneous reassociation of dispersed adult-rat pancreatic-islet cells into aggregates with 3-dimensional architecture typical of native islets. Diabetes. 1987;36:783–90.

70. Kamiya A, Kinoshita T, Ito Y, Matsui T, Morikawa Y, et al. Fetal liver development requires a paracrine action of oncostatin M through the gp130 signal transducer. EMBO J. 1999;18:2127–36.
71. Kamiya A, Kojima N, Kinoshita T, Sakai Y, Miyaijma A. Maturation of fetal hepatocytes in vitro by extracellular matrices and oncostatin M: induction of tryptophan oxygenase. Hepatology. 2002;35:1351–9.
72. Matsui T, Kinoshita T, Hirano T, Yokota T, Miyajima A. STAT3 down-regulates the expression of cyclin D during liver development. J Biol Chem. 2002;277:36167–73.
73. Banas A, Yamamoto Y, Teratani T, Ochiya T. Stem cell plasticity: Learning from hepatogenic differentiation strategies. Dev Dyn. 2007;236:3228–41.
74. Lee OK, Kuo TK, Chen WM, Lee KD, Hsieh SL, et al. Isolation of multipotent mesenchymal stem cells from umbilical cord blood. Blood. 2004;103:1669–75.
75. Arany S, Catalan MA, Roztocil E, Ovitt CE. Ascl3 knockout and cell ablation models reveal complexity of salivary gland maintenance and regeneration. Dev Biol. 2011;353:186–93.
76. Courtois G, Morgan JG, Campbell LA, Fourel G, Crabtree GR. Interaction of a liver-specific nuclear factor with the fibrinogen and alpha-1-antitrypsin promoters. Science. 1987;238:688–92.
77. Cereghini S, Blumenfeld M, Yaniv M. A liver-specific factor essential for albumin transcription differs between differentiated and dedifferentiated rat hepatoma-cells. Genes Dev. 1988;2:957–74.
78. Ishiyama T, Kano J, Minami Y, Iijima T, Morishita Y, et al. Expression of HNFs and C/EBP alpha is correlated with immunocytochemical differentiation of cell lines derived from human hepatocellular carcinomas, hepatoblastomas and immortalized hepatocytes. Cancer Sci. 2003;94:757–63.
79. Gao N, LeLay J, Vatamaniuk MZ, Rieck S, Friedman JR, et al. Dynamic regulation of Pdx1 enhancers by Foxa1 and Foxa2 is essential for pancreas development. Genes Dev. 2008;22:3435–48.
80. Xu CH, Lu XW, Chen EZ, He ZY, Uyunbilig B, et al. Genome-wide roles of Foxa2 in directing liver specification. J Mol Cell Biol. 2012;4:420–2.
81. Gao N, White P, Doliba N, Golson ML, Matschinsky FM, et al. Foxa2 controls vesicle docking and insulin secretion in mature beta cells. Cell Metab. 2007;6:267–79.
82. Carrasco M, Delgado I, Soria B, Martin F, Rojas A. GATA4 and GATA6 control mouse pancreas organogenesis. J Clin Invest. 2012;122:3504–15.
83. Haumaitre C, Barbacci E, Jenny M, Ott MO, Gradwohl G, et al. Lack of TCF2/vHNF1 in mice leads to pancreas agenesis. Proc Natl Acad Sci U S A. 2005;102:1490–5.
84. Watt AJ, Garrison WD, Duncan SA. HNF4: a central regulator of hepatocyte differentiation and function. Hepatology. 2003;37:1249–53.
85. Pierreux CE, Vanhorenbeeck V, Jacquemin P, Lemaigre FP, Rousseau GG. The transcription factor hepatocyte nuclear factor-6/Onecut-1 controls the expression of its paralog Onecut-3 in developing mouse endoderm. J Biol Chem. 2004;279:51298–304.
86. Du A, Hunter CS, Murray J, Noble D, Cai CL, et al. Islet-1 is required for the maturation, proliferation, and survival of the endocrine pancreas. Diabetes. 2009;58:2059–69.
87. Pin CL, Rukstalis JM, Johnson C, Konieczny SF. The bHLH transcription factor Mist1 is required to maintain exocrine pancreas cell organization and acinar cell identity. J Cell Biochem. 2001;155:519–30.
88. Imai Y, Patel HR, Hawkins EJ, Doliba NM, Matschinsky FM, et al. Insulin secretion is increased in pancreatic islets of neuropeptide Y-deficient mice. Endocrinology. 2007;148:5716–23.
89. Gu C, Stein GH, Pan N, Goebbels S, Hornberg H, et al. Pancreatic beta cells require NeuroD to achieve and maintain functional maturity. Cell Metab. 2010;11:298–310.
90. Magenheim J, Klein AM, Stanger BZ, Ashery-Padan R, Sosa-Pineda B, et al. Ngn3(+) endocrine progenitor cells control the fate and morphogenesis of pancreatic ductal epithelium. Dev Biol. 2011;359:26–36.
91. Papizan JB, Singer RA, Tschen SI, Dhawan S, Friel JM, et al. Nkx2.2 repressor complex regulates islet beta-cell specification and prevents beta-to-alpha-cell reprogramming. Genes Dev. 2011;25:2291–305.
92. Taylor BL, Liu FF, Sander M. Nkx6.1 is essential for maintaining the functional state of pancreatic beta cells. Cell Rep. 2013;4:1262–75.
93. Flanagan SE, De Franco E, Lango Allen H, Zerah M, Abdul-Rasoul MM, et al. Analysis of transcription factors key for mouse pancreatic development establishes NKX2-2 and MNX1 mutations as causes of neonatal diabetes in man. Cell Metab. 2014;19:146–54.
94. Bonnefond A, Vaillant E, Philippe J, Skrobek B, Lobbens S, et al. Transcription factor gene MNX1 is a novel cause of permanent neonatal diabetes in a consanguineous family. Diabetes Metab. 2013;39:276–80.
95. Hart AW, Mella S, Mendrychowski J, van Heyningen V, Kleinjan DA. The developmental regulator Pax6 is essential for maintenance of islet cell function in the adult mouse pancreas. PLoS One. 2013;8:e54173.
96. Jennings RE, Berry AA, Kirkwood-Wilson R, Roberts NA, Hearn T, et al. Development of the human pancreas from foregut to endocrine commitment. Diabetes. 2013;62:3514–22.
97. Gu G, Dubauskaite J, Melton DA. Direct evidence for the pancreatic lineage: NGN3+ cells are islet progenitors and are distinct from duct progenitors. Development. 2002;129:2447–57.
98. Brissova M, Blaha M, Spear C, Nicholson W, Radhika A, et al. Reduced PDX-1 expression impairs islet response to insulin resistance and worsens glucose homeostasis. Am J Physiol Endocrinol Metab. 2005;288:E707–14.

99. Yoshitomi H, Zaret KS. Endothelial cell interactions initiate dorsal pancreas development by selectively inducing the transcription factor Ptf1a. Development. 2004;131:807–17.
100. Weedon MN, Cebola I, Patch AM, Flanagan SE, De Franco E, et al. Recessive mutations in a distal PTF1A enhancer cause isolated pancreatic agenesis. Nat Genet. 2014;46:61–4.
101. Kawaguchi Y. Sox9 and programming of liver and pancreatic progenitors. J Clin Invest. 2013;123:1881–6.
102. Mead TJ, Wang Q, Bhattaram P, Dy P, Afelik S, et al. A far-upstream (−70 kb) enhancer mediates Sox9 auto-regulation in somatic tissues during development and adult regeneration. Nucleic Acids Res. 2013;41:4459–69.
103. Kanai-Azuma M, Kanai Y, Gad JM, Tajima Y, Taya C, et al. Depletion of definitive gut endoderm in Sox17-null mutant mice. Development. 2002;129:2367–79.
104. Tran SD, Liu YN, Xia DS, Maria OM, Khalili S, et al. Paracrine effects of bone marrow soup restore organ function, regeneration, and repair in salivary glands damaged by irradiation. PLoS One. 2013;8:e61632.
105. Caplan AI, Dennis JE. Mesenchymal stem cells as trophic mediators. J Cell Biochem. 2006;98:1076–84.
106. Fang D, Hu S, Liu Y, Quan VH, Seuntjens J, et al. Identification of the active components in Bone Marrow Soup: a mitigator against irradiation-injury to salivary glands. Sci Rep. 2015;5:16017.
107. Entesarian M, Dahlqvist J, Shashi V, Stanley CS, Falahat B, et al. FGF10 missense mutations in aplasia of lacrimal and salivary glands (ALSG). Eur J Hum Genet. 2007;15:379–82.
108. Scheckenbach K, Balz V, Wagenmann M, Hoffmann TK. An intronic alteration of the fibroblast growth factor 10 gene causing ALSG-(aplasia of lacrimal and salivary glands) syndrome. BMC Med Genet. 2008;9:114.
109. Lombaert IMA, Brunsting JF, Wierenga PK, Kampinga HH, De Haan G, et al. Keratinocyte growth factor prevents radiation damage to salivary glands by expansion of the stem/progenitor pool. Stem Cells. 2008;26:2595–601.
110. Panman L, Andersson E, Alekseenko Z, Hedlund E, Kee N, et al. Transcription factor-induced lineage selection of stem-cell-derived neural progenitor cells. Cell Stem Cell. 2011;8:663–75.
111. Jones S. An overview of the basic helix-loop-helix proteins. Genome Biol. 2004;5:226.
112. Fairman R, Beran-Steed RK, Anthony-Cahill SJ, Lear JD, Stafford 3rd WF, et al. Multiple oligomeric states regulate the DNA binding of helix-loop-helix peptides. Proc Natl Acad Sci U S A. 1993;90:10429–33.
113. Shih HP, Seymour PA, Patel NA, Xie R, Wang A, et al. A gene regulatory network cooperatively controlled by Pdx1 and Sox9 governs lineage allocation of foregut progenitor cells. Cell Rep. 2015;13:326–36.
114. Kawakami M, Ishikawa H, Tachibana T, Tanaka A, Mataga I. Functional transplantation of salivary gland cells differentiated from mouse early ES cells in vitro. Hum Cell. 2013;26:80–90.
115. Zhou Q, Brown J, Kanarek A, Rajagopal J, Melton DA. In vivo reprogramming of adult pancreatic exocrine cells to beta-cells. Nature. 2008;455:627–U630.
116. Lima MJ, Muir KR, Docherty HM, Drummond R, McGowan NWA, et al. Suppression of epithelial-to-mesenchymal transitioning enhances ex vivo reprogramming of human exocrine pancreatic tissue toward functional insulin-producing beta-like cells. Diabetes. 2013;62:2821–33.
117. Li L, Yi ZH, Seno M, Kojima I. Activin A and betacellulin – effect on regeneration of pancreatic beta-cells in neonatal streptozotocin-treated rats. Diabetes. 2004;53:608–15.
118. Huang PY, He ZY, Ji SY, Sun HW, Xiang D, et al. Induction of functional hepatocyte-like cells from mouse fibroblasts by defined factors. Nature. 2011;475:386–U142.
119. Yu B, He ZY, You P, Han QW, Xiang D, et al. Reprogramming fibroblasts into bipotential hepatic stem cells by defined factors. Cell Stem Cell. 2013;13:328–40.
120. Ahlgren U, Pfaff SL, Jessell TM, Edlund T, Edlund H. Independent requirement for ISL1 in formation of pancreatic mesenchyme and islet cells. Nature. 1997;385:257–60.
121. Edlund H. Pancreatic organogenesis – developmental mechanisms and implications for therapy. Nat Rev Genet. 2002;3:524–32.
122. Matsuoka TA, Kaneto H, Miyatsuka T, Yamamoto T, Yamamoto K, et al. Regulation of MafA expression in pancreatic beta-cells in db/db mice with diabetes. Diabetes. 2010;59:1709–20.
123. Harmon JS, Bogdani M, Parazzoli SD, Mak SS, Oseid EA, et al. Beta-Cell-specific overexpression of glutathione peroxidase preserves intranuclear MafA and reverses diabetes in db/db mice. Endocrinology. 2009;150:4855–62.
124. Butler AE, Robertson RP, Hernandez R, Matveyenko AV, Gurlo T, et al. Beta cell nuclear musculoaponeurotic fibrosarcoma oncogene family A (MafA) is deficient in type 2 diabetes. Diabetologia. 2012;55:2985–8.
125. Guo S, Dai C, Guo M, Taylor B, Harmon JS, et al. Inactivation of specific beta cell transcription factors in type 2 diabetes. J Clin Invest. 2013;123:3305–16.
126. Talchai C, Xuan S, Lin HV, Sussel L, Accili D. Pancreatic beta cell dedifferentiation as a mechanism of diabetic beta cell failure. Cell. 2012;150:1223–34.
127. Ghaye AP, Bergemann D, Tarifeno-Saldivia E, Flasse LC, Von Berg V, et al. Progenitor potential of nkx6.1-expressing cells throughout zebrafish life and during beta cell regeneration. BMC Biol. 2015;13:70.
128. Chan JY, Luzuriaga J, Bensellam M, Biden TJ, Laybutt DR. Failure of the adaptive unfolded protein response in islets of obese mice is linked with abnormalities in beta-cell gene expression and progression to diabetes. Diabetes. 2013;62:1557–68.

129. Li H, Arber S, Jessell TM, Edlund H. Selective agenesis of the dorsal pancreas in mice lacking homeobox gene Hlxb9. Nat Genet. 1999;23:67–70.
130. Conrad E, Stein R, Hunter CS. Revealing transcription factors during human pancreatic beta cell development. Trends Endocrinol Metab. 2014;25:407–14.
131. Hamasaki A, Yamada Y, Kurose T, Ban N, Nagashima K, et al. Adult pancreatic islets require differential pax6 gene dosage. Biochem Biophys Res Commun. 2007;353:40–6.
132. Wen JH, Chen YY, Song SJ, Ding J, Gao Y, et al. Paired box 6 (PAX6) regulates glucose metabolism via proinsulin processing mediated by prohormone convertase 1/3 (PC1/3). Diabetologia. 2009;52:504–13.
133. Ding J, Gao Y, Zhao J, Yan H, Guo SY, et al. Pax6 haploinsufficiency causes abnormal metabolic homeostasis by down-regulating glucagon-like peptide 1 in mice. Endocrinology. 2009;150:2136–44.
134. Habener JF, Kemp DM, Thomas MK. Minireview: transcriptional regulation in pancreatic development. Endocrinology. 2005;146:1025–34.
135. Fujimoto K, Polonsky KS. Pdx1 and other factors that regulate pancreatic beta-cell survival. Diabetes Obes Metab. 2009;11:30–7.
136. Spence JR, Lange AW, Lin SCJ, Kaestner KH, Lowy AM, et al. Sox17 regulates organ lineage segregation of ventral foregut progenitor cells. Dev Cell. 2009;17:62–74.
137. Kamiya A, Inoue Y, Gonzalez FJ. Role of the hepatocyte nuclear factor 4 alpha in control of the pregnane X receptor during fetal liver development. Hepatology. 2003;37:1375–84.
138. Li JX, Ning G, Duncan SA. Mammalian hepatocyte differentiation requires the transcription factor HNF-4 alpha. Genes Dev. 2000;14:464–74.
139. Jin JL, Iakova P, Jiang YJ, Lewis K, Sullivan E, et al. Transcriptional and translational regulation of C/EBP beta-HDAC1 protein complexes controls different levels of p53, SIRT1, and PGC1 alpha proteins at the early and late stages of liver cancer. J Biol Chem. 2013;288:14451–62.
140. Schrem H, Klempnauer J, Borlak J. Liver-enriched transcription factors in liver function and development. Part II: the C/EBPs and D site-binding protein in cell cycle control, carcinogenesis, circadian gene regulation, liver regeneration, apoptosis, and liver-specific gene regulation. Pharmacol Rev. 2004;56:291–330.
141. Flodby P, Barlow C, Kylefjord H, Ahrlund-Richter L, Xanthopoulos KG. Increased hepatic cell proliferation and lung abnormalities in mice deficient in CCAAT/enhancer binding protein alpha. J Biol Chem. 1996;271:24753–60.
142. Nakamura T, Ueno T, Sakamoto M, Sakata R, Torimura T, et al. Suppression of transforming growth factor-beta results in upregulation of transcription of regeneration factors after chronic liver injury. J Hepatol. 2004;41:974–82.
143. Fausto N. Liver regeneration. J Hepatol. 2000;32:19–31.
144. Koniaris LG, McKillop IH, Schwartz SI, Zimmers TA. Liver regeneration. J Am Coll Surg. 2003;197:634–59.
145. Crissey MAS, Leu JI, De Angelis RA, Greenbaum LE, Scearce LM, et al. Liver-specific and proliferation-induced deoxyribonuclease I hypersensitive sites in the mouse insulin-like growth factor binding protein-1 gene. Hepatology. 1999;30:1187–97.
146. Tomizawa M, Garfield S, Xanthopoulos KG. Hepatocytes deficient in CCAAT enhancer binding protein alpha (C/EBP alpha) exhibit both hepatocyte and biliary epithelial cell character. Hepatology. 1998;28:294a.
147. Yamasaki H, Sada A, Iwata T, Niwa T, Tomizawa M, et al. Suppression of C/EBP alpha expression in periportal hepatoblasts may stimulate biliary cell differentiation through increased Hnf6 and Hnf1b expression. Development. 2006;133:4233–43.
148. Antoniou A, Raynaud P, Cordi S, Zong YW, Tronche F, et al. Intrahepatic bile ducts develop according to a new mode of tubulogenesis regulated by the transcription factor SOX9. Gastroenterology. 2009;136:2325–33.
149. Bullard T, Koek L, Roztocil E, Kingsley PD, Mirels L, et al. Ascl3 expression marks a progenitor population of both acinar and ductal cells in mouse salivary glands. Dev Biol. 2008;320:72–8.
150. Yoshida S, Ohbo K, Takakura A, Takebayashi H, Okada T, et al. Sgn1, a basic helix-loop-helix transcription factor delineates the salivary gland duct cell lineage in mice. Dev Biol. 2001;240:517–30.
151. Aure MH, Konieczny SF, Ovitt CE. Salivary gland homeostasis is maintained through acinar cell self-duplication. Dev Cell. 2015;33:231–7.
152. Gagliardi A, Mullin NP, Tan ZY, Colby D, Kousa AI, et al. A direct physical interaction between Nanog and Sox2 regulates embryonic stem cell self-renewal. EMBO J. 2013;32:2231–47.
153. Knox SM, Lombaert IM, Reed X, Vitale-Cross L, Gutkind JS, et al. Parasympathetic innervation maintains epithelial progenitor cells during salivary organogenesis. Science. 2010;329:1645–7.
154. Pereira L, Yi F, Merrill BJ. Repression of Nanog gene transcription by Tcf3 limits embryonic stem cell self-renewal. Mol Cell Biol. 2006; 26:7479–91.
155. Wray J, Kalkan T, Gomez-Lopez S, Eckardt D, Cook A, et al. Inhibition of glycogen synthase kinase-3 alleviates Tcf3 repression of the pluripotency network and increases embryonic stem cell resistance to differentiation. Nat Cell Biol. 2011;13:838–45.
156. Kelly KF, Ng DY, Jayakumaran G, Wood GA, Koide H, et al. Beta-catenin enhances Oct-4 activity and reinforces pluripotency through a TCF-independent mechanism. Cell Stem Cell. 2011;8:214–27.
157. Yi F, Pereira L, Hoffman JA, Shy BR, Yuen CM, et al. Opposing effects of Tcf3 and Tcf1 control Wnt

stimulation of embryonic stem cell self-renewal. Nat Cell Biol. 2011;13:762–U383.
158. Atlasi Y, Noori R, Gaspar C, Franken P, Sacchetti A, et al. Wnt signaling regulates the lineage differentiation potential of mouse embryonic stem cells through Tcf3 down-regulation. PLoS Genet. 2013;9:e1003424.
159. Wu CI, Hoffman JA, Shy BR, Ford EM, Fuchs E, et al. Function of Wnt/beta-catenin in counteracting Tcf3 repression through the Tcf3-beta-catenin interaction. Development. 2012;139:2118–29.
160. Marmorstein LY, Kinev AV, Chan GKT, Bochar DA, Beniya H, et al. A human BRCA2 complex containing a structural DNA binding component influences cell cycle progression. Cell. 2001;104:247–57.
161. Lin JW, Biankin AV, Horb ME, Ghosh B, Prasad NB, et al. Differential requirement for ptf1a in endocrine and exocrine lineages of developing zebrafish pancreas. Dev Biol. 2004;274:491–503.
162. Cano DA, Soria B, Martin F, Rojas A. Transcriptional control of mammalian pancreas organogenesis. Cell Mol Life Sci. 2014;71:2383–402.
163. Garg V, Kathiriya IS, Barnes R, Schluterman MK, King IN, et al. GATA4 mutations cause human congenital heart defects and reveal an interaction with TBX5. Nature. 2003;424:443–7.
164. Watt AJ, Zhao R, Li JX, Duncan SA. Development of the mammalian liver and ventral pancreas is dependent on GATA4. BMC Dev Biol. 2007;7:37.
165. Lee CS, Sund NJ, Vatamaniuk MZ, Matschinsky FM, Stoffers DA, et al. Foxa2 controls Pdx1 gene expression in pancreatic beta-cells in vivo. Diabetes. 2002;51:2546–51.
166. Lee CS, Sund NJ, Behr R, Herrera PL, Kaestner KH. Foxa2 is required for the differentiation of pancreatic alpha-cells. Dev Biol. 2005;278:484–95.
167. Zaret KS, Carroll JS. Pioneer transcription factors: establishing competence for gene expression. Genes Dev. 2011;25:2227–41.
168. Le Lay J, Kaestner KH. The fox genes in the liver: from organogenesis to functional integration. Physiol Rev. 2010;90:1–22.
169. Jeon J, Correa-Medina M, Ricordi C, Edlund H, Diez JA. Endocrine cell clustering during human pancreas development. J Histochem Cytochem. 2009;57:811–24.
170. Rubio-Cabezas O, Jensen JN, Hodgson MI, Codner E, Ellard S, et al. Permanent neonatal diabetes and enteric anendocrinosis associated with biallelic mutations in NEUROG3. Diabetes. 2011;60:1349–53.
171. Gradwohl G, Dierich A, LeMeur M, Guillemot F. Neurogenin3 is required for the development of the four endocrine cell lineages of the pancreas. Proc Natl Acad Sci U S A. 2000;97:1607–11.
172. Johnson CL, Kowalik AS, Rajakumar N, Pin CL. Mist1 is necessary for the establishment of granule organization in serous exocrine cells of the gastrointestinal tract. Mech Dev. 2004;121:261–72.
173. Pin CL, Bonvissuto AC, Konieczny SF. Mist1 expression is a common link among serous exocrine cells exhibiting regulated exocytosis. Anat Rec. 2000;259:157–67.
174. Direnzo D, Hess DA, Damsz B, Hallett JE, Marshall B, et al. Induced Mist1 expression promotes remodeling of mouse pancreatic acinar cells. Gastroenterology. 2012;143:469–80.
175. Rukstalis JM, Kowalik A, Zhu LQ, Lidington D, Pin CL, et al. Exocrine specific expression of Connexin32 is dependent on the basic helix-loop-helix transcription factor Mist1. J Cell Sci. 2003;116:3315–25.
176. Pan FC, Wright C. Pancreas organogenesis: from bud to plexus to gland. Dev Dyn. 2011;240: 530–65.
177. McDonald E, Li JM, Krishnamurthy M, Fellows GF, Goodyer CG, et al. SOX9 regulates endocrine cell differentiation during human fetal pancreas development. Int J Biochem Cell Biol. 2012;44:72–83.
178. Selman K, Kafatos FC. Transdifferentiation in the labial gland of silk moths: is DNA required for cellular metamorphosis? Cell Differ. 1974;3:81–94.
179. Eguchi G, Okada TS. Differentiation of lens tissue from the progeny of chick retinal pigment cells cultured in vitro: a demonstration of a switch of cell types in clonal cell culture. Proc Natl Acad Sci U S A. 1973;70:1495–9.
180. Araki M, Okada TS. Differentiation of lens and pigment cells in cultures of neural retinal cells of early chick embryos. Dev Biol. 1977;60:278–86.
181. Bronnerfraser M. Transdifferentiation – flexibility in cell-differentiation – Okada, Ts. Science. 1992;256:1054–5.
182. Eberhard D, Tosh D. Transdifferentiation and metaplasia as a paradigm for understanding development and disease. Cell Mol Life Sci. 2008;65:33–40.
183. Gruh IMU. Transdifferentiation of stem cells: a critical view. In: Martin U, editor. Engineering of stem cells. Heidelberg: Springer; 2009. p. 73–106.
184. Burke ZD, Tosh D. Therapeutic potential of transdifferentiated cells. Clin Sci (Lond). 2005;108:309–21.
185. Torper O, Pfisterer U, Wolf DA, Pereira M, Lau S, et al. Generation of induced neurons via direct conversion in vivo. Proc Natl Acad Sci U S A. 2013;110:7038–43.
186. Merrell AJ, Stanger BZ. Adult cell plasticity in vivo: de-differentiation and transdifferentiation are back in style. Nat Rev Mol Cell Biol. 2016;17:413–145.
187. Burke ZD, Thowfeequ S, Peran M, Tosh D. Stem cells in the adult pancreas and liver. Biochem J. 2007;404:169–78.
188. Kume S. Stem-cell-based approaches for regenerative medicine. Dev Growth Differ. 2005;47:393–402.
189. Dabeva MD, Hurston E, Shafritz DA. Transcription factor and liver-specific mRNA expression in facultative epithelial progenitor cells of liver and pancreas. Am J Pathol. 1995;147:1633–48.
190. McLin VA, Zorn AM. Organogenesis: making pancreas from liver. Curr Biol. 2003;13:R96–8.
191. Horb ME, Shen CN, Tosh D, Slack JMW. Experimental conversion of liver to pancreas. Curr Biol. 2003;13:105–15.

192. Kojima H, Nakamura T, Fujita Y, Kishi A, Fujimiya M, et al. Combined expression of pancreatic duodenal homeobox 1 and islet factor 1 induces immature enterocytes to produce insulin. Diabetes. 2002;51:1398–408.
193. von Mach MA, Hengstler JG, Brulport M, Eberhardt M, Schormann W, et al. In vitro cultured islet-derived progenitor cells of human origin express human albumin in severe combined immunodeficiency mouse liver in vivo. Stem Cells. 2004;22:1134–41.
194. Lardon J, De Breuck S, Rooman I, Van Lommel L, Kruhoffer M, et al. Plasticity in the adult rat pancreas: Transdifferentiation of exocrine to hepatocyte-like cells in primary culture. Hepatology. 2004;39:1499–507.
195. Kawaguchi Y, Cooper B, Gannon M, Ray M, MacDonald RJ, et al. The role of the transcriptional regulator Ptf1a in converting intestinal to pancreatic progenitors. Nat Genet. 2002;32:128–34.
196. Ferber S, Halkin A, Cohen H, Ber I, Einav Y, et al. Pancreatic and duodenal homeobox gene 1 induces expression of insulin genes in liver and ameliorates streptozotocin-induced hyperglycemia. Nat Med. 2000;6:568–72.
197. Yi F, Liu GH, Izpisua Belmonte JC. Rejuvenating liver and pancreas through cell transdifferentiation. Cell Res. 2012;22:616–9.
198. Kaneto H, Nakatani Y, Miyatsuka T, Matsuoka T, Matsuhisa M, et al. PDX-1/VP16 fusion protein, together with NeuroD or Ngn3, markedly induces insulin gene transcription and ameliorates glucose tolerance. Diabetes. 2005;54:1009–22.
199. Hisatomi Y, Okumura K, Nakamura K, Matsumoto S, Satoh A, et al. Flow cytometric isolation of endodermal progenitors from mouse salivary gland differentiate into hepatic and pancreatic lineages. Hepatology. 2004;39:667–75.
200. Rothova M, Thompson H, Lickert H, Tucker AS. Lineage tracing of the endoderm during oral development. Dev Dyn. 2012;241:1183–91.
201. Okumura K, Nakamura K, Hisatomi Y, Nagano K, Tanaka Y, et al. Salivary gland progenitor cells induced by duct ligation differentiate into hepatic and pancreatic lineages. Hepatology. 2003;38:104–13.
202. Denny PC, Denny PA. Dynamics of parenchymal cell division, differentiation, and apoptosis in the young adult female mouse submandibular gland. Anat Rec. 1999;254:408–17.
203. Konings AW, Coppes RP, Vissink A. On the mechanism of salivary gland radiosensitivity. Int J Radiat Oncol Biol Phys. 2005;62:1187–94.
204. Patel VN, Hoffman MP. Salivary gland development: a template for regeneration. Semin Cell Dev Biol. 2014;25:52–60.
205. Nanduri LS, Maimets M, Pringle SA, van der Zwaag M, van Os RP, et al. Regeneration of irradiated salivary glands with stem cell marker expressing cells. Radiother Oncol. 2011;99:367–72.
206. Rock JR, Onaitis MW, Rawlins EL, Lu Y, Clark CP, et al. Basal cells as stem cells of the mouse trachea and human airway epithelium. Proc Natl Acad Sci U S A. 2009;106:12771–5.
207. Arnold K, Sarkar A, Yram MA, Polo JM, Bronson R, et al. Sox2(+) adult stem and progenitor cells are important for tissue regeneration and survival of mice. Cell Stem Cell. 2011;9:317–29.
208. Nelson J, Manzella K, Baker OJ. Current cell models for bioengineering a salivary gland: a mini-review of emerging technologies. Oral Dis. 2013;19:236–44.
209. Maimets M, Rocchi C, Bron R, Pringle S, Kuipers J, et al. Long-term in vitro expansion of salivary gland stem cells driven by Wnt signals. Stem Cells Rep. 2016;6:150–62.
210. Yaniv A, Neumann Y, David R, Stiubea-Cohen R, Orbach Y, et al. Establishment of immortal multipotent rat salivary progenitor cell line toward salivary gland regeneration. Tissue Eng Part C Methods. 2011;17:69–78.
211. Sutton, M.T. & Bonfield, T.L. Stem cells: innovations in clinical applications. Stem Cells Int 2014, 516278 (2014).
212. Shamoon H, Duffy H, Fleischer N, Engel S, Saenger P, et al. The effect of intensive treatment of diabetes on the development and progression of long-term complications in insulin-dependent diabetes-mellitus. N Engl J Med. 1993;329:977–86.
213. Bouwens L, Houbracken I, Mfopou JK. The use of stem cells for pancreatic regeneration in diabetes mellitus. Nat Rev Endocrinol. 2013;9:598–606.
214. Gennero L, Roos MA, Sperber K, Denysenko T, Bernabei P, et al. Pluripotent plasticity of stem cells and liver repopulation. Cell Biochem Funct. 2010;28:178–89.
215. Miyajima A, Tanaka M, Itoh T. Stem/progenitor cells in liver development, homeostasis, regeneration, and reprogramming. Cell Stem Cell. 2014;14:561–74.
216. Schwartz RE, Reyes M, Koodie L, Jiang YH, Blackstad M, et al. Multipotent adult progenitor cells from bone marrow differentiate into functional hepatocyte-like cells. J Clin Investig. 2002;109:1291–302.
217. Reyes M, Lund T, Lenvik T, Aguiar D, Koodie L, et al. Purification and ex vivo expansion of postnatal human marrow mesodermal progenitor cells (Retracted article. See vol. 113, pg. 2370, 2009). Blood. 2001;98:2615–25.
218. Lu LL, Liu YJ, Yang SG, Zhao QJ, Wang X, et al. Isolation and characterization of human umbilical cord mesenchymal stem cells with hematopoiesis-supportive function and other potentials. Haematologica. 2006;91:1017–26.
219. Banas A, Teratani T, Yamamoto Y, Tokuhara M, Takeshita F, et al. Adipose tissue-derived mesenchymal stem cells as a source of human hepatocytes. Hepatology. 2007;46:219–28.
220. Talens-Visconti R, Bonora A, Jover R, Mirabet V, Carbonell F, et al. Human mesenchymal stem cells from adipose tissue: Differentiation into hepatic lineage. Toxicol In Vitro. 2007;21:324–9.

221. Planat-Benard V, Silvestre JS, Cousin B, Andre M, Nibbelink M, et al. Plasticity of human adipose lineage cells toward endothelial cells – physiological and therapeutic perspectives. Circulation. 2004;109:656–63.
222. Seo MJ, Suh SY, Bae YC, Jung JS. Differentiation of human adipose stromal cells into hepatic lineage in vitro and in vivo. Biochem Biophys Res Commun. 2005;328:258–64.
223. Lange C, Bassler P, Lioznov MV, Bruns H, Kluth D, et al. Liver-specific gene expression in mesenchymal stem cells is induced by liver cells. World J Gastroenterol. 2005;11:4497–504.
224. Rai RM, Yang SQ, McClain C, Karp CL, Klein AS, et al. Kupffer cell depletion by gadolinium chloride enhances liver regeneration after partial hepatectomy in rats. Am J Physiol. 1996;270:G909–18.
225. Sato Y, Araki H, Kato J, Nakamura K, Kawano Y, et al. Human mesenchymal stem cells xenografted directly to rat liver are differentiated into human hepatocytes without fusion. Blood. 2005; 106:756–63.
226. Chamberlain J, Yamagami T, Colletti E, Theise ND, Desai J, et al. Efficient generation of human hepatocytes by the intrahepatic delivery of clonal human mesenchymal stem cells in fetal sheep. Hepatology. 2007;46:1935–45.
227. Meier RPH, Muller YD, Morel P, Gonelle-Gispert C, Buhler LH. Transplantation of mesenchymal stem cells for the treatment of liver diseases, is there enough evidence? Stem Cell Res. 2013; 11:1348–64.
228. Bae SH. Clinical application of stem cells in liver diseases. Korean J Hepatol. 2008;14:309–17.
229. Russo FP, Alison MR, Bigger BW, Amofah E, Florou A, et al. The bone marrow functionally contributes to liver fibrosis. Gastroenterology. 2006;130:1807–21.
230. Khalili S, Liu YN, Kornete M, Roescher N, Kodama S, et al. Mesenchymal stromal cells improve salivary function and reduce lymphocytic infiltrates in mice with Sjogren's-like disease. PLoS One. 2012;7:e38615.
231. Coppes RP, Zeilstra LJW, Kampinga HH, Konings AWT. Early to late sparing of radiation damage to the parotid gland by adrenergic and muscarinic receptor agonists. Br J Cancer. 2001;85:1055–63.
232. Konings AWT, Faber H, Cotteleer F, Vissink A, Coppes RP. Secondary radiation damage as the main cause for unexpected volume effects: a histopathologic study of the parotid gland. Int J Radiat Oncol Biol Phys. 2006;64:98–105.
233. Jeong J, Baek H, Kim YJ, Choi Y, Lee H, et al. Human salivary gland stem cells ameliorate hyposalivation of radiation-damaged rat salivary glands. Exp Mol Med. 2013;45:e58.
234. Lombaert IMA, Brunsting JF, Wierenga PK, Faber H, Stokman MA, et al. Rescue of salivary gland function after stem cell transplantation in irradiated glands. PLoS One. 2008;3:e2063.
235. Kojima H, Fujimiya M, Matsumura K, Younan P, Imaeda H, et al. NeuroD-betacellulin gene therapy induces islet neogenesis in the liver and reverses diabetes in mice. Nat Med. 2003;9:596–603.
236. Nakatsuji N, Nakajima F, Tokunaga K. HLA-haplotype banking and iPS cells. Nat Biotechnol. 2008;26:739–40.
237. Lanza RP, Hayes JL, Chick WL. Encapsulated cell technology. Nat Biotechnol. 1996;14:1107–11.
238. Dean SK, Yulyana Y, Williams G, Sidhu KS, Tuch BE. Differentiation of encapsulated embryonic stem cells after transplantation. Transplantation. 2006;82:1175–84.
239. Selmani Z, Naji A, Gaiffe E, Obert L, Tiberghien P, et al. HLA-G is a crucial immunosuppressive molecule secreted by adult human mesenchymal stem cells. Transplantation. 2009;87:S62–6.
240. Jacobs SA, Roobrouck VD, Verfaillie CM, Van Gool SW. Immunological characteristics of human mesenchymal stem cells and multipotent adult progenitor cells. Immunol Cell Biol. 2013;91:32–9.

Part III

Bioengineering of Salivary Glands

Current Cell Models for Bioengineering Salivary Glands

7

Olga J. Baker

Abstract

Saliva is critical for sustaining oral health. Patients with salivary gland hypofunction (symptomatically, xerostomia) have difficulties in basic oral functions such as chewing, tasting, and swallowing foods. Additionally, they suffer from caries, periodontal disease, and a variety of microbial infections. Salivary flow has been significantly improved through the use of gene therapies and secretory agonists, thereby mitigating dry mouth symptoms. However, scientific advancements in the area of clinically applied implants are needed to more fully restore compromised salivary gland function. In this book chapter, we will evaluate the advantages and limitations of commonly used salivary cell lines. Furthermore, we will summarize ongoing studies on current salivary cell isolation methods and cell culture techniques using different biomaterials. Then, we will describe the use of stem, progenitor, and acinar cells as candidates for salivary gland regeneration and bioengineering studies. Finally, we will highlight the use of scaffolds for growing salivary glands in vitro. Together, these studies represent the state of the art and trends in the emerging field of salivary gland bioengineering.

7.1 Introduction

Hyposalivation contributes to tooth decay, periodontitis, and microbial infections. Additionally, it impairs activities of daily living such as speaking, chewing, and swallowing [22]. Hyposalivation is associated with the following conditions: (a) Sjögren's syndrome (SS), an autoimmune disease with a prevalence between 0.1 and 3 % in the United States [48]; (b) radiotherapy for head and neck cancer, accounting for up to 3 % of all cancers in the United States [35]; (c) side effects of commonly used medications [3, 76]; and (d) developmental disorders affecting ectodermal tissues and organs [59].

Common treatments for hyposalivation include the use of secretory agonists and saliva substitutes, both of which provide only temporary

O.J. Baker, DDS, PhD
School of Dentistry, The University of Utah,
383 Colorow Dr Room 289A, Salt Lake City,
UT 84108-1201, USA
e-mail: olga.baker@hsc.utah.edu

S. Cha (ed.), *Salivary Gland Development and Regeneration*, DOI 10.1007/978-3-319-43513-8_7

relief [75]. Experimental treatments such as use of adenoviral [9] and ultrasound-assisted transfection of aquaporin-1 [77] or stem cells [54] offer the possibility of significant improvement. Alternatively, salivary gland transplantation remains an attractive option if other treatments using a native salivary gland prove ineffective, unreliable, or fail to maintain treatment gains. However, transplantation of a naturally occurring salivary gland is problematic because finding donors, as with any organ donation, is a complex and unpredictable process [78] and the chances of rejection by the host are high [17].

Despite the difficulties involved in salivary gland implantation, alternative approaches provide new possibilities for making this treatment viable. In this chapter, current trends in the emerging field of salivary gland bioengineering involving cell lines, primary cells, and stem/progenitor cells will be summarized. Note that particular variations on some strategies are too numerous to cover individually; however, no significant strategy has been omitted, and multiple examples within each category are provided.

7.2 Commonly Used Salivary Cell Lines

Normal cells are unable to replicate beyond several rounds of proliferation due to progressive telomere shortening with each successive round. Specifically, cellular senescence (i.e., naturally occurring loss of a cell's ability to divide and grow) results when the telomeres reach a critically reduced length and DNA is damaged [14]. An immortalized cell line is a population of cells from a multicellular organism which would normally not proliferate indefinitely but, due to mutation, has evaded normal cellular senescence and undergoes continuous division [49]. Immortalized cells are different from stem cells, which can also divide indefinitely but form a normal part of the development of a multicellular organism (i.e., are not attributable to mutation or intentional in vitro modification). Immortalized cell lines can be derived from tumors, such as HeLa cells that were obtained from human cervical cancer cells [26]. However, many of the current immortalized cell lines are intentionally created by introduction of a virus that greatly extends the number of viable cell replications and allows for the possibility that a mutation might occur that could result in a truly immortalized cell. For instance, overexpression of the large T antigen of the simian virus 40 (SV40) represses the retinoblastoma (Rb) and p53 genes (both of which are critical controllers of the cell cycle) [62]. Other viral genes include those from the human papilloma virus family, which also target Rb and p53 [8]. Another method to induce cell immortalization involves the use of the gene telomerase (hTERT), which extends the DNA sequence of telomeres and allows for infinite cell divisions [44]. The use of cell lines to bioengineer a salivary gland would appear to offer an attractive alternative to development and implantation of an artificial salivary gland. However, no currently available cell line fully recapitulates the morphological and functional features of the native salivary acinar cells; moreover, some are tumorigenic [62]. Consequently, we believe cell lines are currently not suitable for bioengineering and/or implantation and that their use should be limited to understanding basic physiological mechanisms in native glands. Below we evaluate the advantages and limitations of commonly used salivary cell lines (see Table 7.1).

7.2.1 HSY

Human parotid gland (HSY) is an epithelial cell line derived from the acinar-intercalated duct region of the human parotid gland [81]. These cells are cuboidal in shape, have papillary infoldings as well as microvilli on their free border, and exhibit low levels of secretory granules. HSY cells maintain polarized monolayer organization [81], which is critical for engineering a gland capable of fluid secretion. They respond to muscarinic and β-adrenergic autonomic agonists to increase intracellular free calcium concentration ($[Ca^{2+}]_i$) and cyclic AMP ($[cAMP]_i$), respectively [61], features that are essential for saliva secretion in vivo [73]. It has been proposed that salivary intercalated ducts

Table 7.1 Commonly used immortalized salivary cell lines

Cell line	Source	Cell polarity	Amylase expression	Response to secretory agonist	Reference
HSY	Human parotid adenocarcinoma	+	+	+	Yanagawa et al. [81]
HSG	Human submandibular gland	+	+	+	Shirasuna et al. [69]
SMIE	Rat submandibular gland	+	–	+	He et al. [28, 29]
RSMT-A5	Rat submandibular gland	–	–	–	Brown [12]
SMG-C6	Rat submandibular gland	+	Unknown	Unknown	Quissell et al. [65]
SMG-C10	Rat submandibular gland	+	Unknown	Unknown	Quissell et al. [65]
Par-C10	Rat parotid gland	+	–	+	Quissell et al. [66]
Par-C5	Rat parotid gland	+	–	–	Quissell et al. [66]
GManSV	tdTomato mouse submandibular gland	Unknown	Unknown	Unknown	Furukawa et al. [24]

Modified from Nelson et al. [56]

function as the reservoir for progenitor cells in salivary glands [55]. More recently, FGF10-induced ERK1/2 phosphorylation in HSY cells was noted, indicating that they respond to growth factors linked to salivary gland morphogenesis [80]. Given that HSY cells have similar morphological features to those of intercalated duct cells, it would be interesting to study whether non-transfected HSY cells may behave like progenitor stem cells when transplanted in vivo. If this were the case, they could be used for differentiation studies (with the aim of developing an artificial salivary gland), though we believe such applications have not yet been explored.

7.2.2 HSG

Human submandibular gland (HSG) is a neoplastic intercalated duct cell line derived from an irradiated human submandibular gland (SMG) [69]. Histologically, HSG cells can be presented either in cuboidal or conical shape and have easily visible desmosomes with sporadic tight junction (TJ) complexes [69]. They appear capable of fluid and protein secretion due to the presence of intercellular connections, microvilli, and protein synthesis machinery.

HSG cells have been used as an in vitro model for salivary gland secretion, morphology, and regeneration [31, 37, 38, 45, 70]. Moreover, several features indicate that they are a potential source for developing an artificial salivary gland. Specifically, HSG cells differentiate into acinar structures and express amylase when cultured on Matrigel [31]. Additionally, they have an innate capacity to increase $[Ca^{2+}]_i$ in response to muscarinic and purinergic agonists [52]. Furthermore, HSG cells can be modulated by regulators of apoptosis, thereby providing researchers with a shutoff mechanism for tumorigenic cells in vivo [23]. Finally, they are regulated by growth factor receptors (e.g., EGFR) that may be used to promote repair or regeneration of salivary tissue [67].

Despite the many advantages of using HSG for salivary gland bioengineering, there are significant drawbacks as well. Specifically, HSG cells grown on plastic appear unable to form TJs. Interestingly, they have been noted to express TJs and aquaporins when grown on permeable supports coated with Matrigel [45]. However, future studies will be necessary to understand the barrier properties of this model (i.e., characterization of TJ morphology and in-depth analysis of monolayer permeability). Finally, a challenge presented by the use of HSG is the frequent contamination of this cell line with HeLa cells [18], which may be overcome through proper control protocols, as discussed below.

7.2.3 SMIE

Rat submandibular gland acinar epithelial (SMIE) is an immortalized epithelial cell line derived from rat SMG [29]. This cell line was originally named RSMG and later renamed to SMIE because of the adenovirus (12S E1A gene product) used to immortalize the cells [28]. SMIE cells were established to study polarized functions in salivary epithelium due to their ability to form TJs when grown on permeable supports [28, 51]. Structurally, these cells resemble salivary glandular epithelium with immature lumens [28]. SMIE cells display selective barrier function and fluid transport [28, 51] and have also been shown to secrete luciferase, when transfected with a pGL3-EGFSP construct [1]. These studies indicate that SMIE cells can be used to modulate fluid and protein secretion for future bioengineering applications.

7.2.4 RSMT-A5

Rat submandibular duct epithelial (RSMT-A5) also known as (A5) cell line was derived through transformation of rat SMG cells by way of treatment with 3-methylcholanthrene [12]. This cell line displays a ductal epithelium phenotype and expresses a high density of α1-andrenergic receptors with metabolic behavior similar to smooth muscle cells [30]. Recent studies demonstrated that A5 cells were able to uptake nanoparticles [27]. Taken together, these results indicate that A5 cells are useful for receptor characterization and signaling studies; however, they might not be suitable for the study of protein secretion due to transfection difficulties [1].

7.2.5 SMG-C6 and SMG-C10

Rat submandibular gland epithelial (SMG-C) cell lines were isolated through transfection of a replication-defective simian virus 40 (SV40) genome into rat SMG acinar cells [65]. Only two of the formed clones, termed SMG-C6 and SMG-C10, were found to be both well differentiated and of epithelial origin [65]. Structurally, these cell lines polarize due to their ability of form TJs and desmosomes [65]. Additionally, secretory features (i.e., domes, granules, and canaliculi) are observed within them [65]. Functionally, SMG-C6 respond to muscarinic and purinergic agonists by increasing $[Ca^{2+}]_i$ [42]. Furthermore, both SMG-C6 and SMG-C10 respond to β-adrenergic agonists by increasing $[cAMP]_i$ [42]. Of the two cell lines, SMG-C6 seems to be better differentiated than SMG-C10 due to a greater quantity of secretory cellular structures and a more stable $[Ca^{2+}]_i$ release [65]. Moreover, SMG-C6 and SMG-C10 lines develop a high transepithelial resistance when grown on collagen-coated polycarbonate filters [16].

In addition to the signaling processes detailed above, additional functions for SMG-C6 and SMG-C10 have been observed. Specifically, both cell lines are excellent models to study sodium channels and expression of the epithelial sodium channel protein (ENaC), given their ability to modulate sodium transport when grown in a culture medium lacking glucocorticoids or mineralocorticoids [74]. Studies using SMG-C10 cells also indicated that the vanilloid receptor 4 (TRPV4) was functionally connected to aquaporin-5 volume [6]. More recently, a molecule involved in the regulation of energy metabolism and inflammatory responses (adiponectin) was found to promote fluid secretion in SMG-C6 cells [21]. The above studies indicate SMG-C6 and SMG-C10 are potential candidates for understanding regulation of cell volume and secretory function in salivary gland bioengineering. Finally, studies using SMG-C6 cells demonstrated that apoptosis could be modulated through a Fas-mediated pathway [2], indicating their potential for in vivo transplantation studies without the risk of uncontrolled cellular growth.

7.2.6 Par-C10 and Par-C5

Following development of the SMG-C6 and SMG-C10 cell lines, another study was done to isolate cells from a rat parotid gland [66]. Similar to the SMG-C cell line development, parotid salivary cells were transfected with an origin-defective SV40 plasmid [66]. Morphology and receptor-mediated $[Ca^{2+}]_i$ responses were used as a screening technique to monitor cell differentiation [66]. The rat parotid (Par-C5 and Par-C10) cell lines were found to exhibit a significant elevation of $[Ca^{2+}]_i$ in response to cholinergic, muscarinic, and α_1-adrenergic agonists [41]. These cell lines demonstrated an increase of $[cAMP]_i$ in response to α_1-adrenergic agonists [66] as well as increases of $[Ca^{2+}]_i$ in response to M3R muscarinic agonists [10]. No functional amylase expression has been observed in Par-C10 cells when grown either on plastic or growth factor-reduced (GFR) Matrigel, although an interesting study on the Par-C 3-9 clones reported a 16-fold increase in amylase content following incubation with rat serum [83]. However, further studies are needed to consistently demonstrate improved amylase production in these cell lines.

The Par-C10 cell line has been widely studied, given its ability to form polarized monolayers, which makes it a great model for studying barrier function and ion transport in salivary epithelium [72]. Specifically, ion secretion in Par-C10 cells has been well characterized, thereby establishing that it is regulated by basolateral α_1-adrenergic and muscarinic cholinergic receptors as well as apical $P2Y_2$ receptors [72]. Furthermore, Par-C10 cells express Na^+/H^+ exchangers, Na^+-HCO_3^- cotransporters, and anion exchange proteins on their basolateral surfaces [20]. These proteins, which regulate transepithelial transport, are sensitive to changes in both $[Ca^{2+}]_i$ and $[cAMP]_i$ [20]. Par-C10 single cells also are capable of forming salivary spheres when grown on Matrigel. Under these conditions, Par-C10 acinar-like spheres expressed TJs, aquaporins, ion transporters, and muscarinic receptor 3 [7]. These features make Par-C10 acinar-like spheres an intriguing model to characterize cell volume regulation and ion secretion in salivary epithelium. Moreover, recent studies demonstrated that recombinant adenovirus vectors can modify Par-C10 cells [11], thereby making them useful not only for bioengineering purposes but also as a gene therapy model.

7.2.7 Fluorescent Salivary Cell Lines

Recent studies generated a mesenchymal stem cell line from transgenic mice overexpressing the red fluorescent protein tdTomato (tdTomato mice). Specifically, they immortalized submandibular gland-derived stem cells with the SV40 large T antigen mouse submandibular (GManSV) [24]. These cells exhibited high cell migration rates, a spindle-shaped fibroblastic morphology, and expression of mesenchymal stem cell markers. Moreover, they retained multipotent stem cell characteristics, as evidenced by their ability to differentiate into both osteogenic and adipogenic lineages [24]. Taken together, these results indicate that GManSV cells are useful both for imaging and regeneration studies.

7.2.8 Caution When Using Cell Lines

We bear a responsibility as researchers to verify the correct use of cell lines and avoid misidentification by regularly checking to see that they correspond to their original sources. To that end, short tandem repeat profiling (a method used to compare specific loci on DNA from two or more samples) offers an excellent solution. Additionally, clear guidelines for authentication testing, documentation of cell line provenance, and ongoing validation will help ensure that human cell lines are effective and representative models for biomedical research [15].

7.3 Use of Primary Salivary Cells

Primary cells are taken directly from living tissue (e.g., during biopsies) and established for growth in vitro. These cells have undergone very few population doublings and therefore represent well the main functional components of the tissue from which they were derived. As such, the primary cells represent and offer a good option for creating an artificial salivary gland because they closely resemble native tissue (see Table 7.2).

Primary salivary cells for in vitro studies are obtained through cell isolation of salivary glands (SMG, parotid, sublingual, and minor salivary

Table 7.2 Commonly used human and mouse primary salivary cells

Source	Cell	Amylase expression	Response to secretory agonist	References
Human submandibular gland	+	Unknown	+	Tran et al. [71], Pradhan-Bhatt et al. [63, 64]
Human minor salivary glands	+	Unknown	+	Jang et al. [33]
Human parotid gland	+	+	Unknown	Joraku et al. [34]
Mouse parotid gland	+	+	+	McCall et al. [50]
Mouse submandibular gland	+	+	+	Leigh et al. [40], Maruyama et al. [47]

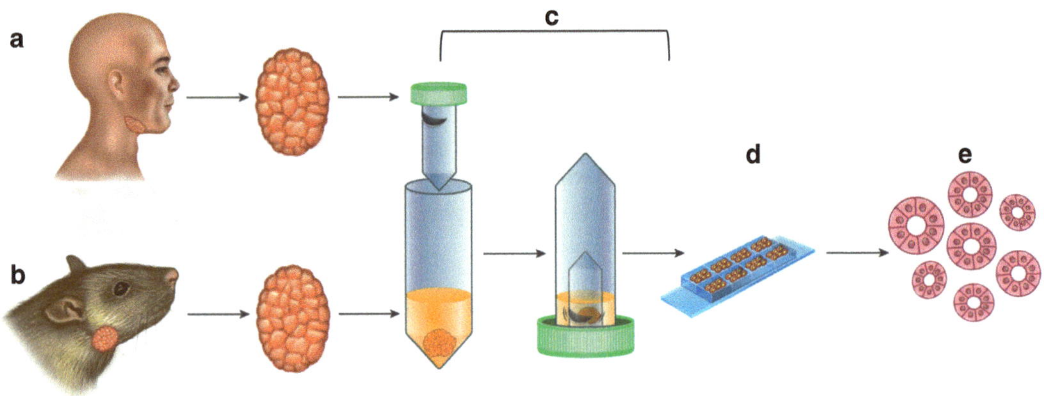

Fig. 7.1 Salivary cell isolation and tissue culture. (**a**) Human and (**b**) mouse submandibular glands are homogenized in a solution containing dispersion enzymes (e.g., hyaluronidase and collagenase) using a (**c**) tissue dissociator. Following dissociation, cells are centrifuged for 5 min at 150 × g and supernatant is removed. Cells are resuspended and plated on (**d**) eight-well chambers mounted on a cover glass and filled with various scaffolds (e.g., laminin-111 or hydrogels) within which salivary cell clusters organize into (**e**) salivary spheres with hollow lumens

glands) using enzymes and mechanical dissociation, by which a cell dispersion of predominantly acinar and ductal cells is obtained and plated on multiple substrates (e.g., permeable supports or Matrigel) (see Fig. 7.1). The limitations of this method include the presence of multiple cell types (i.e., a cell dispersion contains acinar, ductal, progenitor, stem, and myoepithelial cells), slow growth, dedifferentiation, and a finite life span [40, 46, 47, 54]. However, ongoing studies aim to improve the quality of primary cells by optimizing the cell isolation methods and using biomaterials that better mimic the extracellular environment in which cells must be grown.

7.3.1 Salivary Cell Monolayers

A model for secretion studies involves the growth of human SMG on permeable supports, also known as transwells. SMG cells grown under these conditions form cell monolayers with key features indicative of a functional gland (e.g., tight junctions, microvilli, and secretory granules) as well as barrier function regulation [71]. More recently, human minor salivary glands isolated with an explant technique were grown on collagen-coated permeable supports, and protein markers for progenitor and acinar cells were expressed. Importantly, as the calcium concentration increased within the growth medium, these cells acquired a polarized acinar-like phenotype (i.e., increased expression of α-amylase) and demonstrated intact secretion and barrier function [33]. The results shown above indicate that human salivary cells grown as monolayers mimic acinar functioning and are useful for studying fluid and electrolyte secretion; however, further study is needed to develop bioengineering applications (e.g., providing for formation into a branching pattern, as noted in functional salivary glands).

7.3.2 Salivary Spheres

Single human parotid cells can be grown on collagen and Matrigel to form salivary spheres with hollow lumens. Under these conditions, cells exhibit markers of acinar differentiation, including α-amylase, aquaporin-5, and apical TJ expression [34]. Human SMG cells also form salivary spheres when grown on hyaluronic acid hydrogels. Under these conditions, cells express TJ proteins and α-amylase [63] and respond to neurotransmitters as well [64].

Murine salivary cells form spheres with hollow lumens when grown on Matrigel or fibrin hydrogels. Previous studies showed that fibrin hydrogels polymerized with growth factors (i.e., EGF and IGF-1) induced salivary gland differentiation, as indicated by increased levels of α-amylase expression and response to salivary secretory agonists [50]. Further studies using both parotid and SMG cells demonstrated that SMG cell clusters formed more organized and larger structures than were formed by parotid gland cell clusters. However, both SMG and parotid gland cell clusters maintained α-amylase expression, presence of secretory granules, TJs, and agonist-induced secretory responses over time [40]. These results indicate that mouse SMG cell clusters are more promising for the development of a bioengineered salivary gland than parotid gland cell clusters, as they form more organized and functional spheres. Moreover, we recently demonstrated that conditioned medium (from mesenchymal stem cells) enhanced cell organization and multi-lumen formation [47]. These studies indicate that soluble signals secreted by these human mesenchymal stem cells promote formation of a glandular shape. The ability of such mammalian salivary cells to form salivary spheres has been useful for understanding cell assembly mechanisms, secretory function, and cell behavior in culture. However, further studies are needed to better characterize the various structural (i.e., acinar, ductal, or myoepithelial) and functional (i.e., serous and mucous) cell types present in these salivary spheres during time in culture.

7.4 Salivary Gland-Derived Stem and Progenitor Cells

Cells capable of growing more specialized cells (e.g., stem cells and progenitor cells) offer significant possibilities for salivary gland treatment but present serious challenges as well. Stem cells are undifferentiated but can develop into specialized cells and reproduce indefinitely to produce more cells with the same properties [32]. Similarly, progenitor cells are early descendants of stem cells that can also differentiate to form one or more kinds of cells; however, they cannot divide and reproduce indefinitely and are more limited in the kinds of cells they can form [79]. Previous studies indicated that intercalated ducts of salivary glands are enriched with both stem and progenitor cells capable of differentiating and replacing damaged tissues, thereby contributing to the maintenance of acinar cells. Additionally, recent studies indicated that the primary mechanism for maintaining salivary glands is through duplication of acinar cells [5]. These results indicate that all of these cells (i.e., stem, progenitor, and acinar cells) are candidates for the study of salivary gland regeneration and, as such, may have bioengineering applications.

7.4.1 Markers of Stem and Progenitor Cells

As indicated above, salivary glands contain progenitor cells, and multiple markers have been used to characterize the various progenitor cell populations within them. One such marker is *Ascl3*, a transcription factor essential for tissue development and differentiation. Specifically, transplanted *Ascl3*-expressing cells were shown to induce differentiation and tissue repair in salivary glands and actively promote regeneration [68]. Moreover, *Ascl3* was found to be expressed in ductal cells and demilune caps of SMG [13]. These studies indicate *Ascl3*-expressing cells contribute to mature salivary gland tissue maintenance and are likely useful for salivary gland bioengineering studies.

K5 (a structural protein that forms cytoskeleton intermediate filaments) shows some potential as a marker for salivary gland progenitor cells. Specifically, cells expressing K5 (also known as $K5^+$ cells) have been found in ducts of adult salivary glands, which contain progenitor cells [19]. Likewise, $K5^+$ cells give rise to acinar and ductal cells shortly after birth [36]. However, in order to isolate progenitor cells, co-localization of K5 with other cell markers is necessary [43], indicating that K5 expression alone may not be a satisfactory marker for salivary progenitor cells.

Another marker for progenitor cells is Sox2, a member of the Sox family of transcription factors which has been shown to play key roles in many stages of mammalian development [39]. Recent studies showed that an absence of $Sox2^+$ cells in mice resulted in compromised epithelial integrity (including salivary glands) and death [4]. These results indicate that $Sox2^+$ progenitor cells are important for tissue regeneration and survival of mice.

Integrins such as $\alpha_6\beta_1$ have been found to be markers for salivary progenitor cells in rats [58], and cells expressing them were used to establish an immortalized cell line of rat salivary progenitor cells [82]. This cell line is capable of differentiating into both acinar- and ductal-like structures and has the ability to be modulated when grown on Matrigel-based scaffolds; however, cells grown under these conditions display uncontrolled growth. Consequently, further studies are needed to determine whether the acinar- and ductal-like structures generated from this cell line respond to salivary secretory agonists for purposes of growth suppression.

Ductal cells in salivary glands also have been shown to express several stem cell markers (i.e., CD24, CD49f, CD133, and c-Kit^+) [55]. Particularly, c-Kit was definitively established as a stem cell marker and therefore gained the highest priority; however, flow cytometric analysis of cells obtained from SMG indicated that only 0.058 % of salivary cells expressed c-Kit [55]. As such, c-Kit appears not to be an ideal marker for stem cell isolation in salivary glands.

7.4.2 Stem and Progenitor Cells In Vivo

Bone marrow-derived mesenchymal stem cells (BM-MSC) can be used for stem cell usage. Recent studies cocultured BM-MSC with primary mouse SMG cells, which led to transdifferentiation of BM-MSC into a salivary-like phenotype, as indicated by the expression of α-amylase, muscarinic type 3 receptor, aquaporin-5, and cytokeratin 19. These studies successfully identified proteins involved in the process of BM-MSC differentiating into salivary gland epithelial cells; however, further analyses are necessary to determine the function of these factors in mesenchymal stem cell reprogramming [60].

The use of embryonic salivary cells is an alternative option for restoring salivary glands in vivo, as they are a good source of progenitor and stem cells. Specifically, previous studies showed that mouse embryonic salivary cells (i.e., submandibular, sublingual, and parotid glands) grown in an organ culture can be transplanted in vivo. However, significant issues appear to limit the utility of this strategy, including a diminished gland size and a brief period of survival for animal subjects following implantation [57].

Finally, several studies have shown that stem cells derived from postnatal salivary cells can be used to restore damaged salivary glands in vivo [54, 55]. Unfortunately, as mentioned above, salivary gland specimens express low numbers of stem and progenitor cells (e.g., c-Kit^+ cells), thereby limiting their utility for clinical purposes. In order to generate enough stem/progenitor cells for transplantation, recent studies using mouse salivary glands expanded the number of stem cells ex vivo. Specifically, salivary gland sphere-derived single cells were differentiated in vitro into distinct lobular or ductal/lobular organoids containing multiple salivary gland cell lineages [53]. This study indicates that functional salivary gland stem cells can now be purified and expanded ex vivo from single salivary cells; however, follow-up studies using human cells will be needed to translate these findings for clinical work.

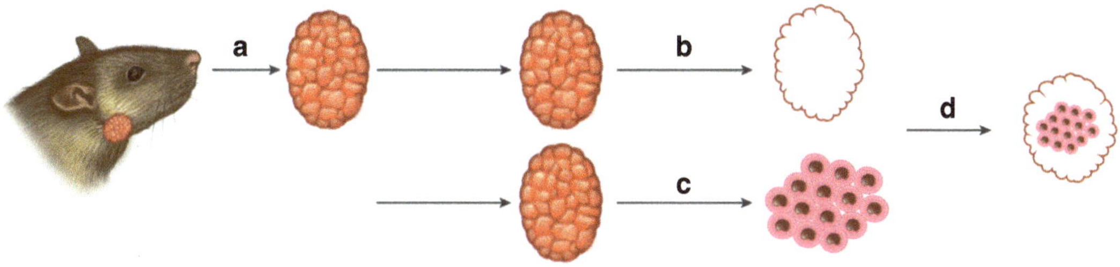

Fig. 7.2 Decellularization of a submandibular gland. (**a**) Two submandibular glands from a single rat are dissected. (**b**) Cells from the first gland are completely removed, and the remaining extracellular matrix is kept for use as a scaffold. (**c**) Primary cells are isolated in the second gland and (**d**) combined with the extracellular matrix, then cultured for 14 days, to generate a salivary structure in vitro

7.5 Scaffolds

Salivary cells can be grown on various substrates (e.g., permeable supports, collagen, Matrigel, hydrogels, etc.), as detailed above. Recently, a study showed that rat SMG can be decellularized by detergent immersion, thereby removing all cells from the tissue and leaving only a scaffold composed of extracellular matrix proteins (Fig. 7.2). Using this scaffold as a support, primary SMG cells were reseeded and cultured in vitro. Results from this study demonstrated that recellularized structures express salivary differentiation markers in vitro [25], thereby offering a promising option for salivary gland bioengineering if secretory function can be demonstrated in future studies.

Conclusion

Construction of an artificial salivary gland is an attractive alternative to repair or regeneration of native glands in patients with hyposalivation. To that end, issues related to the cells to be grown (i.e., longevity, differentiation, function) and the environment in which they are to be cultured (i.e., ability to produce differentiated structures, biocompatibility, and limitation of tumorigenicity) must be resolved.

Acknowledgments This work was supported by the NIH-NIDCR grants R01DE022971, R01DE021697, R01DE021697S1, and R01DE021697S2.

References

1. Aframian DJ, Amit D, David R, Shai E, Deutsch D, Honigman A, Panet A, Palmon A. Reengineering salivary gland cells to enhance protein secretion for use in developing artificial salivary gland device. Tissue Eng. 2007;13:995–1001.
2. Aiba-Masago S, Masago R, Vela-Roch N, Talal N, Dang H. Fas-mediated apoptosis in a rat acinar cell line is dependent on caspase-1 activity. Cell Signal. 2001;13:617–24.
3. Aliko A, Wolff A, Dawes C, Aframian D, Proctor G, Ekstrom J, Narayana N, Villa A, Sia YW, Joshi RK, Mcgowan R, Beier Jensen S, Kerr AR, Lynge Pedersen AM, Vissink A. World Workshop on Oral Medicine VI: clinical implications of medication-induced salivary gland dysfunction. Oral Surg Oral Med Oral Pathol Oral Radiol. 2015;120:185–206.
4. Arnold K, Sarkar A, Yram MA, Polo JM, Bronson R, Sengupta S, Seandel M, Geijsen N, Hochedlinger K. Sox2(+) adult stem and progenitor cells are important for tissue regeneration and survival of mice. Cell Stem Cell. 2011;9:317–29.
5. Aure MH, Konieczny SF, Ovitt CE. Salivary gland homeostasis is maintained through acinar cell self-duplication. Dev Cell. 2015;33:231–7.
6. Aure MH, Roed A, Galtung HK. Intracellular Ca2+ responses and cell volume regulation upon cholinergic and purinergic stimulation in an immortalized salivary cell line. Eur J Oral Sci. 2010;118:237–44.
7. Baker OJ, Schulz DJ, Camden JM, Liao Z, Peterson TS, Seye CI, Petris MJ, Weisman GA. Rat parotid gland cell differentiation in three-dimensional culture. Tissue Eng Part C Methods. 2010;16(5):1135–44.
8. Band V, Zajchowski D, Kulesa V, Sager R. Human papilloma virus DNAs immortalize normal human mammary epithelial cells and reduce their growth factor requirements. Proc Natl Acad Sci U S A. 1990;87:463–7.

9. Baum BJ, Alevizos I, Zheng C, Cotrim AP, Liu S, Mccullagh L, Goldsmith CM, Burbelo PD, Citrin DE, Mitchell JB, Nottingham LK, Rudy SF, Van Waes C, Whatley MA, Brahim JS, Chiorini JA, Danielides S, Turner RJ, Patronas NJ, Chen CC, Nikolov NP, Illei GG. Early responses to adenoviral-mediated transfer of the aquaporin-1 cDNA for radiation-induced salivary hypofunction. Proc Natl Acad Sci U S A. 2012;109:19403–7.
10. Bockman CS, Bradley ME, Dang HK, Zeng W, Scofield MA, Dowd FJ. Molecular and pharmacological characterization of muscarinic receptor subtypes in a rat parotid gland cell line: comparison with native parotid gland. J Pharmacol Exp Ther. 2001;297:718–26.
11. Bori E, Racz G, Burghardt B, Demeter I, Hegyesi O, Varga G, Foldes A. ParC-10 cells for modelling parotid gland tissue reorganization. Fogorv Sz. 2014;107:99–105.
12. Brown AM. In vitro transformation of submandibular gland epithelial cells and fibroblasts of adult rats by methylcholanthrene. Cancer Res. 1973;33:2779–89.
13. Bullard T, Koek L, Roztocil E, Kingsley PD, Mirels L, Ovitt CE. Ascl3 expression marks a progenitor population of both acinar and ductal cells in mouse salivary glands. Dev Biol. 2008;320:72–8.
14. Campisi J, d'Adda di Fagagna F. Cellular senescence: when bad things happen to good cells. Nat Rev Mol Cell Biol. 2007;8:729–40.
15. Capes-Davis A, Reid YA, Kline MC, Storts DR, Strauss E, Dirks WG, Drexler HG, Macleod RA, Sykes G, Kohara A, Nakamura Y, Elmore E, Nims RW, Alston-Roberts C, Barallon R, Los GV, Nardone RM, Price PJ, Steuer A, Thomson J, Masters JR, Kerrigan L. Match criteria for human cell line authentication: where do we draw the line? Int J Cancer. 2013;132:2510–9.
16. Castro R, Barlow-Walden L, Woodson T, Kerecman JD, Zhang GH, Martinez JR. Ion transport in an immortalized rat submandibular cell line SMG-C6. Proc Soc Exp Biol Med. 2000;225:39–48.
17. Chong AS, Alegre ML. The impact of infection and tissue damage in solid-organ transplantation. Nat Rev Immunol. 2012;12:459–71.
18. Committee I. C. L. A. Database of cross-contaminated or misidentified cell lines [Online]. 2014. Available: http://iclac.org/databases/cross-contaminations/2014.
19. Coppes RP, Stokman MA. Stem cells and the repair of radiation-induced salivary gland damage. Oral Dis. 2011;17:143–53.
20. Demeter I, Szucs A, Hegyesi O, Foldes A, Racz GZ, Burghardt B, Steward MC, Varga G. Vectorial bicarbonate transport by Par-C10 salivary cells. J Physiol Pharmacol. 2009;60(Suppl 7):197–204.
21. Ding C, Li L, Su YC, Xiang RL, Cong X, Yu HK, Li SL, Wu LL, Yu GY. Adiponectin increases secretion of rat submandibular gland via adiponectin receptors-mediated AMPK signaling. PLoS One. 2013;8:e63878.
22. Epstein JB, Beier Jensen S. Management of Hyposalivation and Xerostomia: Criteria for Treatment Strategies. Compend Contin Educ Dent. 2015;36:600–3.
23. Fukuda M, Tanaka S, Suzuki S, Kusama K, Kaneko T, Sakashita H. Cimetidine induces apoptosis of human salivary gland tumor cells. Oncol Rep. 2007;17:673–8.
24. Furukawa S, Kuwajima Y, Chosa N, Satoh K, Ohtsuka M, Miura H, Kimura M, Inoko H, Ishisaki A, Fujimura A, Miura H. Establishment of immortalized mesenchymal stem cells derived from the submandibular glands of tdTomato transgenic mice. Exp Ther Med. 2015;10:1380–6.
25. Gao Z, Wu T, Xu J, Liu G, Xie Y, Zhang C, Wang J, Wang S. Generation of Bioartificial Salivary Gland Using Whole-Organ Decellularized Bioscaffold. Cells Tissues Organs. 2014;200:171–80.
26. Gey GO, Coffman WD, Kubicek MT. Tissue culture studies of the proliferative capacity of cervical carcinoma and normal epithelium. Cancer Res. 1952;12:264–5.
27. Goldman EB, Zak A, Tenne R, Kartvelishvily E, Levin-Zaidman S, Neumann Y, Stiubea-Cohen R, Palmon A, Hovav AH, Aframian DJ. Biocompatibility of tungsten disulfide inorganic nanotubes and fullerene-like nanoparticles with salivary gland cells. Tissue Eng Part A. 2015;21:1013–23.
28. He X, Kuijpers GA, Goping G, Kulakusky JA, Zheng C, Delporte C, Tse CM, Redman RS, Donowitz M, Pollard HB, Baum BJ. A polarized salivary cell monolayer useful for studying transepithelial fluid movement in vitro. Pflugers Arch. 1998;435:375–81.
29. He XJ, Frank DP, Tabak LA. Establishment and characterization of 12S adenoviral E1A immortalized rat submandibular gland epithelial cells. Biochem Biophys Res Commun. 1990;170:336–43.
30. He XJ, Wu XZ, Brown AM, Wellner RB, Baum BJ. Characteristics of alpha 1-adrenergic receptors in a rat salivary cell line, RSMT-A5. Gen Pharmacol. 1989;20:175–81.
31. Hoffman MP, Kibbey MC, Letterio JJ, Kleinman HK. Role of laminin-1 and TGF-beta 3 in acinar differentiation of a human submandibular gland cell line (HSG). J Cell Sci. 1996;109(Pt 8):2013–21.
32. Hsu YC, Fuchs E. A family business: stem cell progeny join the niche to regulate homeostasis. Nat Rev Mol Cell Biol. 2012;13:103–14.
33. Jang SI, Ong HL, Gallo A, Liu X, Illei G, Alevizos I. Establishment of functional acinar-like cultures from human salivary glands. J Dent Res. 2015;94:304–11.
34. Joraku A, Sullivan CA, Yoo J, Atala A. In-vitro reconstitution of three-dimensional human salivary gland tissue structures. Differentiation. 2007;75:318–24.
35. Knecht R. Current treatment concepts in head and neck cancer: Highlights of the 2015 Annual Meeting of the American Society for Clinical Oncology. HNO. 2015;63:605.
36. Knox SM, Lombaert IM, Reed X, Vitale-Cross L, Gutkind JS, Hoffman MP. Parasympathetic innervation

maintains epithelial progenitor cells during salivary organogenesis. Science. 2010;329:1645–7.
37. Kurokawa R, Kyakumoto S, OTA M. Autocrine growth factor in defined serum-free medium of human salivary gland adenocarcinoma cell line HSG. Cancer Res. 1989;49:5136–42.
38. Lam K, Zhang L, Bewick M, Lafrenie RM. HSG cells differentiated by culture on extracellular matrix involves induction of S-adenosylmethione decarboxylase and ornithine decarboxylase. J Cell Physiol. 2005;203:353–61.
39. Lefebvre V, Dumitriu B, Penzo-Mendez A, Han Y, Pallavi B. Control of cell fate and differentiation by Sry-related high-mobility-group box (Sox) transcription factors. Int J Biochem Cell Biol. 2007;39:2195–214.
40. Leigh NJ, Nelson JW, Mellas RE, Mccall AD, Baker OJ. Three-dimensional cultures of mouse submandibular and parotid glands: a comparative study. J Tissue Eng Regen Med. 2014. http://onlinelibrary.wiley.com/doi/10.1002/term.1952/abstract;jsessionid=08D6882890A70711C3E23CA9FAB89BD1.f04t02
41. Liu X, Mork AC, Sun X, Castro R, Martinez JR, Zhang GH. Regulation of Ca(2+) signals in a parotid cell line Par-C5. Arch Oral Biol. 2001;46:1141–9.
42. Liu XB, Sun X, Mork AC, Dodds MW, Martinez JR, Zhang GH. Characterization of the calcium signaling system in the submandibular cell line SMG-C6. Proc Soc Exp Biol Med. 2000;225:211–20.
43. Lombaert IM, Knox SM, Hoffman MP. Salivary gland progenitor cell biology provides a rationale for therapeutic salivary gland regeneration. Oral Dis. 2011;17:445–9.
44. Lu MH, Liao ZL, Zhao XY, Fan YH, Lin XL, Fang DC, Guo H, Yang SM. hTERT-based therapy: a universal anticancer approach (Review). Oncol Rep. 2012;28:1945–52.
45. Maria OM, Maria O, Liu Y, Komarova SV, Tran SD. Matrigel improves functional properties of human submandibular salivary gland cell line. Int J Biochem Cell Biol. 2011a;43:622–31.
46. Maria OM, Zeitouni A, Gologan O, Tran SD. Matrigel improves functional properties of primary human salivary gland cells. Tissue Eng Part A. 2011b;17:1229–38.
47. Maruyama CL, Leigh NJ, Nelson JW, Mccall AD, Mellas RE, Lei P, Andreadis ST, Baker OJ. Stem cell-soluble signals enhance multilumen formation in SMG cell clusters. J Dent Res. 2015;94:1610–7.
48. Maslinska M, Przygodzka M, Kwiatkowska B, Sikorska-Siudek K. Sjogren's syndrome: still not fully understood disease. Rheumatol Int. 2015;35:233–41.
49. Masters JR. Human cancer cell lines: fact and fantasy. Nat Rev Mol Cell Biol. 2000;1:233–6.
50. Mccall AD, Nelson JW, Leigh NJ, Duffey ME, Lei P, Andreadis ST, Baker OJ. Growth factors polymerized within fibrin hydrogel promote amylase production in parotid cells. Tissue Eng Part A. 2013;19(19–20):2215–25.
51. Michikawa H, Fujita-Yoshigaki J, Sugiya H. Enhancement of barrier function by overexpression of claudin-4 in tight junctions of submandibular gland cells. Cell Tissue Res. 2008;334:255–64.
52. Nagy K, Szlavik V, Racz G, Ovari G, Vag J, Varga G. Human submandibular gland (HSG) cell line as a model for studying salivary gland Ca2+ signalling mechanisms. Acta Physiol Hung. 2007;94:301–13.
53. Nanduri LS, Baanstra M, Faber H, Rocchi C, Zwart E, de Haan G, Van OS R, Coppes RP. Purification and ex vivo expansion of fully functional salivary gland stem cells. Stem Cell Rep. 2014;3:957–64.
54. Nanduri LS, Lombaert I, van der Zwaag M, Faber H, Brunsting JF, van OS RP, Coppes RP. Salisphere derived c-Kit< sup>+ cell transplantation restores tissue homeostasis in irradiated salivary gland. Radiother Oncol. 2013a;108:458–63.
55. Nanduri LS, Maimets M, Pringle SA, van der Zwaag M, van OS RP, Coppes RP. Regeneration of irradiated salivary glands with stem cell marker expressing cells. Radiother Oncol. 2011;99:367–72.
56. Nelson J, Manzella K, Baker OJ. Current cell models for bioengineering a salivary gland: a mini-review of emerging technologies. Oral Dis. 2013;19:236–44.
57. Ogawa M, Oshima M, Imamura A, Sekine Y, Ishida K, Yamashita K, Nakajima K, Hirayama M, Tachikawa T, Tsuji T. Functional salivary gland regeneration by transplantation of a bioengineered organ germ. Nat Commun. 2013;4:1–10.
58. Okumura K, Nakamura K, Hisatomi Y, Nagano K, Tanaka Y, Terada K, Sugiyama T, Umeyama K, Matsumoto K, Yamamoto T, Endo F. Salivary gland progenitor cells induced by duct ligation differentiate into hepatic and pancreatic lineages. Hepatology. 2003;38:104–13.
59. Pagnan NA, Visinoni AF. Update on ectodermal dysplasias clinical classification. Am J Med Genet A. 2014;164A:2415–23.
60. Park YJ, Koh J, Gauna AE, Chen S, Cha S. Identification of regulatory factors for mesenchymal stem cell-derived salivary epithelial cells in a co-culture system. PLoS One. 2014;9:e112158.
61. Patton LL, Pollack S, Wellner RB. Responsiveness of a human parotid epithelial cell line (HSY) to autonomic stimulation: muscarinic control of K+ transport. In Vitro Cell Dev Biol. 1991;27A:779–85.
62. Pipas JM. SV40: cell transformation and tumorigenesis. Virology. 2009;384:294–303.
63. Pradhan S, Liu C, Zhang C, Jia X, Farach-Carson MC, Witt RL. Lumen formation in three-dimensional cultures of salivary acinar cells. Otolaryngol Head Neck Surg. 2010;142:191–5.
64. Pradhan-Bhatt S, Harrington DA, Duncan RL, Jia X, Witt RL, Farach-Carson MC. Implantable three-dimensional salivary spheroid assemblies demonstrate fluid and protein secretory responses to neurotransmitters. Tissue Eng Part A. 2013;19:1610–20.
65. Quissell DO, Barzen KA, Gruenert DC, Redman RS, Camden JM, Turner JT. Development and characterization of SV40 immortalized rat submandibular acinar cell lines. In Vitro Cell Dev Biol Anim. 1997;33:164–73.

66. Quissell DO, Barzen KA, Redman RS, Camden JM, Turner JT. Development and characterization of SV40 immortalized rat parotid acinar cell lines. In Vitro Cell Dev Biol Anim. 1998;34:58–67.
67. Ratchford AM, Baker OJ, Camden JM, Rikka S, Petris MJ, Seye CI, Erb L, Weisman GA. P2Y2 nucleotide receptors mediate metalloprotease-dependent phosphorylation of epidermal growth factor receptor and ErbB3 in human salivary gland cells. J Biol Chem. 2010;285:7545–55.
68. Rugel-Stahl A, Elliott ME, Ovitt CE. Ascl3 marks adult progenitor cells of the mouse salivary gland. Stem Cell Res. 2012;8:379–87.
69. Shirasuna K, Sato M, Miyazaki T. A neoplastic epithelial duct cell line established from an irradiated human salivary gland. Cancer. 1981;48:745–52.
70. Soltoff SP, Hedden L. Isoproterenol and cAMP block ERK phosphorylation and enhance [Ca2+]i increases and oxygen consumption by muscarinic receptor stimulation in rat parotid and submandibular acinar cells. J Biol Chem. 2010;285:13337–48.
71. Tran SD, Wang J, Bandyopadhyay BC, Redman RS, Dutra A, Pak E, Swaim WD, Gerstenhaber JA, Bryant JM, Zheng C, Goldsmith CM, Kok MR, Wellner RB, Baum BJ. Primary culture of polarized human salivary epithelial cells for use in developing an artificial salivary gland. Tissue Eng. 2005;11:172–81.
72. Turner JT, Redman RS, Camden JM, Landon LA, Quissell DO. A rat parotid gland cell line, Par-C10, exhibits neurotransmitter-regulated transepithelial anion secretion. Am J Physiol. 1998;275:C367–74.
73. Turner RJ, Sugiya H. Understanding salivary fluid and protein secretion. Oral Dis. 2002;8:3–11.
74. Vasquez MM, Mustafa SB, Choudary A, Seidner SR, Castro R. Regulation of epithelial Na+ channel (ENaC) in the salivary cell line SMG-C6. Exp Biol Med (Maywood). 2009;234:522–31.
75. Villa A, Connell CL, Abati S. Diagnosis and management of xerostomia and hyposalivation. Ther Clin Risk Manag. 2015a;11:45–51.
76. Villa A, Wolff A, Aframian D, Vissink A, Ekstrom J, Proctor G, Mcgowan R, Narayana N, Aliko A, Sia YW, Joshi RK, Jensen SB, Kerr AR, Dawes C, Pedersen AM. World Workshop on Oral Medicine VI: a systematic review of medication-induced salivary gland dysfunction: prevalence, diagnosis, and treatment. Clin Oral Investig. 2015b;19:1563–80.
77. Wang Z, Zourelias L, Wu C, Edwards PC, Trombetta M, Passineau MJ. Ultrasound-assisted non-viral gene transfer of AQP1 to the irradiated minipig parotid gland restores fluid secretion. Gene Ther. 2015;22(9):739–49.
78. Watson CJ, Dark JH. Organ transplantation: historical perspective and current practice. Br J Anaesth. 2012;108(Suppl 1):i29–42.
79. Weissman IL, Anderson DJ, Gage F. Stem and progenitor cells: origins, phenotypes, lineage commitments, and transdifferentiations. Annu Rev Cell Dev Biol. 2001;17:387–403.
80. Yamada A, Futagi M, Fukumoto E, Saito K, Yoshizaki K, Ishikawa M, Arakaki M, Hino R, Sugawara Y, Ishikawa M, Naruse M, Miyazaki K, Nakamura T, Fukumoto S. Connexin 43 is necessary for salivary gland branching morphogenesis and FGF10-induced ERK1/2 phosphorylation. J Biol Chem. 2016;291:904–12.
81. Yanagawa T, Hayashi Y, Nagamine S, Yoshida H, Yura Y, Sato M. Generation of cells with phenotypes of both intercalated duct-type and myoepithelial cells in human parotid gland adenocarcinoma clonal cells grown in athymic nude mice. Virchows Arch B Cell Pathol Incl Mol Pathol. 1986;51:187–95.
82. Yaniv A, Neumann Y, David R, Stiubea-Cohen R, Orbach Y, Lang S, Rotter N, DVIR-Ginzberg M, Aframian DJ, Palmon A. Establishment of immortal multipotent rat salivary progenitor cell line toward salivary gland regeneration. Tissue Eng Part C Methods. 2011;17:69–78.
83. Zhu Y, Aletta JM, Wen J, Zhang X, Higgins D, Rubin RP. Rat serum induces a differentiated phenotype in a rat parotid acinar cell line. Am J Physiol. 1998;275:G259–68.

Matrix Biology of the Salivary Gland: A Guide for Tissue Engineering

8

Mariane Martinez, Danielle Wu,
Mary C. Farach-Carson, and Daniel A. Harrington

Abstract

Salivary glands produce saliva needed to carry out daily functions such as initiation of food digestion, lubrication of the oral cavity, and prevention of oral diseases. Xerostomia, or dry mouth due to hyposalivation, can occur in individuals with Sjögren's syndrome or in patients who receive radiation therapy to treat head and neck cancer. This chapter will focus on the bioengineering approaches in salivary gland regeneration that seek to restore salivary function in patients suffering from xerostomia. A brief description of salivary gland function, structure, and development will be provided first as this information is vital to inform any salivary gland tissue engineering efforts. Additionally, examples of salivary gland-derived stem/progenitor cells that are used in various salivary gland regeneration models will be introduced along with a brief description of each utility as source material for tissue engineering. Lastly, we will review matrices for three-dimensional cell culture, including decellularized native extracellular matrix scaffolds, Matrigel®, and scaffolds containing biologically derived natural polymers, polysaccharides, and biologically active protein fragments. Together, these elements will provide a current view of the state-of-the-art clinical approaches to relieve xerostomia using three-dimensional culture techniques.

M. Martinez • D. Wu, PhD
D.A. Harrington, PhD (✉)
Department of BioSciences, Rice University, Houston, TX, USA
e-mail: mariane.martinez@rice.edu; daniellewu@rice.edu; daniel.a.harrington@rice.edu

M.C. Farach-Carson, PhD
Departments of BioSciences and Bioengineering, Rice University, Houston, TX, USA
e-mail: farachca@rice.edu

8.1 Motivation for Engineering a Neo-Salivary Gland

Over 50,000 Americans suffer from head and neck cancer annually [1]. The majority of those cancer patients with locally invasive forms of the disease will receive a standard treatment that includes radiation therapy (RT). Despite the broad availability and use of intensity-modulated radiotherapy (IMRT) as an RT delivery method,

S. Cha (ed.), *Salivary Gland Development and Regeneration*, DOI 10.1007/978-3-319-43513-8_8

and its high spatial precision, RT still too often leads to drastic adverse morphological changes in salivary gland structure, including ductal metaplasia, periductal fibrosis, and necrotic loss of salivary acinar cells [2, 3]. Consequently, ~64 % of long-term head and neck cancer survivors who were treated with conventional two-dimensional (2D) RT suffer from moderate to severe RT-induced xerostomia, or dry mouth, which greatly reduces their quality of life due to hyposalivation [4]. Current treatments for xerostomia include cholinergic sialagogues, artificial saliva substitutes, strict oral hygiene, and other largely palliative treatments. However, these strategies are only temporary, and many produce several undesirable side effects including, but not limited to, nausea, vomiting, and/or excessive sweating [5]. Hence, they are often abandoned by patients, as a majority of patients fail to return to their dental follow-up appointments after completion of radiation therapy [6].

For these reasons, a tissue-engineered salivary gland, created with patient-derived cells and surgically grafted into the parotid of the xerostomic patient after successful cancer therapy, offers an alternative treatment for this syndrome. Inspiration for designing a salivary neotissue is found in the organization and composition of the native tissue. This chapter considers the cellular and extracellular components of salivary glands and their potential for implementation in the creation of an engineered tissue replacement.

8.2 Salivary Gland Structure and Development

8.2.1 Salivary Gland Structure and Function

The salivary system in humans consists of three major bilateral salivary glands: the parotid, submandibular, and sublingual glands. Parotid glands are the largest and are located inferior and anterior to the ear. Submandibular and sublingual glands are located inferior to the tongue and floor of the oral cavity. Additionally, there are numerous minor serous fluid and mucous-producing salivary glands lining the oral cavity [7]. All of these salivary glands are responsible for producing the salivary components that initiate digestion of food, lubricate the oral cavity, facilitate swallowing, and maintain the dental flora to avoid dental diseases [8].

Functional salivary glands have a complex cellular organization that allows saliva to be produced constitutively and in greater amounts upon demand. A tightly structured and extensively routed network ensures that saliva is unidirectionally secreted into the oral cavity. Proximal functional units in salivary glands consist of acinar cells (serous and/or mucinous cells) that produce proteins and fluid transported in saliva. The spherically organized pyramidal acinar cells are highly interconnected to each other and to the surrounding myoepithelial cells via cell-cell adhesions (see Sect. 8.2.2.1). Stellate-shaped myoepithelial cells are postulated to be responsible for contracting the acinus and forcing saliva out into an elaborate system of ducts leading into the oral cavity [9–11]. Functional acini secrete salivary contents into their lumens that merge into a hierarchical transport system composed of intercalated, striated, and excretory ductal regions, which both transport and continue to modify the composition of saliva as it exits the gland [12].

Polarized acinar cells comprise the acinar units that selectively secrete salivary components into the lumens of the glands. Polarized, or asymmetrically organized, acinar cells are interconnected on their lateral surfaces by cell-cell junctions including adherens junctions, tight junctions, gap junctions, and desmosomes and are anchored on their basal surface to the basement membrane through integrin-mediated binding [13–20]. Acinar cells also interact with myoepithelial cells on their basal sides. Organelle positioning is distinct in polarized cells; in salivary acinar cells, the nucleus is positioned at the base of the cell, near the basement membrane and far away from the lumen. The smooth and extensive rough endoplasmic reticulum (ER) and the Golgi apparatus that synthesize and package salivary contents for transport are positioned between the nucleus and the secretory lumen. The many secretory vesicles

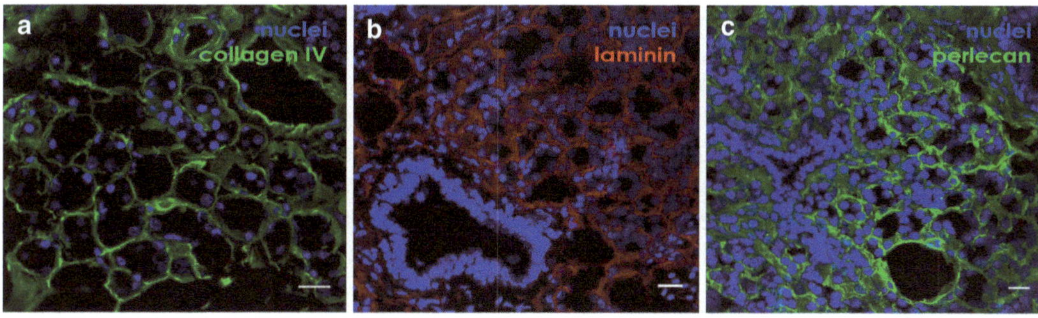

Fig. 8.1 Basement membrane proteins assist in the organization and function of acinar and ductal compartments in human parotid glands. Immunohistochemistry is used to identify critical extracellular matrix proteins in human salivary gland tissue including (**a**) collagen IV (*green*), (**b**) laminin (*red*), and (**c**) perlecan/HSPG2 (*green*) that surround acini and intercalated ducts. Scale bars are 20 μm, and nuclei (*blue*) are counterstained with 4',6-diamidino-2-phenylindole (DAPI)

that release salivary contents into the lumen are in close proximity to the apical membrane [21–23]. This polarized organization and unique nature of secretory vesicles have been described as key factors for proper secretion, which can be lost in the selective destruction of the salivary acinar units during radiation [24–26].

A well-organized basement membrane is crucial for maintaining cell polarity. The basement membrane is a specialized network of extracellular matrix (ECM) proteins mainly composed of collagen IV, laminin, perlecan/HSPG2 (Fig. 8.1), and entactin/nidogen [27, 28]. Each of the acini and ductal cells in the salivary glands is separated from underlying connective tissue by a meshwork of laminin heterotrimers and collagen IV, which are spot-welded together by entactin [29]. The ~800-million-year-old heparan sulfate proteoglycan 2, perlecan/HSPG2, plays a unique role in the border functions of the basement membrane and as a growth factor supply depot important for wound healing [30, 31]. Additional functions of the basement membrane include promoting salivary cell organization, providing mechanical support to the gland, providing guidance cues during salivary development and hence regeneration, and storing heparin-binding (HB) factors that can be actively released in case of immediate need.

Salivary glands produce and secrete saliva constitutively at a relatively constant rate of 0.3 mL/min or on demand at higher levels of about 7 mL/min during a meal when they are stimulated by the surrounding autonomic nerves [32, 33]. Cholinergic neurotransmitters released by the surrounding parasympathetic innervation stimulate the glands to secrete salivary fluid, while adrenergic neurotransmitters released by the surrounding sympathetic innervation stimulate the glands to secrete salivary proteins (see 8.2.2). Additionally, autonomic innervation is required for proper development of the salivary glands during embryogenesis and in postnatal maturation, as discussed in the following section.

8.2.2 Salivary Gland Development

Much of what is known of salivary gland development was discovered through the study of developmental processes in a mouse submandibular gland model and is summarized briefly below [34–36]. Salivary gland development consists of five distinct stages: prebud, initial bud, pseudoglandular, canalicular, and terminal bud. During the prebud stage at embryonic day 11.5 (E11.5), the ectodermal epithelium adjacent to the tongue thickens, marking the location of the base of the salivary gland. The thickened epithelium continues to proliferate and grow, forming an initial bud that extends into the neighboring neural crest mesenchyme (E12.5). Simultaneously, the base of the bud invaginates to create the duct of the developing salivary bud. Branching morphogenesis of the salivary gland commences during the pseudoglandular stage (E13.5), yielding a multilobed structure with multiple salivary buds connected to the base of the gland via epithelial branches. As branching morphogenesis continues during the canalicular stage (E15.5), these epithelial branches begin to develop a lumen via apoptosis of inner epithelial cells, migration of outer epithelial cells, and proliferation of epithelial cells at the tips of the branches [36, 37]. Salivary gland differentiation begins and the majority of ductal lumens are formed by the terminal bud stage (E17.5), but the remaining undifferentiated cells and unformed lumens will be differentiated and formed postnatally, respectively.

Aside from being instrumental for the physiology of the mature gland, nerves also are essential during salivary gland development. Salivary glands are innervated by postganglionic parasympathetic and sympathetic nerve fibers that stimulate fluid-rich and protein-rich secretions [12, 32, 38]. Parasympathetic postganglionic nerve fibers originate from the submandibular ganglion for the SMG, while they originate from the otic ganglion for the parotid gland. Sympathetic nerve fibers innervating all glands originate from the superior cervical ganglion, using the external carotid plexus as a guide to branch off to the salivary glands [32, 38, 39]. Sympathetic denervation in neonatal rats led to a reduction in acinar cell size and in the number of granules produced, suggesting that proper sympathetic innervation is important for acinar cell maturation [40]. Studies performed with fetal mice showed that the parasympathetic ganglia from the submandibular ganglion grow in parallel and are directed by the developing SMG epithelium [41, 42]. Lastly, disruption of parasympathetic innervation in mouse embryonic SMG explant cultures reduced the number of cytokeratin 5 (K5) expressing progenitor cells, as well as a reduction in branching morphogenesis of the organ [43]. The epithelial progenitor population needed for organogenesis was rescued by stimulation of the muscarinic M1 receptor (M1), via acetylcholine treatment, and was dependent on epithelial growth factor (EGF) receptor signaling. Together, these studies indicate that innervation contributes significantly to proper branching morphogenesis and differentiation of the salivary gland.

Studies of embryonic salivary gland development have identified important factors needed for successful development of a functional salivary gland: growth factors (GFs) (i.e., EGF and fibroblast growth factors (FGFs)), innervation, vascularization, and epithelial and mesenchymal cell-derived ECM (discussed further in the following section). To advance the field of salivary tissue replacement, these critical factors should be strategically incorporated into the design of the bioengineered organ, regardless of whether or not they are integrated in vitro into the scaffold for organ culture or added in vivo after implantation of the scaffold. Moreover, to fully recapitulate the events in development, the addition of key morphogenetic factors needs to be temporally and spatially controlled, as they would be during native gland development. For example, the introduction of certain ECM components like fibronectin, which is critical for cleft formation needed during branching morphogenesis, and FGFs needed for bud elongation and cleft formation should only be introduced during the onset of branching and not during initial clustering and assembly. The ideal salivary gland tissue engineering scaffold would support these temporospatially distinguished functions and promote stable long-term ingrowth of the host's native vascularization and innervation as the neogland develops.

8.2.2.1 Importance of the ECM During Salivary Gland Development

Successful salivary gland development depends on the temporospatial signaling provided by the development and maturation of the underlying basement membrane that separates the basal surface of the glandular acinar and myoepithelial cells from the underlying connective tissue. The basement membrane is dynamic and is actively remodeled throughout the development, homeostasis, and wound healing. Under normal conditions, the basement membrane maintains tissue homeostasis of the salivary organ and provides direction and orientation for secretion.

Proteases, such as matrix metalloproteinases (MMPs), partially cleave the basement membrane to loosen the matrix and allow for acinar growth. Moreover, in addition to providing the extra space needed for tissue expansion, turnover releases a number of locally sequestered growth factors needed to further stimulate gland development [37, 44, 45]. One study using mouse embryonic SMG explants showed the dynamics of the basement membrane during branching morphogenesis [44]. Salivary epithelial cells at the tip of the buds digested the basement membrane through the activation of MMPs, creating perforations through which they could migrate. The basement membrane is seen to move inward toward the clefts, forming the ducts that appear during development. Furthermore, interactions between the developing salivary epithelium and the basement membrane are vital for the growth of an initial salivary epithelial bud, branching morphogenesis of that bud, and differentiation and polarization of the salivary parenchymal cells including acinar and ductal cells.

The ECM is crucial for the formation of a multilobed secretory organ such as the salivary gland through branching morphogenesis, and in the case of salivary acinus, the ECM is secreted both by the epithelial cells, particularly the myoepithelial cells, and the neighboring mesenchymal cells [46–48]. Studies in which interactions between the salivary epithelial cells and the surrounding basement membrane proteins were decreased inhibited branching morphogenesis, demonstrating the need for a well-organized basement membrane to guide gland development [46, 49, 50]. Fibronectin, a glycoprotein in the ECM that is arranged into fibrils, is present early on during gland development and establishes cleft formation during branching morphogenesis [46, 51, 52]. Fibronectin expression in developing mouse SMGs was assessed via laser microdissection and RT-PCR of the salivary tissue, allowing the tissue to be divided into cell populations of cleft-derived epithelial cells, bud-derived epithelial cells, cleft-neighboring mesenchyme, and bud-neighboring mesenchyme [46]. Fibronectin expression was high in both the cleft-neighboring and bud-neighboring mesenchyme. Unexpectedly, fibronectin expression was observed in salivary epithelial cells, with increased expression of fibronectin localized to the epithelial cells closest to where the cleft will form. This fibronectin deposition was accompanied by a decrease in E-cadherin expression [46].

To test fibronectin's role in branching morphogenesis, the protein was targeted with specific blocking polyclonal or monoclonal antibodies that could reduce its interactions with epithelial cells in the developing mouse SMG [46]. Branching morphogenesis was partially inhibited by anti-fibronectin antibody treatment, shown by a dose-dependent reduction in bud formation. A similar decrease in branching was observed in early salivary buds treated with siRNA against fibronectin and was rescued by exogenous fibronectin in a concentration-dependent manner if it was added to the organ culture. Treatment with antibodies targeting integrins α_5 and β_1 had the greatest negative effect on branching morphogenesis, suggesting an important role for the fibronectin receptor $\alpha_5\beta_1$ in fibronectin-mediated branching morphogenesis.

Interstitial collagens also are important to initiate branching morphogenesis during the early development of salivary glands [47, 53, 54]. Using radiolabeled amino acids, soluble tropocollagen was shown to be secreted by mesenchymal cells and selectively polymerized into collagen fibrils by salivary epithelium from embryonic SMG rudiments [47]. Studies in which collagen synthesis was inhibited, or the protein was degraded with collagenases, reduced the extent of branching morphogenesis of the SMG in the ex vivo embryonic model [54–58]. Inhibition of collagen synthesis and secretion by treatment with L-azetidine-2-carboxylic acid (LACA) or α,α'-dipyridyl drastically reduced branching of the embryonic developing SMGs [57]. However, inhibition of collagen cross-linking by treatment with β-aminopropionitrile had no effect on the morphogenetic activity of the SMGs treated. Collagenase treatment of developing salivary glands, whether for a short or long period, inhibited branching morphogenesis of the gland [58]. Expression of collagens I, III, IV, and V was analyzed by immunohistochemical techniques [51, 53]. Collagens I, III, IV, and V were found to be present throughout the salivary mesenchyme; however, collagen III accumulated distinctly at the indented sites and clefts of the salivary bud. Although collagen IV was present in the mesenchyme, it was localized in higher concentrations in the surrounding basement membrane layer [51]. Collagenase derived from bovine dental pulp that specifically cleaves collagen I and III was shown to inhibit branching morphogenesis of the developing salivary gland, and this effect was reversible with the addition of a collagenase inhibitor (bovine serum albumin or fetal calf serum) [56].

Laminin is an abundant component of the basement membrane that also plays an important role in salivary gland development. The heterotrimer is made up of one of each subunit chain: α (1–5), β (1–3), and γ (1–3), which co-assembles into a polymeric network lying directly under the epithelium and/or endothelium [27]. The α chain of laminin uniquely contributes five LG domains at the C-terminal end, which make up the globular, or G, domain that has many cell-binding sites enabling cell-laminin adhesion [59]. Cellular receptors for laminin mediate the biological responses by the salivary epithelium, and those include integrins α_3, α_6, β_1, and α-dystroglycan [37, 50, 60–63]. Laminin glycoproteins mainly are localized to the basement membrane surrounding the developing mouse SMG, but expression of laminin chain isoforms varies during development and continues to be expressed up to the time of maturation of the gland [37, 51, 62, 64]. Laminins α_1 and α_5 are expressed throughout the basement membrane of developing mouse SMGs during branching morphogenesis and are localized to the basement membrane surrounding the ducts in the branched organ; specifically, α_1 is restricted to the only the excretory ducts [49, 65]. Laminin α_2 was shown to be expressed at lower levels early during development at E13 and at higher levels in the basement membrane surrounding matured acini [66]. Laminin α_3 is present in the early stages of development (mouse) around the stalk of the bud and growing ducts and in later stages in the basement membrane surrounding the excretory duct and myoepithelial cells [67].

Antibodies against the E3 domain of the laminin α_1 chain inhibited branching morphogenesis and led to the formation of a discontinuous basement membrane [49]. Targeting integrin α_6 with antibodies also disturbed branching morphogenesis of the embryonic SMG, but did not disrupt the formation of a continuous basement membrane. Furthermore, this study suggests that the E3 domain from laminin α_1 chain is important for the formation of a well-organized, continuous basement membrane. Kadoya et al. tested the activity of various sequences from laminin α_1 and α_2 chains, and a specific sequence (RKRLQVQLSIRT) in laminin α_1 LG4 domain was shown to be important for salivary gland development [66]. Embryonic mouse SMG glands failed to undergo branching morphogenesis or form a continuous basement membrane when treated with AG-73 peptide. Furthermore, the homologous sequence to AG-73, MG-73, in laminin α_2 chain had no effect on branching morphogenesis of the developing gland, suggesting that laminin α_2 may be important for terminal bud differentiation and not for branching morphogenesis. Targeting laminin α_5 activity in the developing SMG with A5G77f peptide derived from globular domain LG4 inhibited branching morphogenesis, but did not inhibit basement membrane formation [65]. The salivary rudiments treated with A5G77f contained terminal epithelial buds with cleft formation, but they failed to form elongated ducts, suggesting that laminin α_5 cell adhesion is important for ductal elongation.

Perlecan/HSPG2 is a heparan sulfate proteoglycan present in the basement membrane surrounding various epithelia, including the salivary gland epithelium. Successful salivary gland growth and development relies on the presence of perlecan because it serves as a depot for the broad category of heparin-binding (HB) GFs that enable cell proliferation and differentiation [30, 37, 68, 69]. When properly synthesized by the cell, perlecan is modified within domain I and V with long glycosaminoglycan chains, primarily heparan sulfate and some chondroitin sulfate. The presence of heparan sulfate enables its growth factor binding and delivery function [30, 70–72]. Important GFs for salivary gland development delivered by perlecan's heparan sulfate chains include FGF-1, FGF-2, FGF-9, FGF-10, HB-EGF, vascular endothelial growth factor (VEGF), and platelet-derived growth factor (PDGF) for vascularization [30]. Furthermore, these growth factors, often in the presence of heparan sulfate, stimulate FGFRs, and this activation in turn triggers the biological processes that occur during salivary gland development. Thus, the contribution of perlecan as a key co-receptor providing heparan sulfate for growth factor presentation remains very important [69, 73–76]. Through the use of the ex vivo embryonic mouse SMG model, FGF-7 and FGF-10 have been shown to play a positive role in stimulating salivary epithelial bud growth. Embryonic mouse SMGs treated with FGF-7 and FGF-10 had larger buds when compared to the control SMGs, while those treated with FGF-2 had a decrease in the number of buds [73]. Additionally, FGF-7 and FGF-10 have differential effects on ductal formation within a heparan sulfate-rich ECM, driving ductal branching or elongation, respectively [69].

Aside from its role in branching morphogenesis, the ECM also is critical for the establishment and stabilization of polarity of the salivary parenchymal cells. Salivary epithelial cells asymmetrically organize their cytoplasmic and membrane-bound proteins during cytodifferentiation, leading to a polarized epithelial cell with an apical membrane facing the lumen separated from the basolateral membrane. Polarization of epithelial cells involves both cell-cell and cell-ECM adhesion, but how exactly each one aids in establishment of polarity is still unclear. This asymmetric organization, or polarity, of salivary acinar and ductal cells is required for the proper function of the salivary gland, i.e., the directional secretion of salivary proteins and fluid. It is essential that these polarized functions be recreated in any engineered gland designed to restore directional secretion of saliva.

Polarized epithelial cells are physically interconnected by cell-cell adhesions including tight junctions, adherens junctions, desmosomes, and gap junctions, all of which aid in establishing and maintaining cell polarity [77]. Tight junctions are composed of cytoplasmic scaffolding proteins zonula occludins, and transmembrane proteins claudins, occludins, and junctional adhesion molecules [14, 78]. Tight junctions are responsible for forming an epithelial barrier to prevent diffusion of solutes via the paracellular space, allowing for the formation of a transepithelial ion gradient [78]. Additionally, tight junctions prevent the fluid movement of integral membrane proteins between the apical surface and the basolateral membrane [79]. Adherens junctions are composed of transmembrane protein cadherins and cytoplasmic scaffolding proteins including catenins [80]. In polarized cells, catenins associate with cadherins to form the complex that links with F-actin, and these F-actin-linked complexes are directly linked by cadherins between the cells in the epithelial sheet, forming a continuous actin belt that strengthens the epithelium [81]. Desmosomes are another form of cell-adhesions that use cadherins to interconnect neighboring epithelial cells [82]. Desmosomal cadherins desmoglein and desmocollin are the transmembrane proteins, and their N-terminal domains interact with each other in the intercellular space to make the cell-cell adhesions that provide mechanical strength to the epithelial sheet [83]. Cytoplasmic proteins plakoglobin and plakophilin are part of the scaffolding complex that connects the cadherins to the intermediate filament-binding protein desmoplakin, forming cell-cell adhesions linked to the cytoskeletal intermediate filaments [84]. Lastly, polarized epithelial cells are physically and chemically interconnected by gap junctions that are composed of connexin (Cx) proteins that form cell-cell channels that allow the transport of ions and small molecules from the cytoplasm of one cell to another adjacent cell [20]. Connexin (Cx) proteins mediate coordinated intercellular signaling during secretion [19], and they also play a role in the proper development of the salivary gland [85].

Asymmetrically organized salivary epithelial cells also are connected to the underlying basement membrane via integrin-mediated binding in hemidesmosomes and focal adhesions, which also play a role in establishing and maintaining epithelial cell polarity. Focal adhesions and hemidesmosomes are composed of transmembrane integrin receptors which bind to the ECM outside of the cell and are linked to a scaffold complex that is associated with the actin and keratin intermediate filament cytoskeleton, respectively [86, 87]. Some of the proteins that make up the intracellular scaffold complex for focal adhesions include talin, vinculin, focal adhesion kinase, and α-actinin. The intracellular scaffold of hemidesmosomes is composed of an inner plaque that connects the adhesion site to the intermediate filaments and an outer plaque that lies parallel to the cell membrane [88]. Formation of these integrin-mediated adhesion sites along with cell-cell adhesion sites activates members of the Rho family small guanosine triphosphatases (RhoGTPases), such as Rac1, Cdc42, and RhoA, which mediate the reorganization of the cytoskeletal filaments [77, 89, 90]. This cytoskeletal reorganization is vital for the asymmetrical positioning of organelles and polarity complexes within the polarized epithelial cells [13, 77]. Because of the importance of cell-matrix adhesions on cell polarity, a scaffold for three-dimensional (3D) organ culture should provide integrin-binding sites to the cells to activate the players needed for cytoskeletal and organelle reorganization in vitro.

Polarity complexes, primarily characterized to date in ductal-like cells, are crucial to the establishment and maintenance of epithelial cell polarity, and they include the Crumbs (CRB) complex, the partitioning-defective (PAR) complex, and the Scribble complex [91].The CRB complex consists of the transmembrane homologous CRB proteins, and its cytoplasmic domain is associated with protein associated with lin seven (PALS-1) [77, 91, 92]. PALS-1 links with PALS-1-associated tight junction protein (PATJ) via one of its L27 domains [77]. PATJ then connects the CRBs complex to tight junctions by interaction with its PDZ domains, localizing the polarity complex to the apicolateral side of the membrane. The PAR complex is made up of PDZ-domain-containing scaffold proteins Par3 and Par6 and serine/threonine kinase atypical protein kinase C (aPKC) [14, 77, 93, 94]. Par3 and Par6 both have the ability to bind to aPKC and to each other. The PAR complex binds to tight junctions via Par3, creating an apical-basal border in polarized epithelial cells [14, 77]. Additionally, Par6 in this polarity complex associates with RhoGTPase Cdc42, and together this complex enables actin cytoskeletal reorganization and organelle translocation [95, 96]. The Scribble complex contains the conserved proteins scribble, Discs large (Dlg), and lethal giant larvae (Lgl), altogether making a polarity complex that defines the basolateral membrane of a polarized cell [77, 97]. The presence of noncoding mir200c, a key regulator of polarity in epithelial cells, in the developing submandibular end bud lends further support to the notion that the salivary gland uses many of the same mechanisms to establish polarity as do other polarized epithelia [98]. It remains unclear whether the secretory acinar cells of the salivary gland polarize precisely in the same way as do the ductal cells, but it is likely many of the same proteins and complexes are involved.

Together, these cell adhesion and polarity complexes regulate the asymmetric cell organization needed for proper function of the variety of epithelial cells found in the salivary gland. However, characterization of these protein complexes in native salivary epithelial tissue remains to be investigated, particularly with respect to the differences in the acinar and ductal regions. When the exact nature of the cell adhesion and polarity complexes present in the various salivary regions is determined, then these polarity complex proteins can serve as markers for correct salivary epithelial assembly and polarization in a 3D in vitro organ culture model suitable for tissue repair.

8.3 Salivary Gland-Derived Stem/Progenitor Cells Used for Regeneration Models

Engineering a functional organ in vitro requires all of the cell types, and their distinct functions, present in that native organ. Those various cell types can be isolated and maintained in various differentiated states, or stem/progenitor cells can be induced to differentiate into the various cell types needed. Much work has been performed to locate, isolate, and characterize salivary stem/progenitor cells for tissue engineering purposes. This section will focus on some of the efforts performed by multiple research groups to isolate and characterize salivary stem/progenitor cells and determine their utility in gland regeneration.

8.3.1 Salisphere-Derived c-Kit+ Cells for Salivary Gland Regeneration

A salisphere cell culture method was developed to culture salivary stem cells that can be used for salivary gland function restoration [99]. The method entails mechanically and enzymatically dissociating salivary tissue and culturing the isolated cells on nonadherent plates to drive cell aggregation into salispheres. The salisphere cell culture technique allowed for fluorescence-activated cell sorting (FACS)-mediated isolation of c-Kit+/Sca-1+/Musashi-1+ cells that could differentiate into acinar and ductal cell types [99]. Salivary c-Kit+ stem cells transplanted into irradiated mouse salivary beds restored salivary tissue morphology and function after 90 days compared to the irradiated-only control. To verify that the same population of stem cells was present in human salivary glands, salivary c-Kit+ stem cells were isolated from non-tumorigenic parotid and SMG tissue and were cultured via the salisphere method [100]. Human-derived c-Kit+ cells also had the ability to self-renew and differentiate when cultured in vitro in Matrigel®.

A study by Nanduri et al. further characterized stem cell marker expression of murine cells derived from the salisphere culture system [101]. Salivary stem cell populations expressing c-Kit, CD133, CD49f, and CD24 were isolated, and stem cell populations expressing each marker were quantified using FACS. The expression level of CD49f and CD24 was high in the salisphere-derived stem cells at day 0, with positive expression in more than 50 % of cells, while the expression level of CD133 (6 %) and c-Kit (0.058 %) was low. Expression levels of CD133, CD49f, and CD24 remained constant at day 3 of salisphere culture, but the expression of c-Kit increased by tenfold (0.65 %). Salivary restoration (measured by salivary flow rate) in the mouse-irradiated model was achieved most efficiently with c-Kit+ cells, as it only required an injection of 300 c-Kit+ cells, while the rest of the stem cell populations required 10,000–134,000 cells to achieve salivary restoration.

A subsequent study assessed the ability of a stem cell population positive for c-Kit, CD24, and/or CD49f to restore salivary function in irradiated salivary glands [102]. Although all of the stem cell populations (c-Kit+, c-Kit+/CD24+, c-Kit+/CD49f+, and c-Kit+/CD24+/CD49f+) restored some percentage of salivary flow in irradiated salivary glands, the stem cell population expressing c-Kit, CD24, and CD49f restored salivary flow rate in irradiated salivary glands to an average of >50 % of the salivary flow rate of the untreated control. Restoration of salivary gland tissue morphology and cell phenotype by transplantation of c-Kit+/CD24+/CD49f+ cells was shown by a reduction in fibrosis, reduction in oversized blood vessels, and increased expression of differentiated ductal cell markers and ductal stem markers. Lastly, another lab showed that CD49f and Thy-1 expressing salivary progenitor cells isolated from adult human salivary glands could differentiate into pancreatic cell phenotypes when cultured as spheres, showing their potential to transdifferentiate into other cell types [103].

8.3.2 Isolation of Adult Stem Cells from Human Salivary Tissue

Researchers also have reported other methods for the isolation and maintenance of salivary-derived cells with stem/progenitor markers [104, 105]. Our own group has isolated primary human salivary stem/progenitor cells (hS/PCs) from healthy parotid and SMG tissue that can organize into salivary structures and that display self-renewal and extended proliferation capabilities [106–109]. Specifically, hS/PCs cultured in 3D hyaluronic acid-based hydrogels retain self-renewing properties that allow them to continue to proliferate for over 48 days (further discussed in polysaccharide Sect. 8.4.3.2) [108]. These hS/PCs express c-Kit, K5, and K14 and can be differentiated to display salivary and ductal phenotypes (data in press and [108]).

Rotter et al. reported a population of stem cells isolated from adult human parotid tissue using enzymatic digestion methods [104]. Mesenchymal stem cell (MSC) marker expression was assessed with flow cytometry analysis and showed that these isolated cells expressed CD29, CD44, CD73, and CD90. Analysis of these cells for hematopoietic stem cell marker expression showed that these cells are negative for CD34, CD45, and CD133, indicating that they are not of hematopoietic origin. Salivary gland-derived MSCs were cultured in various tissue-specific differentiation media to demonstrate their ability to differentiate into adipogenic, osteogenic, or chondrogenic cell phenotypes. Importantly, these authors do not suggest that these cells demonstrate any salivary-specific function or ability to differentiate into acinar, ductal, or myoepithelial phenotypes.

Another lab also reported the isolation of MSCs from adult human salivary tissue that could be used for salivary regenerative purposes [110]. In this case, human SMGs were processed enzymatically and isolated cells were cultured on collagen I-coated tissue culture flasks. Human salivary gland stem cells (hSGSCs) were cultured for longer periods of up to 5 weeks, and then assessed for mesenchymal and hematopoietic stem cell marker expression via FACS. Isolated hSGSCs were positive for MSC markers CD44, CD49f, CD90, and CD105, but not for hematopoietic stem cell markers CD34 and CD45. These MSCs could be differentiated into adipogenic, osteogenic, and chondrogenic cell types. Sequential treatments with differentiation media (containing EGF/hepatocyte growth factor (HGF) or solely EGF) resulted in amylase expression in a small fraction of these cells. Transplantation of hSGSCs into radiation-damaged rat salivary beds partially restored salivary flow and normalized tissue morphology and reduced the number of cells undergoing apoptosis. This study suggests some salivary gland regenerative potential of MSCs derived from human salivary tissue, but this approach has many challenges to overcome before such use can be achieved.

8.4 3D Scaffolds Used in Salivary Gland Regeneration Models

Tissue engineering of a functional organ requires a 3D biomimetic scaffold that allows cells to organize into their native structure. There have been a myriad of scaffolds generated and utilized for 3D cell culture for tissue engineering purposes. Specifically, studies using 3D scaffolds for salivary cell culture have been extremely informative for future salivary gland tissue engineering efforts; these studies will be reviewed in this section and are outlined in Table 8.1.

8.4.1 Decellularized Native ECM Scaffolds for Regenerative Purposes

Organ donor shortage is a significant problem in the United States. There are currently over 123,000 people on the waitlist to receive an organ, and more than 6500 people will die waiting each year [118]. One solution to the shortage of donated organs is to utilize allogeneic and xenogeneic organs to create ECM scaffolds for organ regeneration. Decellularization of organs via physical, chemical, and/or enzymatic methods leaves behind a natural scaffold composed of the ECM proteins and glycosaminoglycans (reviewed in [119, 120]). Published decellularization methods vary greatly depending on the chemical agents, physical forces, organ type, and organ species used (Table 8.2) [113, 119, 120, 140, 151–156]. In general, mechanical/physical methods, chemical methods, and/or enzymatic methods can be applied when decellularizing organs, and these methods can be used in combination for more efficient decellularization depending on tissue type (refer to [119, 120] for a more detailed review of techniques).

Efforts to decellularize various organs including the heart, lungs, kidneys, and trachea, among

Table 8.1 Scaffolds used for tissue engineering purposes

Type of matrix	Examples
Decellularized ECM scaffold	Decellularized mouse, rat, porcine, and human lungs [111, 112] Decellularized and recellularized rat submandibular gland [113]
Matrigel®/GFR Matrigel®	Matrigel® used for culturing primary salivary gland cells [114] Growth factor-reduced Matrigel® [115]
Biologically derived polymers/ natural polymers	Proteins: Collagen Laminin Fibrin [115] Polysaccharides: Hyaluronic acid Agarose Dextran Chitosan Protein/polysaccharide hybrid polymers: Collagen/HA Laminin/cellulose Gelatin/chitosan Fibrin/alginate Fibrin/agarose
Protein fragments/peptides	Laminin peptides integrated into agarose gels [116] Collagen-derived degradable peptides (PQ) [117] Collagen I Collagen IV Perlecan domain IV peptide [106] Fibronectin-derived RGD peptide

others, in the past decade have paved the way for future advancements in decellularization of other organs [113, 136, 140, 151–156]. An example of a highly branched organ of high relevance to salivary glands, with respect to tissue structure and organization, is the lung. Like salivary gland development, lung development depends on branching morphogenesis of the primary buds formed to develop the fully branched, differentiated organ [157–159].

Ott et al. developed a low-concentration sodium dodecyl sulfate (SDS)-based perfusion decellularization protocol for decellularizing rat lungs to yield ECM scaffolds for lung regeneration purposes [141]. Perfusion decellularization of rat lungs created acellular ECM scaffolds that maintained the native tissue architecture including vasculature, airways, and alveoli. Decellularized lung ECM scaffolds were recellularized with human umbilical cord endothelial cells (HUVECs) and fetal rat lung epithelial cells to restore vasculature and the lung parenchyma. Recellularized rat lungs were perfused with blood and cultured under physiological conditions with the use of a bioreactor, and were orthotopically transplanted into a lung resection rat model. Regenerated rat lungs were connected to the recipient's pulmonary artery and vein to restore blood flow and were ventilated via the recipient's airway. Although the regenerated lung functionally provided gas exchange in vivo for up to 6 h before respiratory failure, there were observable complications including pulmonary secretions and interstitial edema in the graft. Additionally, the lung was not completely regenerated as there were areas of the lung tissue lacking surfactants, containing squamous epithelium instead of mature secretory cells, and possible type II cell hyperplasia. A follow-up study showed that integrating improved graft preservation and postoperative weaning protocols allowed for enhanced in vivo gas exchange in the rat lung resection model [111]. However, the recellularized scaffolds again failed to be completely regenerated and led to fibrosis surrounding the graft and infection.

Table 8.2 Decellularization methods

	Method	Mechanism of action	Examples; references
Physical methods	Freeze/thaw cycles	Disrupts cell membrane through the formation of ice crystals	[121–126]
	Sonication	Disrupts cell membrane through the use of applied sound energy	[122, 127–129]
	Orbital shaking conditions	Typically used to enhance the removal of cell particles from ECM	[130–132]
	Pressure	Pressure applied to tissue can aid in bursting cells	[133–135]
Chemical methods	Nonionic detergents (Triton X-100)	Permeabilizes the cell membrane without denaturing proteins	[124, 128, 133, 136–139]
	Ionic detergents (sodium dodecyl sulfate, sodium deoxycholate, Triton-X-200)	Permeabilizes the cell membrane while also denaturing proteins	[111–113, 133, 136–142]
	Zwitterionic detergents (CHAPS, sulfobetaines)	Permeabilizes the cell membrane with little or no denaturing of proteins	[112, 139, 142, 143]
	Hypotonic and hypertonic solutions	Alternating between low concentrations and high concentrations of NaCl solution to disrupt cell membranes	[125, 144–146]
	Alcohols (isopropanol, ethanol, methanol, glycerol)	Disrupts cell membranes but also fixes tissue; the remaining ECM is partially cross-linked	[147–149]
	Acids/bases (peracetic acid, acetic acid, sodium hydroxide)	Disinfects while also removing nucleic acids and hydrolyzing the ECM proteins, especially collagen Damages ECM as it degrades it to a higher extent	[112, 140, 150]
	Chelating agents (EDTA, EGTA)	Takes up ions necessary for cell-cell and cell-ECM binding	[125, 126, 149]
Enzymatic methods	Enzymes targeting proteins (trypsin, collagenase, lipase)	Typically used in combination with other methods as it removes cellular proteins and ECM proteins, depending on the enzyme type	[125, 126, 149]
	Enzymes targeting nucleotides (nucleases)	Typically used in combination with other decellularization methods because it degrades nucleic acids left behind from cell lysis	[125, 126, 132, 146]

Other lung decellularization examples contain the use of Triton X/sodium deoxycholate (SDC)-based and 3-[(3-cholamidopropyl)dimethylammonio]-1-propanesulfonate (CHAPS)-based detergents [137, 138, 140, 143, 153]. Mouse lungs have been successfully decellularized with Triton X/SDC-based decellularization method and have been recellularized with different cell types [137, 138]. Mouse lungs decellularized with the detergent solution via the trachea, and the right ventricle generated lung ECM scaffolds that maintained the tissue architecture and ECM integrity including an intact vasculature network and ECM proteins collagen I, collagen IV, laminin, and fibronectin, respectively [138]. However, scaffolds decellularized by this method contained higher amounts of remaining contaminating cell cytoskeletal proteins. Lung scaffolds were recellularized with bone marrow-derived mouse MSCs, and although cell attachment was successful, MSCs failed to differentiate into differentiated lung epithelial cells. Price et al. integrated the use of a ventilator to their decellularization method to apply physiological levels of mechanical stretch [137]. Decellularized ECM scaffolds in this study also retained their native tissue architecture including airways, alveoli, and blood vessel network. The remaining ECM contained normal levels of collagens, but lower levels of elastin, laminin, and GAGs. These scaffolds were reseeded with fetal lung cells that were localized to alveolar areas and expressed cytokeratin 18 and pro-Sp-C, a pulmonary-associated surfactant protein.

Other studies using the Triton X/SDC-based decellularization method translated this technology to clinically relevant, human-sized lungs [140, 153]. Price et al. followed up their previous lung decellularization work by developing an automated system for organ decellularization via perfusion that allows for the control of airway and vascular perfusion pressures [153]. Porcine lungs were decellularized via perfusion with the following solutions in sequential order: deionized water (2 h), Triton X (5 h), sodium deoxycholate (5 h), NaCl (5 h), DNase (2 h), and PBS (5 h). This automated method reduced decellularization time from days to a day and made decellularization of the lung more consistent when compared to a manual decellularization. Decellularization of the lung with this automated system yielded decellularized lung ECM that maintained the structural integrity of the tissue, removed α-galactose, reduced levels of DNA, and retained expression of collagen I, collagen IV, elastin, fibronectin, vitronectin, and laminin. Weymann et al. similarly used Triton X/SDC detergent-based perfusion decellularization methods to successfully generate an acellular lung ECM scaffold that retained the three-dimensional tissue architecture [140]. Acellular ECM scaffolds decellularized by this method had reduced DNA content compared to native tissue, maintained expression of collagen I, elastin, and GAGs, and had comparable biomechanical integrity.

CHAPS-based detergent solution has been used in the successful decellularization of rat lungs [112, 143]. Using this method, Petersen et al. generated acellular lung ECM scaffolds that maintained the three-dimensional branched tissue architecture and could be recellularized with lung epithelial and rat lung microvascular endothelial cells [143]. DNA content was reduced by 99 %, and cellular proteins' major histocompatibility complex (MHC) I, MHC II, and β-actin were removed from acellular scaffolds generated with this CHAPs-based decellularization method. ECM proteins' collagen, laminin, and elastin, at lower levels, were preserved in the acellular scaffolds, while most of the GAGs were depleted. However, in a study by the Ott Lab that used SDS, Triton X/SDC, and CHAPs-based decellularization methods on rat lungs, the CHAPS-based method led to acellular scaffolds with higher levels of DNA content and cytoplasmic proteins and lower levels of collagen and laminin peptides [112]. Based on this study, this group chose the SDS-based detergent decellularization method to decellularize clinically relevant porcine and human lungs. Acellular porcine and lung ECM scaffolds had a reduced DNA and cytoplasmic protein content, with preservation of ECM components' elastin, collagen IV, fibronectin, laminin, and GAGs. Additionally, the human ECM scaffolds were biocompatible with small airway epithelial cells (SAECs), pulmonary alveolar epithelial cells (PAECs), and human umbilical vein endothelial cells (HUVECs), shown by cell adhesion onto the scaffold and cell viability after 5 days of culture. The whole human lung ECM scaffold, and not just tissue slices, also was successfully seeded with human PAECs that were distributed via the airways by gravity and were cultured under constant pressure for 4 days with no visible tissue damage.

Another study also used the three different published detergent-based decellularization methods (SDS [111, 141], Triton X/SDC [137, 138, 153], and CHAPS [143]) on mouse lungs to compare the effectiveness of each decellularization method and the ability to recellularize each scaffold [139]. ECM scaffolds processed by the three different protocols were assessed for preservation of the overall tissue architecture by histological methods. The remaining ECM scaffolds from SDS and Triton X/SDC better maintained the native lung tissue architecture when compared to ECM scaffolds from CHAPS. Decellularized ECM from all three methods was analyzed for protein retention by performing immunohistochemistry, mass spectrometry, and western blotting. Lung ECM decellularized with SDS and CHAPS had higher retention of collagen I, collagen IV, and fibronectin when compared to Triton X/SDC. Tissue decellularized with SDS and Triton X/SDC had higher retention of laminin compared to CHAPS. Decellularization methods using Triton X/SDC and CHAPS were more effective at removing nuclear proteins, but less effective at removing cytoplasmic proteins when compared to the method using SDS. Proteolytic activity was measured in each decellularized tissue treated with the various detergent-based methods, and methods using SDS and CHAPS had the lowest proteolytic activity after 24 h. Remaining ECM from all three detergent-based methods were recellularized using bone marrow-derived mouse mesenchymal stem cells and C10 mouse lung epithelial cells. Notably, there were no observable differences in cell attachment, proliferation, and apoptosis between the varying detergent-based methods.

Lessons learned from decellularization of the lung studies are highly relevant and applicable to any attempts to successfully decellularize the salivary glands. Based on the successful generation of an acellular lung ECM scaffold that has the native three-dimensional tissue architecture preserved with the use of SDS-based decellularization method, this method seems to be the most useful in an organ with similar tissue architecture to the lung. However, decellularizing the salivary gland poses challenges when compared to the lungs. The vasculature network surrounding the salivary gland is too small to use for perfusion of detergent solution. Another option is to perfuse the detergent solution through the salivary ductal system, which also is quite small and delicate when attempting to cannulate, making it technically difficult to connect to an external perfusion system.

One attempt to translate this SDS-based method to the salivary gland was reported by Gao et al. [113]. Rat SMGs were successfully collected and decellularized via a previously published method that employed detergent solutions containing SDS for 32 h, Triton X-100 for 2 h, and DNase I for 1 h [119, 155]. However, in this study, perfusion decellularization techniques were unsuccessful because of the difficult anatomy. After the decellularization process was completed, the remaining ECM was assessed through histology, immunohistochemistry, scanning electron microscopy (SEM), and DNA and protein quantification analysis, showing that the ECM retained its natural structure and comparable levels of ECM proteins that include collagen I, collagen IV, laminin, and fibronectin. Decellularized ECM scaffolds were recellularized with primary rat SMG cells injected into the main duct of the gland and were maintained in suspension in a rotary cell culture system. Recellularized scaffolds were assessed by histology and immunohistochemistry, which showed that SMG cells remained in the scaffold after injection and expressed cell-cell adhesion markers (E-cadherin and occludin), aquaporin 5 (AQP-5), and α-amylase. This study was useful in showing the possibilities of using decellularization and recellularization techniques to salivary gland regeneration purposes. However, the salivary glands of humans are much larger organs than those of rats. An animal source such as the pig that retains a similar facial anatomy to that of humans would be more ideal for generating an organ scaffold that could restore salivary function in humans.

8.4.2 Matrigel®/GFR Matrigel®

Matrigel®, a basement membrane extract derived from Engelbreth-Holm-Swarm (EHS) mouse sarcoma cells, is commonly used for 3D cell culture because it contains growth factors and basement membrane proteins found in the ECM, which greatly enhance the growth and viability of cells that may not grow well in GF-depleted matrices. Matrigel® often is used for cell culture to assess the biological activity of those cells in a growth factor and basement membrane replete matrix. However, commercial Matrigel® typically is not considered a controlled and fully characterized matrix as it contains various growth factors and basement membrane proteins, and their concentrations vary from batch to batch. Although Matrigel® is a useful matrix for studying the biological activity of cells, it is not ideal for regeneration or tissue engineering purposes as it is not biocompatible with humans because it is rodent derived. Despite this, the results of studies using Matrigel® for salivary cell culture are insightful because they can provide a reference scaffold for comparison when trying to generate and assess the usefulness of a customized, biocompatible scaffold with ECM proteins/peptides.

Rat parotid gland cells (Par-C10) are salivary acinar cells isolated from rat parotid glands and transformed with simian virus 40 [160]. Baker et al. have cultured Par-C10s on Matrigel® and demonstrated their arrangement into acinar-like spheres that expressed tight junction proteins, M3 receptor, and AQP-3, AQP-5 and were responsive to carbachol treatment [161]. Other studies have used the human submandibular gland (HSG) cell line in combination with Matrigel®/basement membrane extract to study the in vitro cell organization and phenotype of salivary cells. Human submandibular gland (HSG) cells, a cell line derived from intercalated ducts of irradiated SMGs (now known to be contaminated with HeLa, see below), have been shown to form a 3D reticular network resembling acini-like structures connected to duct-like structures, with the HSG cells taking on a differentiated acinar cell phenotype, i.e., well-developed Golgi apparatus, microvilli-like projections from the apical surface, the presence of granules, amylase production, and a decrease in cell division [162, 163]. Hoffman et al. further investigated the role of basement membrane components on acinar development and cytodifferentiation, showing that laminin-1 and transforming growth factor-β_3 are important for acinar cell differentiation [164]. In another study, HSG cells cultured on Matrigel® or laminin-1 and treated with transforming growth factor-α had a synergistic increase in α-amylase expression, which was mediated through protein kinase C (PKC) and ERK1/2 [165]. A study performed by Maria et al. showed that 3D spheroids formed on Matrigel® expressed α-amylase, tight junction proteins, AQP-5, CD44, and CD166 [114]. HSG cells grown in Matrigel® formed acini-like structures that displayed reduced cell proliferation and increased in cell apoptosis [166]. Although decreased cell proliferation tends to accompany cell differentiation, further characterization of these salivary structures for acinar phenotypic markers is needed to fully show that they are differentiated structures. Additionally, the HSG cell line is listed on the International Cell Line Authentication Committee's (ICLAC) Database of Cross-Contaminated or Misidentified Cell Lines (version 7.1, 2013-08-22) as likely to be contaminated by HeLa cells [167]. This is additionally reported by the European Collection of Cell Cultures (ECACC) and the Japanese Collection of Research Bioresources (JCRB) Cell Bank. For these reasons, HSG cell lines are no longer a reliable model for salivary cell function.

Although cell lines are convenient to use because of their hardiness and extended passaging capability, they are not always representative of the cell phenotype/activity from the native tissue. Additionally, immortalized cell lines are not useful for regenerative or tissue engineering purposes as they can lead to tumor formation as a result of their uncontrolled proliferation subsequent to immortalization. Primary human salivary cells grown on top of Matrigel® ("2.5D" cell culture) formed acini-like spheroids that expressed tight junction proteins, AQP-5, α-amylase, CD44, and CD166 [168]. In another study, primary human SMG cells were isolated from fresh salivary tissue that was processed with a dissociation buffer to yield single cells that could be passaged on tissue culture plates [169].

These primary human SMG cells cultured on basement membrane extract (BME) formed acinotubular structures, which is believed to be differentiated because of decreased proliferation and increased expression of α-amylase, occludin, claudin-1, and claudin-3 [144]. Our group showed that primary human salivary stem/progenitor cells (hS/PCs) cultured on Matrigel® self-assembled into 3D acini-like structures [106]. Salivary cells cultured on Matrigel® expressed E-cadherin, AQP-5, K19, and α-amylase, showing that these acini-like structures are at least partially differentiated. Additionally, the cells grown on Matrigel® formed stress fibers and had localized phospho-FAK expression at focal adhesion sites. The components present in Matrigel® that stimulate salivary-derived hS/PCs, or other salivary stem/progenitor cells, to differentiate into the multiple salivary epithelial cell types include laminin, perlecan/HSPG2, and collagen IV. Integration of these motifs into a scaffold used for tissue engineering can thus provide a defined, scalable, and bioactive support for salivary gland replacement.

8.4.3 Biologically Derived Polymers/Natural Polymers

As previously mentioned, Matrigel® contains many components from the ECM, and, thus, it is not easy to determine the connection between biological responses of cells grown on Matrigel® and a specific protein component(s). Growing cells on matrices with limited numbers of precisely defined components helps to identify a connection between the activity of cells and an individual ECM component. Matrices can be made up of biologically derived whole proteins, polysaccharides, peptides, or combinations of these. As the size of some ECM proteins are very large, it is difficult to integrate multiple purified proteins into one scaffold. For this reason, recent work has focused on mining small peptide sequences or recombinant protein subdomains with specific biological activity that can be easily integrated into tissue engineering scaffolds or hydrogels [170].

8.4.3.1 Salivary Cell Culture Utilizing Whole Proteins in the Matrix

Collagen I gels have been used frequently as a matrix for culture of both epithelial and mesenchymal cells. Collagen I is a heterotrimer composed of two α_1 chains and one α_2 chain that organize into triple-helical fibers [171]. This fibrous protein polymer is the most abundant protein in connective tissue, providing multiple integrin-binding sites to promote cell migration, phospho-FAK activation, phosphoinositol-3-kinase (PI3-kinase), and mitogen-activated protein kinase (MAPK) cascades [27, 172].

Mouse-derived SMG salivary gland cells grown on rat tail-derived collagen I gels formed ductal-like structures that retained expression of EGF when cultured with media containing testosterone, triiodothyronine, and hydrocortisone [173]. Rat SMG-derived salivary (RSMG-1) cells cultured under serum-free conditions in collagen I gels underwent branching morphogenesis and formed branched structures when treated with HGF [174]. In another study, human primary parotid gland cells were isolated using tissue explant culture methods and used at early passages [175]. These primary parotid gland cells organized into acini-like and ductal-like structures when grown on collagen I/GFR Matrigel® gels [175]. These salivary structures expressed α-amylase, AQP-5, and tight junction proteins occludin, claudin-1, and ZO-1. Burford-Mason et al. reported the use of collagen I gels for culturing rat SMG glands as a model for the pathobiology of salivary glands [176]. In this study, rat SMG organoids cultured in collagen retained their secretory activity, and each salivary cell type (acinar, ductal, and myoepithelial) retained their phenotypic marker expression for some time in culture.

Although collagen I provides integrin-binding sites for cell attachment and activation of signaling cascades, it is not the most appropriate ECM component to use for salivary gland regeneration as it is most commonly associated with connective tissue, wound healing, scarring, and fibrosis. A scaffold with components from the basement membrane, which directly surrounds the salivary epithelial cells, should be more effective for providing a scaffold that recreates the native environment and signals and enables regeneration or tissue engineering based reconstruction of the salivary gland.

Fibrin hydrogels provide another example of how intact proteins are used for salivary gland cell culture. Fibrin, a nonglobular protein derived from fibrinogen, polymerizes into non-soluble fibers that play a role in blood clot formation [177]. Fibrin polymers also can be cross-linked to form hydrogels for 3D cell culture. Mouse parotid gland-derived salivary cells cultured on fibrin-based hydrogel polymerized with growth factors (EGF and insulin-like growth factor-1) and combined with growth factor-reduced (GFR) Matrigel® formed 3D salivary spheroids [115].

Laminin is a major component of the basement membrane that directs salivary gland development. Specifically, Cantara et al. developed poly(lactic-*co*-glycolic acid) nanofiber scaffolds that were functionalized with laminin-111 [178]. Two salivary cell lines were used in this study: (1) immortalized mouse ductal SMG epithelial cells (SIMS) and (2) immortalized rat SMG acinar epithelial cells (SMGC10). Both SIMS and SMGC10 cells grown on laminin-111-functionalized nanofibers showed enhanced formation of tight junctions, shown by immunofluorescence for ZO-1 and occludins at the apical-basal membrane border. Moreover, laminin-111-modified nanofibers increased cell proliferation, without reducing cell viability. This study showed that a matrix for salivary gland tissue engineering could be functionalized with basement membrane components, especially laminin-111, to promote the polarization and differentiation of salivary epithelial cells. In particular, this study utilized the whole laminin-111 isoform in the scaffold for cell culture, but peptides with the biologically active sequences also can be synthesized and incorporated into a variety of cell culture matrices.

8.4.3.2 Use of Polysaccharides in Tissue Engineering Scaffolds

Polysaccharides, including cellulose, alginate, agarose, and hyaluronic acid (HA), often are used to generate scaffolds for 3D cell culture. Cellulose is a linear polymer of 1,4-β-linked D-glucose units commonly found in the cell wall of plant cells and produced by some microbial organisms; its derivatives have been used for microbial and mammalian cell culture [179–181]. Agarose and alginate also are polysaccharides, derived from natural biological sources, generally cell walls of algae. These materials are readily commercially available, and both are frequently used in the laboratory for a variety of biological applications. Both are largely considered as nonadhesive, "blank slates" when used with many mammalian cell types, as neither contains the common adhesive proteins (e.g. fibronectin, vitronectin) nor other adhesion motifs found in a complete ECM. Yet, these materials preserve much of the mechanical properties (viscoelasticity and low elastic modulus) found in native ECM. Agarose powders can be dissolved with heat and cooled to form aqueous hydrogels, while alginates are readily soluble and can be triggered to gel with ionic gradients (Ca^{2+} is most common). Cells seeded onto these materials, or encapsulated within them, tend to aggregate into multicellular spheroids for survival. The nonadhesive nature of these polysaccharide hydrogels preferentially supports cell-cell interactions, because there are no cell-ECM interactions available from these substrates.

HA-based systems have seen a significant increase in interest over recent years, as researchers have developed the tools to purify and functionalize the base polymer. HA is a linear polysaccharide composed of the repeating disaccharide unit β-1,4-D glucuronic acid-β-1,3-*N*-acetyl-D-glucosamine [182, 183]. HA is a ubiquitously expressed glycosaminoglycan that is found throughout the body as an essential component of ECM. Unlike many other hydrogel systems, HA is inherently bioactive, as many cells have HA receptors for interaction with the base material. Moreover, cells express HA receptors CD44 and hyaluronan-mediated motility receptor (RHAMM), allowing them to interact with the surrounding HA polymers and activate cellular signaling processes [184, 185]. Carboxylic acids along each saccharide unit offer preferred reactive sites for selectively functionalizing the polymer. Many orthogonal chemistries exist for initiating cross-linking and hydrogel network formation; however these must be cytocompatible, especially for cells embedded within the pre-gelled matrix [186]. HA is ideal for tissue engineering purposes as it is nonimmunogenic, biocompatible and biodegradable and can be chemically functionalized with biologically active peptides from the ECM [183, 186]. Also, as HA is negatively charged and highly polar, therefore, hydrogels composed of HA are very hydrophilic, hydrated gels with variable swelling ratios [187].

The Prestwich Laboratory has published extensively on the preparation and application of various HA systems, and their work has resulted in commercially available preparations as the HyStem® product line. Our lab has used the commercially available HyStem® hydrogel composed of thiol-modified hyaluronic acid cross-linked with polyethylene glycol diacyrlate (PEGDA) for culture of salivary cells. Primary human salivary stem/progenitor-like cells (hS/PCs) cultured on 2.5D or in 3D HA-based HyStem® hydrogels formed acini-like structures that expressed cell-cell adhesion markers (β-catenin, E-cadherin, ZO-1), cholinergic M3-muscarinic receptors, and β-adrenergic receptors [107, 108]. Salivary structures formed were proliferative up to 48 days in culture and were responsive to isoproterenol and norepinephrine stimulation [108]. HA hydrogels seeded with hS/PCs were implanted in a parotid gland three-fourths resection model, allowing the hydrogels to be directly in contact with the salivary bed [109]. Implanted acini-like spheroids expressed HA receptors CD168/RHAMM and CD44, also a progenitor marker, and retained progenitor marker CD44 expression after at least 1 week of being implanted in the rat salivary bed [109].

HyStem® materials are sold by BioTime and have received FDA 510(k) approval for marketing in use as a wound matrix (K134037 for BioTime's *Premvia*™ product). It has been assessed for ISO 10993 biocompatibility tests, EN 13276-1 tests for primary wound dressings, and human cytocompatibility tests. *Premvia*™ is approved for syringe-based delivery within tunneled or undermined wounds, or over superficial wounds. HyStem's *Renevia*™ product is similarly based on the HyStem platform and is under evaluation in clinical trials in Spain as an injectable hydrogel and cell carrier to treat facial wasting in HIV+ patients. *Renevia*™ has passed Phase I clinical trials for safety in humans as an injectable in the retro-auricular area. Other licensees of this same technology target veterinary wound repair applications, and other human wound repair or topical applications (e.g., BakerDVM's *Remend*® corneal repair, wound repair, and eye lubrication drops). The progression of these and other biomaterials through FDA approval is a critical step in their eventual use in tissue engineering applications.

8.4.4 Protein Fragments/ Biologically Active Peptides

Scaffolds for tissue engineering purposes should be strategically designed to have bioactive components to induce the cellular processes needed for the development, or regeneration, of the desired specific organ. Peptides derived from ECM components can be integrated into scaffolds to promote cell-ECM adhesion and cell-cell adhesion in cell culture. Additionally, peptides can be integrated into hydrogels to aid in polymerization, as in the case of step-growth thiol-ene polymerizations of poly(ethylene glycol) hydrogels [188].

There are several available laminin-derived peptides including IKVAV, AG10, AG32, and AG73/MG73 from the α-chain, and YIGSR from the β chain [189]. HSG formed spheroids that resemble acini when cultured on scaffolds containing AG73 (RKRLQVQLSIRT) or the homologous MG73 (KNRLTIELEVRT); however, these spheroids lacked polarity and lumens [116, 162, 190]. Branching morphogenesis was inhibited in embryonic SMGs cultured on a combination of Matrigel®, laminin, and nidogen when treated with soluble AG73 peptide [189]. Therefore, epithelial cell interaction with AG73 peptide sequence is vital for branching, but it is not sufficient to promote acinar polarity and differentiation.

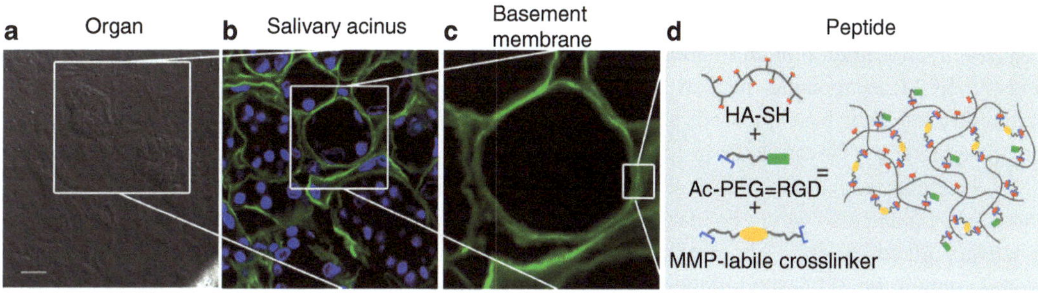

Fig. 8.2 Illustration of multiple organ levels for consideration in salivary gland engineering. (**a**) Examination at the organ level (bright field DIC image, scale bar is 20 μm) of human parotid gland demonstrates organization into acini and ductal systems. (**b**) Closer study of the salivary acinus with an antibody against collagen IV (*green*) and nuclear stain (*blue*) identifies the basement membrane (*BM*) that surrounds each acinar structure. (**c**) Within that BM, a meshwork of protein and glycosaminoglycan components arranges to direct cell polarization. (**d**) On the molecular level, key elements of this supportive BM may be reproduced by employing modular peptides with controlled spatial and/or temporal positioning

Our lab previously discovered a novel peptide (TWSKVGGHLRPGIVQSG) from domain IV of perlecan/HSPG2 that promotes cell adhesion, spreading, and FAK activation of various cell types [191]. Salivary hS/PCs cultured on perlecan domain IV peptide formed acini-like spheroids that secreted α-amylase over 6 days at comparable levels to structures formed on Matrigel® [106]. Additionally, salivary spheroids cultured on perlecan domain IV peptide expressed tight junction protein E-cadherin, water channel protein AQP-5, as well as activated FAK. Other peptides can be integrated into three-dimensional cell culture for tissue engineering to promote cell adhesion and motility, and these include fibronectin-derived RGD peptide and MMP-sensitive, collagen I-derived PQ peptide commonly used as a cleavable cross-linker (Fig. 8.2). Fibronectin is one of the ECM proteins that enables cleft formation during branching morphogenesis, thus, integrating an integrin-binding site from fibronectin may permit clefting of the hS/PC spheroids in our HA-based hydrogel model. Additionally, collagen I is expressed and localized throughout the surrounding salivary mesenchyme. MMPs expressed by the salivary epithelium during development play a role in modulating cleft formation, as inhibiting MMP1 with TIMPs leads to increased cleft formation in embryonic salivary glands, while treatment with exogenous MMP1 diminishes cleft formation [56, 192]. Our lab has integrated these peptides into HA-based scaffolds for prostate xenograft-derived cancer cells and is currently working on translating this scaffold for 3D culture of hS/PCs [193].

8.5 Future Directions

Much has been learned about the important role that ECM plays in salivary gland development, and thus, many efforts have focused on developing biologically active scaffolds with incorporated ECM components. An ideal scaffold for salivary gland engineering would contain cell adhesion sites and degradable cross-linkers that salivary epithelial cells could selectively cleave to make space for growth. Cell adhesion sequences from laminin, fibronectin, and collagen are available for use in tissue engineering scaffolds, but should be used in a manner that is relevant to the development of the organ. For example, cell adhesion sites that are vital for branching morphogenesis should be integrated early on in the scaffold to promote branching of salivary acini-like spheroids. It should be noted that salivary epithelial cells, in the case of hS/PCs grown in HA-based hydrogels, secrete their own ECM after a few days in culture, although it is unclear if it is well organized [106]. Therefore, it is possible that a scaffold for successful salivary gland engineering may need only the initial ECM components to trigger branching morphogenesis, and the ECM components needed for terminal differentiation may be expressed, secreted, and organized later by salivary cells in culture.

A decellularized native ECM scaffold from the salivary gland also would be ideal to generate a salivary gland in vitro. The acellular scaffold would contain the 3D tissue architecture present in the native salivary gland, including a vascular network, branched lobules, and a ductal system. Clinically relevant acellular scaffolds of human size, whether porcine or human derived, would be best to generate optimal decellularization and recellularization methods. However, such scaffolds would need to be efficiently decellularized, leaving behind little to no antigens from the donor that would cause any immunoreaction from the host. Moreover, the decellularized scaffold also must be effectively sterilized to avoid causing any infection at the graft site. Regardless of whether the scaffold is a hydrogel with ECM peptides or a native acellular ECM, the mechanical properties must be equal or similar to the mechanical properties of the ECM of the natural gland.

The cell type used to engineer a salivary gland in vitro for implantation should be chosen wisely. Development of a salivary gland either will require all of the various differentiated salivary cell types (acinar, ductal, and myoepithelial) or a salivary stem/progenitor cell that can differentiate into all of the cell types upon receiving the proper ECM and growth factor/morphogenic cues. In addition, salivary cells for implantation into a human host must be human derived and ideally derived from the patient to receive the graft to avoid host rejection, i.e., an autograft model. Our lab routinely uses isolated healthy, non-tumorigenic salivary stem/progenitor cells from the head and neck cancer patient obtained before radiation therapy. The plan is to use these cells to grow the salivary gland implant during the time that the patient receives radiotherapy, such that it will be ready to transplant back into the patient after therapy has been completed.

Although there have been advances in research aimed toward generating a salivary gland, there is much work remaining to be done to design a scaffold with the proper spatial-temporal cues for complete salivary gland functional restoration. The successful generation of a salivary gland in vitro will not only drastically improve head and neck cancer patients' quality of life, but it would be informative for tissue engineering of other secretory organs.

Acknowledgments This work was supported by NIH R01DE022969 (MCFC, DAH), NIH F32DE024697 (DW), NSF Graduate Research Fellowship Program (GRFP) DGE-1450681 (MM), and private philanthropic donations. The authors would like to thank Dr. Robert L. Witt, Dr. Xinqiao Jia, Dr. Swati Pradhan-Bhatt, as well as all of the salivary team members for many helpful discussions.

References

1. American Cancer Society. Cancer Facts & Figures. 2016.
2. Sullivan CA, Haddad RI, Tishler RB, et al. Chemoradiation-induced cell loss in human submandibularglands.Laryngoscope.2005;115:958–64. doi:10.1097/01.MLG.0000163340.90211.87.
3. AlDuhaiby EZ, Breen S, Bissonnette J-P, et al. A national survey of the availability of intensity-modulated radiation therapy and stereotactic radiosurgery in Canada. Radiat Oncol. 2012;7:18. doi:10.1186/1748-717X-7-18.
4. Wijers OB, Levendag PC, Braaksma MMJ, et al. Patients with head and neck cancer cured by radiation therapy: a survey of the dry mouth syndrome in long-term survivors. Head Neck. 2002;24:737–47. doi:10.1002/hed.10129.
5. Dirix P, Nuyts S, Van den Bogaert W. Radiation-induced xerostomia in patients with head and neck cancer: a literature review. Cancer. 2006;107:2525–34. doi:10.1002/cncr.22302.
6. Toljanic JA, Heshmati RH, Bedard J-F. Dental follow-up compliance in a population of irradiated head and neck cancer patients. Oral Surg Oral Med Oral Pathol Oral Radiol Endod. 2002;93:35–8. doi:http://dx.doi.org/10.1067/moe.2002.116599.
7. Ellis H. Anatomy of the salivary glands. Surg. 2012;30:569–72. doi:10.1016/j.mpsur.2012.09.008.
8. Marsh PD, Do T, Beighton D, DA D. Influence of saliva on the oral microbiota. Periodontol 2000. 2016;70:80–92. doi:10.1111/prd.12098.
9. Emmelin N, Garrett JR, Ohlin P. Neural control of salivary myoepithelial cells. J Physiol. 1968;196:381–396.1.
10. Emmelin N, Gjörstrup P. On the function of myoepithelial cells in salivary glands. J Physiol. 1973;230:185–98.

11. Chitturi RT, Veeravarmal V, Nirmal RM, Reddy BVR. Myoepithelial cells (MEC) of the salivary glands in health and tumours. J Clin Diagn Res. 2015;9: ZE14–8. doi:10.7860/JCDR/2015/11372.5707.
12. Diaz Silva G. Physiology of the salivary glands. Odontol Chil. 1987;35:130–6.
13. Bornens M. Organelle positioning and cell polarity. Nat Rev Mol Cell Biol. 2008;9:874–86. doi:10.1038/nrm2524.
14. Shin K, Fogg VC, Margolis B. Tight junctions and cell polarity. Annu Rev Cell Dev Biol. 2006;22:207–35. doi:10.1146/annurev.cellbio.22.010305.104219.
15. Dbouk HA, Mroue RM, El-Sabban ME, Talhouk RS. Connexins: a myriad of functions extending beyond assembly of gap junction channels. Cell Commun Signal. 2009;7:1–17. doi:10.1186/1478-811X-7-4.
16. Shimono M, Muramatsu T, Hashimoto S, Inoue T. Connexin expression in the salivary glands. Eur J Morphol. 1996;34:197–202. doi:10.1076/ejom.34.3.197.13027.
17. Lourenço SV, Lima DMC, Uyekita SH, et al. Expression of beta-1 integrin in human developing salivary glands and its parallel relation with maturation markers: in situ hybridisation and immunofluorescence study. Arch Oral Biol. 2007;52:1064–71. doi:10.1016/j.archoralbio.2007.05.002.
18. Lourenço SV, Nico MMS, Kapas S. Integrin expression in developing human salivary glands. Oral Surg Oral Med Oral Pathol Oral Radiol Endod. 2005;100:193. doi:10.1016/j.tripleo.2005.05.036.
19. Segawa A, Takemura H, Yamashina S. Calcium signaling in tissue: diversity and domain-specific integration of individual cell response in salivary glands. J Cell Sci. 2002;115:1869–76.
20. Meda P, Pepper MS, Traub O, et al. Differential expression of gap junction connexins in endocrine and exocrine glands. Endocrinology. 1993;133:2371–8. doi:10.1210/endo.133.5.8404689.
21. Castle JD. Protein secretion by rat parotid acinar cells: pathways and regulation. Ann N Y Acad Sci. 1998;842:115–24. doi:10.1111/j.1749-6632.1998.tb09639.x.
22. Gorr S-U, Venkatesh SG, Darling DS. Parotid secretory granules: crossroads of secretory pathways and protein storage. J Dent Res. 2005;84:500–9.
23. Ishikawa Y, Cho G, Yuan Z, et al. Water channels and zymogen granules in salivary glands. J Pharmacol Sci. 2006;100:495–512. doi:10.1254/jphs.CRJ06007X.
24. Nagler R, Marmary Y, Fox PC, et al. Irradiation-induced damage to the salivary glands: the role of redox-active iron and copper. Radiat Res. 1997;147:468–76. doi:10.2307/3579504.
25. Nagler RM, Laufer D. Protection against irradiation-induced damage to salivary glands by adrenergic agonist administration. Int J Radiat Oncol. 1998;40:477–81. doi:10.1016/S0360-3016(97)00574-9.
26. Grundmann O, Mitchell GC, Limesand KH. Sensitivity of salivary glands to radiation: from animal models to therapies. J Dent Res. 2009;88:894–903. doi:10.1177/0022034509343143.
27. LeBleu VS, Macdonald B, Kalluri R. Structure and function of basement membranes. Exp Biol Med (Maywood). 2007;232:1121–9. doi:10.3181/0703-MR-72.
28. Yurchenco PD, Amenta PS, Patton BL. Basement membrane assembly, stability and activities observed through a developmental lens. Matrix Biol. 2004;22:521–38. doi:10.1016/j.matbio.2003.10.006.
29. Skalova A, Leivo I. Basement membrane proteins in salivary gland tumours. Virchows Arch A. 1992;420:425–31. doi:10.1007/BF01600514.
30. Farach-Carson MC, Warren CR, Harrington DA, Carson DD. Border patrol: insights into the unique role of perlecan/heparan sulfate proteoglycan 2 at cell and tissue borders. Matrix Biol. 2014;34:64–79. doi:10.1016/j.matbio.2013.08.004.
31. Warren CR, Kassir E, Spurlin J, et al. Evolution of the perlecan/HSPG2 gene and its activation in regenerating *nematostella vectensis*. PLoS One. 2015;10:e0124578. doi:10.1371/journal.pone.0124578.
32. Proctor GB, Carpenter GH. Regulation of salivary gland function by autonomic nerves. Auton Neurosci. 2007;133:3–18. doi:10.1016/j.autneu.2006.10.006.
33. Humphrey SP, Williamson RT. A review of saliva: normal composition, flow, and function. J Prosthet Dent. 2001;85:162–9. doi:10.1067/mpr.2001.113778.
34. Borghese E. The development in vitro of the submandibular gland and sublingual glands of the *mus musculus*. J Anat. 1950;84:287–302.
35. Borghese E. Explanation experiments on the influence of connective tissue capsule on the development of the epithelial part of the submandibular gland of *Mus musculus*. J Anat. 1950;84:303–18.
36. Tucker AS. Salivary gland development. Semin Cell Dev Biol. 2007;18:237–44. doi:10.1016/j.semcdb.2007.01.006.
37. Patel VN, Rebustini IT, Hoffman MP. Salivary gland branching morphogenesis. Differentiation. 2006;74:349–64. doi:10.1111/j.1432-0436.2006.00088.x.
38. Ferreira JN, Hoffman MP. Interactions between developing nerves and salivary glands. Organogenesis. 2013;9:199–205.
39. Werner Kahle M, Michael Frotscher M. Color atlas of human anatomy. Volume 3, nervous system and sensory organs. Stuttgart/New York: Thieme; 2003.
40. Henriksson R, Carlsöö B, Danielsson Å, et al. Developmental influences of the sympathetic nervous system on rat parotid gland. J Neurol Sci. 1985;71:183–91. doi:10.1016/0022-510X(85)90058-9.
41. Coughlin MD. Early development of parasympathetic nerves in the mouse submandibular gland. Dev Biol. 1975;43:123–39. doi:10.1016/0012-1606(75)90136-0.
42. Coughlin MD. Target organ stimulation of parasympathetic nerve growth in the developing mouse submandibular gland. Dev Biol. 1975;43:140–58. doi:10.1016/0012-1606(75)90137-2.

43. Knox SM, Lombaert IM, Reed X, et al. Parasympathetic innervation maintains epithelial progenitor cells during salivary organogenesis. Science. 2010;329:1645–7. doi:10.1126/science.1192046.
44. Harunaga JS, Doyle AD, Yamada KM. Local and global dynamics of the basement membrane during branching morphogenesis require protease activity and actomyosin contractility. Dev Biol. 2014;394:197–205. doi:10.1016/j.ydbio.2014.08.014.
45. Koyama N, Hayashi T, Kashimata M. Regulation of branching morphogenesis in fetal mouse submandibular gland by signaling pathways activated by growth factors and α6 integrin. J Oral Biosci. 2011;53:298–303. doi:http://dx.doi.org/10.1016/S1349-0079(11)80022-8.
46. Sakai T, Larsen M, Yamada KM. Fibronectin requirement in branching morphogenesis. Nature. 2003;423:876–81.
47. Kallman F, Grobstein C. Source of collagen at epitheliomesenchymal interfaces during inductive interaction. Dev Biol. 1965;11:169–83. doi:10.1016/0012-1606(65)90055-2.
48. Bernfield MR. Collagen synthesis during epitheliomesenchymal interactions. Dev Biol. 1970;22:213–31. doi:10.1016/0012-1606(70)90151-X.
49. Kadoya Y, Kadoya K, Durbeej M, et al. Antibodies against domain E3 of laminin-1 and integrin alpha 6 subunit perturb branching epithelial morphogenesis of submandibular gland, but by different modes. J Cell Biol. 1995;129:521–34. doi:10.1083/jcb.129.2.521.
50. Durbeej M, Talts JF, Henry MD, et al. Dystroglycan binding to laminin alpha1LG4 module influences epithelial morphogenesis of salivary gland and lung in vitro. Differentiation. 2001;69:121–34. doi:10.1046/j.1432-0436.2001.690206.x.
51. Hardman P, Spooner BS. Localization of extracellular matrix components in developing mouse salivary glands by confocal microscopy. Anat Rec. 1992;234:452–9. doi:10.1002/ar.1092340315.
52. Kumagai M, Sato I. Immunolocalization of fibronectin and collagen types I and III in human fetal parotid and submandibular glands. Cells Tissues Organs. 2003;173:184–90.
53. Nakanishi Y, Nogawa H, Hashimoto Y, et al. Accumulation of collagen III at the cleft points of developing mouse submandibular epithelium. Development. 1988;104:51–9.
54. Hayakawa T, Kishi J, Nakanishi Y. Salivary gland morphogenesis: possible involvement of collagenase. Matrix Suppl. 1992;1:344–51.
55. Spooner BS, Thompson-Pletscher HA, Stokes B, Bassett KE. Extracellular matrix involvement in epithelial branching morphogenesis. Dev Biol. 1986;3:225–60.
56. Fukuda Y, Masuda Y, Kishi J, et al. The role of interstitial collagens in cleft formation of mouse embryonic submandibular gland during initial branching. Development. 1988;103:259–67.
57. Spooner BS, Faubion JM. Collagen involvement in branching morphogenesis of embryonic lung and salivary gland. Dev Biol. 1980;77:84–102. doi:10.1016/0012-1606(80)90458-3.
58. Grobstein C, Cohen J. Collagenase: effect on the morphogenesis of embryonic salivary epithelium in vitro. Science (80-). 1965;150:626–8.
59. Durbeej M. Laminins. Cell Tissue Res. 2010;339:259–68. doi:10.1007/s00441-009-0838-2.
60. Kadoya Y, Yamashina S. Distribution of alpha 6 integrin subunit in developing mouse submandibular gland. J Histochem Cytochem. 1993;41:1707–14.
61. Belkin AM, Stepp MA. Integrins as receptors for laminins. Microsc Res Tech. 2000;51:280–301. doi:10.1002/1097-0029(20001101)51:3<280::AID-JEMT7>3.0.CO;2-O.
62. KadoyaY, YamashinaS. Salivary gland morphogenesis and basement membranes. Anat Sci Int. 2005;80:71–9. doi:10.1111/j.1447-073x.2005.00102.x.
63. Miner JH, Yurchenco PD. Laminin functions in tissue morphogenesis. Annu Rev Cell Dev Biol. 2004;20:255–84. doi:10.1146/annurev.cellbio.20.010403.094555.
64. Ekblom P, Lonai P, Talts JF. Expression and biological role of laminin-1. Matrix Biol. 2003;22:35–47. doi:10.1016/S0945-053X(03)00015-5.
65. Kadoya Y, Mochizuki M, Nomizu M, et al. Role for laminin-α5 chain LG4 module in epithelial branching morphogenesis. Dev Biol. 2003;263:153–64. doi:10.1016/S0012-1606(03)00446-9.
66. Kadoya Y, Nomizu M, Sorokin LM, et al. Laminin alpha1 chain G domain peptide, RKRLQVQLSIRT, inhibits epithelial branching morphogenesis of cultured embryonic mouse submandibular gland. Dev Dyn. 1998;212:394–402. doi:10.1002/(SICI)1097-0177(199807)212:3<394::AID-AJA7>3.0.CO;2-C.
67. Kadoya Y, Yamashina S. Localization of laminin-5, HD1/plectin, and BP230 in the submandibular glands of developing and adult mice. Histochem Cell Biol. 1999;112:417–25.
68. Patel VN, Knox SM, Likar KM, et al. Heparanase cleavage of perlecan heparan sulfate modulates FGF10 activity during ex vivo submandibular gland branching morphogenesis. Development. 2007;134:4177–86. doi:10.1242/dev.011171.
69. Patel VN, Likar KM, Zisman-Rozen S, et al. Specific heparan sulfate structures modulate FGF10-mediated submandibular gland epithelial morphogenesis and differentiation. J Biol Chem. 2008;283:9308–17. doi:10.1074/jbc.M709995200.
70. Knox SM, Whitelock JM. Perlecan: how does one molecule do so many things? Cell Mol Life Sci. 2006;63:2435–45. doi:10.1007/s00018-006-6162-z.
71. Farach-Carson MC, Carson DD. Perlecan--a multifunctional extracellular proteoglycan scaffold. Glycobiology. 2007;17:897–905. doi:10.1093/glycob/cwm043.
72. Hopf M, Göhring W, Kohfeldt E, et al. Recombinant domain IV of perlecan binds to nidogens,

laminin-nidogen complex, fibronectin, fibulin-2 and heparin. Eur J Biochem. 1999;259:917–25.
73. Hoffman MP, Kidder BL, Steinberg ZL, et al. Gene expression profiles of mouse submandibular gland development: FGFR1 regulates branching morphogenesis in vitro through BMP- and FGF-dependent mechanisms. Development. 2002;129:5767–78.
74. Ohuchi H, Hori Y, Yamasaki M, et al. FGF10 acts as a major ligand for FGF receptor 2 IIIb in mouse multi-organ development. Biochem Biophys Res Commun. 2000;277:643–9. doi:10.1006/bbrc.2000.3721.
75. De Moerlooze L, Spencer-Dene B, Revest JM, et al. An important role for the IIIb isoform of fibroblast growth factor receptor 2 (FGFR2) in mesenchymal-epithelial signalling during mouse organogenesis. Development. 2000;127:483–92.
76. Steinberg Z, Myers C, Heim VM, et al. FGFR2b signaling regulates ex vivo submandibular gland epithelial cell proliferation and branching morphogenesis. Development. 2005;132:1223–34. doi:10.1242/dev.01690.
77. Roignot J, Peng X, Mostov K. Polarity in mammalian epithelial morphogenesis. Cold Spring Harb Perspect Biol. 2013;5(2) doi:10.1101/cshperspect.a013789.
78. Baker OJ. Tight junctions in salivary epithelium. J Biomed Biotechnol. 2010;2010:278948. doi:10.1155/2010/278948.
79. Trimble WS, Grinstein S. Barriers to the free diffusion of proteins and lipids in the plasma membrane. J Cell Biol. 2015;208:259–71. doi:10.1083/jcb.201410071.
80. Hartsock A, Nelson WJ. Adherens and tight junctions: structure, function and connections to the actin cytoskeleton. Biochim Biophys Acta Biomembr. 2008;1778:660–9. doi:10.1016/j.bbamem.2007.07.012.
81. Harris TJ, Tepass U. Adherens junctions: from molecules to morphogenesis. Nat Rev Mol Cell Biol. 2010; 11:502–14. doi: nrm2927 [pii]\r10.1038/nrm2927.
82. Nekrasova O, Green KJ. Desmosome assembly and dynamics. Trends Cell Biol. 2013;23:537–46. doi:10.1016/j.tcb.2013.06.004.
83. Garrod D, Chidgey M. Desmosome structure, composition and function. Biochim Biophys Acta Biomembr. 2008;1778:572–87. doi:http://dx.doi.org/10.1016/j.bbamem.2007.07.014.
84. Delva E, Tucker DK, Kowalczyk AP. The desmosome. Cold Spring Harb Perspect Biol. 2009;1(2):a002543. doi:10.1101/cshperspect.a002543.
85. Yamada A, Futagi M, Fukumoto E, et al. Connexin 43 is necessary for salivary gland branching morphogenesis and FGF10-induced ERK1/2 phosphorylation. J Biol Chem. 2016;291:904–12. doi:10.1074/jbc.M115.674663.
86. Wozniak MA, Modzelewska K, Kwong L, Keely PJ. Focal adhesion regulation of cell behavior. Biochim Biophys Acta. 2004;1692:103–19. doi:10.1016/j.bbamcr.2004.04.007.
87. Green K, Jones J. Desmosomes and hemidesmosomes: structure and function of molecular components. FASEB J. 1996;10:871–81.
88. Litjens SHM, Sonnenberg A. Hemidesmosomes. In: Encyclopedic reference genomics and proteomics molecular medicine. Berlin/Heidelberg: Springer Berlin Heidelberg; 2006. p. 754–8.
89. Yu W, Datta A, Leroy P, et al. Beta1-integrin orients epithelial polarity via Rac1 and laminin. Mol Biol Cell. 2005;16:433–45. doi:10.1091/mbc.E04-05-0435.
90. Liu KD, Datta A, Yu W, et al. Rac1 is required for reorientation of polarity and lumen formation through a PI 3-kinase-dependent pathway. Am J Physiol Renal Physiol. 2007;293:F1633–40. doi:10.1152/ajprenal.00053.2007.
91. Assémat E, Bazellières E, Pallesi-Pocachard E, et al. Polarity complex proteins. Biochim Biophys Acta Biomembr. 2008;1778:614–30. doi:10.1016/j.bbamem.2007.08.029.
92. Bulgakova NA, Knust E. The Crumbs complex: from epithelial-cell polarity to retinal degeneration. J Cell Sci. 2009;122:2587–96. doi:10.1242/jcs.023648.
93. Goldstein B, Macara IG. The PAR proteins: fundamental players in animal cell polarization. Dev Cell. 2007;13:609–22. doi:10.1016/j.devcel.2007.10.007.
94. Chen J, Zhang M. The Par3/Par6/aPKC complex and epithelial cell polarity. Exp Cell Res. 2013;319:1357–64. doi:http://dx.doi.org/10.1016/j.yexcr.2013.03.021.
95. Iden S, Collard JG. Crosstalk between small GTPases and polarity proteins in cell polarization. Nat Rev Mol Cell Biol. 2008;9:846–59. doi:10.1038/nrm2521.
96. Joberty G, Petersen C, Gao L, Macara IG. The cell-polarity protein Par6 links Par3 and atypical protein kinase C to Cdc42. Nat Cell Biol. 2000;2:531–9. doi:10.1038/35019573.
97. Su W-H, Mruk DD, Wong EWP, et al. Polarity protein complex Scribble/Lgl/Dlg and epithelial cell barriers. Adv Exp Med Biol. 2012;763:149–70.
98. Rebustini IT, Hayashi T, Reynolds AD, et al. miR-200c regulates FGFR-dependent epithelial proliferation via Vldlr during submandibular gland branching morphogenesis. Development. 2012;139:191–202. doi:10.1242/dev.070151.
99. Lombaert IMA, Brunsting JF, Wierenga PK, et al. Rescue of salivary gland function after stem cell transplantation in irradiated glands. PLoS One. 2008;3:e2063. doi:10.1371/journal.pone.0002063.
100. Feng J, van der Zwaag M, Stokman MA, et al. Isolation and characterization of human salivary gland cells for stem cell transplantation to reduce radiation-induced hyposalivation. Radiother Oncol. 2009;92:466–71.
101. Nanduri LSY, Maimets M, Pringle SA, et al. Regeneration of irradiated salivary glands with stem cell marker expressing cells. Radiother Oncol. 2011;99:367–72. doi:10.1016/j.radonc.2011.05.085.

102. Nanduri LSY, Lombaert IM, der Zwaag M v, et al. Salisphere derived c-Kit(+) cell transplantation restores tissue homeostasis in irradiated salivary gland. Radiother Oncol. 2013;108:458–63. doi:10.1016/j.radonc.2013.05.020.
103. Sato A, Okumura K, Matsumoto S, et al. Isolation, tissue localization, and cellular characterization of progenitors derived from adult human salivary glands. Cloning Stem Cells. 2007;9:191–205. doi:10.1089/clo.2006.0054.
104. Rotter N, Oder J, Schlenke P, et al. Isolation and characterization of adult stem cells from human salivary glands. Stem Cells Dev. 2008;17:509–18. doi:10.1089/scd.2007.0180.
105. Wang Y, Shnyra A, Africa C, et al. Activation of the extrinsic apoptotic pathway by TNF-alpha in human salivary gland (HSG) cells in vitro, suggests a role for the TNF receptor (TNF-R) and intercellular adhesion molecule-1 (ICAM-1) in Sjögren's syndrome-associated autoimmune sialadenitis. Arch Oral Biol. 2009;54:986–96. doi:10.1016/j.archoralbio.2009.07.011.
106. Pradhan S, Zhang C, Jia X, et al. Perlecan domain IV peptide stimulates salivary gland cell assembly in vitro. Tissue Eng Part A. 2009;15:3309–20. doi:10.1089/ten.TEA.2008.0669.
107. Pradhan S, Liu C, Zhang C, et al. Lumen formation in three-dimensional cultures of salivary acinar cells. Otolaryngol Head Neck Surg. 2010;142:191–5. doi:10.1016/j.otohns.2009.10.039.
108. Pradhan-Bhatt S, Harrington DA, Duncan RL, et al. Implantable three-dimensional salivary spheroid assemblies demonstrate fluid and protein secretory responses to neurotransmitters. Tissue Eng Part A. 2013;19:1610 20. doi:10.1089/tcn.tca.2012.0301.
109. Pradhan-Bhatt S, Harrington DA, Duncan RL, et al. A novel in vivo model for evaluating functional restoration of a tissue engineered salivary gland. Laryngoscope. 2014;124:456–61. doi:10.1002/lary.24297.
110. Jeong J, Baek H, Kim Y-J, et al. Human salivary gland stem cells ameliorate hyposalivation of radiation-damaged rat salivary glands. Exp Mol Med. 2013;45:e58. doi:10.1038/emm.2013.121.
111. Song JJ, Kim SS, Liu Z, et al. Enhanced in vivo function of bioartificial lungs in rats. Ann Thorac Surg. 2011;92:998–1006. doi:http://dx.doi.org/10.1016/j.athoracsur.2011.05.018.
112. Gilpin SE, Guyette JP, Gonzalez G, et al. Perfusion decellularization of human and porcine lungs: bringing the matrix to clinical scale. J Heart Lung Transplant. 2014;33:298–308. doi:http://dx.doi.org/10.1016/j.healun.2013.10.030.
113. Gao Z, Wu T, Xu J, et al. Generation of bioartificial salivary gland using whole-organ decellularized bioscaffold. Cells Tissues Organs. 2014;200:171–80.
114. Maria OM, Maria O, Liu Y, et al. Matrigel improves functional properties of human submandibular salivary gland cell line. Int J Biochem Cell Biol. 2011;43:622–31. doi:10.1016/j.biocel.2011.01.001.
115. McCall AD, Nelson JW, Leigh NJ, et al. Growth factors polymerized within fibrin hydrogel promote amylase production in parotid cells. Tissue Eng Part A. 2013;19(19–20):2215–25. doi:10.1089/ten.tea.2012.0674.
116. Yamada Y, Hozumi K, Aso A, et al. Laminin active peptide/agarose matrices as multifunctional biomaterials for tissue engineering. Biomaterials. 2012;33:4118–25. doi:10.1016/j.biomaterials.2012.02.044.
117. Miller JS, Shen CJ, Legant WR, et al. Bioactive hydrogels made from step-growth derived PEG-peptide macromers. Biomaterials. 2010;31:3736–43. doi:10.1016/j.biomaterials.2010.01.058.Bioactive.
118. American Transplant Foundation. About transplant: facts and myths. 2016. http://www.americantransplantfoundation.org/about-transplant/facts-and-myths/.
119. Gilbert TW, Sellaro TL, Badylak SF. Decellularization of tissues and organs. Biomaterials. 2006;27:3675–83. doi:10.1016/j.biomaterials.2006.02.014.
120. Crapo PM, Gilbert TW, Badylak SF. An overview of tissue and whole organ decellularization processes. Biomaterials. 2011;32:3233–43. doi:10.1016/j.biomaterials.2011.01.057.
121. Cortiella J, Niles J, Cantu A, et al. Influence of acellular natural lung matrix on murine embryonic stem cell differentiation and tissue formation. Tissue Eng Part A. 2010;16:2565–80. doi:10.1089/ten.tea.2009.0730.
122. Hung S-H, Su C-H, Lee F-P, Tseng H. Larynx decellularization: combining freeze-drying and sonication as an effective method. J Voice. 2013;27:289–94. doi:http://dx.doi.org/10.1016/j.jvoice.2013.01.018.
123. Burk J, Erbe I, Berner D, et al. Freeze-thaw cycles enhance decellularization of large tendons. Tissue Eng Part C Methods. 2014;20:276–84. doi:10.1089/ten.TEC.2012.0760.
124. Gardin C, Ricci S, Ferroni L, et al. Decellularization and delipidation protocols of bovine bone and pericardium for bone grafting and guided bone regeneration procedures. PLoS One. 2015;10(7):e0132344. doi:10.1371/journal.pone.0132344.
125. Xu H, Xu B, Yang Q, et al. Comparison of decellularization protocols for preparing a decellularized porcine annulus fibrosus scaffold. PLoS One. 2014;9(1):e86723. doi:10.1371/journal.pone.0086723.
126. Porzionato A, Sfriso MM, Macchi V, et al. Decellularized omentum as novel biologic scaffold for reconstructive surgery and regenerative medicine. Eur J Histochem. 2013;57:24–30. doi:10.4081/ejh.2013.e4.
127. Azhim A, Syazwani N, Morimoto Y, et al. The use of sonication treatment to decellularize aortic tissues for preparation of bioscaffolds. J Biomater Appl. 2014;29:130–41. doi:10.1177/0885328213517579.
128. Oliveira AC, Garzón I, Ionescu AM, et al. Evaluation of small intestine grafts decellularization methods for corneal tissue engineering. PLoS

One. 2013;8(6):e66538. doi:10.1371/journal.pone.0066538.
129. Sarig U, Au-Yeung GCT, Wang Y, et al. Thick acellular heart extracellular matrix with inherent vasculature: a potential platform for myocardial tissue regeneration. Tissue Eng Part A. 2012;18:2125–37. doi:10.1089/ten.TEA.2011.0586.
130. Barnes CA, Brison J, Michel R, et al. The surface molecular functionality of decellularized extracellular matrices. Biomaterials. 2011;32:137–43. doi:10.1016/j.biomaterials.2010.09.007.
131. Nara S, Chameettachal S, Midha S, et al. Preservation of biomacromolecular composition and ultrastructure of a decellularized cornea using a perfusion bioreactor. RSC Adv. 2016;6:2225–40. doi:10.1039/C5RA20745B.
132. Baiguera S, Del Gaudio C, Kuevda E, et al. Dynamic decellularization and cross-linking of rat tracheal matrix. Biomaterials. 2014;35:6344–50. doi:http://dx.doi.org/10.1016/j.biomaterials.2014.04.070.
133. Struecker B, Hillebrandt KH, Voitl R, et al. Porcine liver decellularization under oscillating pressure conditions: a technical refinement to improve the homogeneity of the decellularization process. Tissue Eng Part C Methods. 2015;21:303–13. doi:10.1089/ten.tec.2014.0321.
134. Faulk DM, Wildemann JD, Badylak SF. Decellularization and cell seeding of whole liver biologic scaffolds composed of extracellular matrix. J Clin Exp Hepatol. 2015;5:69–80. doi:http://dx.doi.org/10.1016/j.jceh.2014.03.043.
135. Momtahan N, Poornejad N, Struk JA, et al. Automation of pressure control improves whole porcine heart decellularization. Tissue Eng Part C Methods. 2015;21(11):1148–61. 00:150714090832003 doi:10.1089/ten.tec.2014.0709.
136. Sullivan DC, Mirmalek-Sani S-H, Deegan DB, et al. Decellularization methods of porcine kidneys for whole organ engineering using a high-throughput system. Biomaterials. 2012;33:7756–64. doi:10.1016/j.biomaterials.2012.07.023.
137. Price AP, England KA, Matson AM, et al. Development of a decellularized lung bioreactor system for bioengineering the lung: the matrix reloaded. Tissue Eng Part A. 2010;16:2581–91. doi:10.1089/ten.tea.2009.0659.
138. Daly AB, Wallis JM, Borg ZD, et al. Initial binding and recellularization of decellularized mouse lung scaffolds with bone marrow-derived mesenchymal stromal cells. Tissue Eng Part A. 2011;18:1–16. doi:10.1089/ten.tea.2011.0301.
139. Wallis JM, Borg ZD, Daly AB, et al. Comparative assessment of detergent-based protocols for mouse lung de-cellularization and re-cellularization. Tissue Eng Part C Methods. 2012;18:420–32. doi:10.1089/ten.tec.2011.0567.
140. Weymann A, Patil NP, Sabashnikov A, et al. Perfusion-decellularization of porcine lung and trachea for respiratory bioengineering. Artif Organs. 2015;39:1024–32. doi:10.1111/aor.12481.
141. Ott HC, Clippinger B, Conrad C, et al. Regeneration and orthotopic transplantation of a bioartificial lung. Nat Med. 2010;16:927–33.
142. Hudson TW, Liu SY, Schmidt CE. Engineering an improved acellular nerve graft via optimized chemical processing. Tissue Eng. 2004;10:1346–58. doi:10.1089/ten.2004.10.1346.
143. Petersen TH, Calle EA, Zhao L, et al. Tissue-engineered lungs for in vivo implantation. Science. 2010;329:538–41. doi:10.1126/science.1189345.
144. Elder BD, Eleswarapu SV, Athanasiou KA. Extraction techniques for the decellularization of tissue engineered articular cartilage constructs. Biomaterials. 2009;30:3749–56. doi:10.1016/j.biomaterials.2009.03.050.
145. Meyer SR, Nagendran J, Desai LS, et al. Decellularization reduces the immune response to aortic valve allografts in the rat. J Thorac Cardiovasc Surg. 2005;130:469–76. doi:10.1016/j.jtcvs.2005.03.021.
146. Choi YC, Choi JS, Kim BS, et al. Decellularized extracellular matrix derived from porcine adipose tissue as a xenogeneic biomaterial for tissue engineering. Tissue Eng Part C Methods. 2012;18:866–76. doi:10.1089/ten.tec.2012.0009.
147. Lumpkins SB, Pierre N, McFetridge PS. A mechanical evaluation of three decellularization methods in the design of a xenogeneic scaffold for tissue engineering the temporomandibular joint disc. Acta Biomater. 2008;4:808–16. doi:http://dx.doi.org/10.1016/j.actbio.2008.01.016.
148. Gratzer PF, Harrison RD, Woods T. Matrix alteration and not residual sodium dodecyl sulfate cytotoxicity affects the cellular repopulation of a decellularized matrix. Tissue Eng. 2006;12:2975–83. doi:10.1089/ten.2006.12.ft-234.
149. Sawkins MJ, Bowen W, Dhadda P, et al. Hydrogels derived from demineralized and decellularized bone extracellular matrix. Acta Biomater. 2013;9:7865–73. doi:10.1016/j.actbio.2013.04.029.
150. Saghizadeh M, Winkler MA, Kramerov AA, et al. A simple alkaline method for decellularizing human amniotic membrane for cell culture. PLoS One. 2013;8(11):e79632. doi:10.1371/journal.pone.0079632.
151. Ott HC, Matthiesen TS, Goh S-K, et al. Perfusion-decellularized matrix: using nature's platform to engineer a bioartificial heart. Nat Med. 2008;14:213–21.
152. Robertson MJ, Dries-Devlin JL, Kren SM, et al. Optimizing recellularization of whole decellularized heart extracellular matrix. PLoS One. 2014;9:e90406. doi:10.1371/journal.pone.0090406.
153. Price AP, Godin LM, Domek A, et al. Automated decellularization of intact, human-sized lungs for tissue engineering. Tissue Eng Part C Methods. 2015;21:94–103. doi:10.1089/ten.tec.2013.0756.
154. Moroni F, Mirabella T. Decellularized matrices for cardiovascular tissue engineering. Am J Stem Cells. 2014;3:1–20.

155. Traphagen SB, Fourligas N, Xylas JF, et al. Characterization of natural, decellularized and reseeded porcine tooth bud matrices. Biomaterials. 2012;33:5287–96. doi:10.1016/j.biomaterials.2012.04.010.
156. Macchiarini P, Jungebluth P, Go T, et al. Clinical transplantation of a tissue-engineered airway. Lancet. 2008;372:2023–30. doi:10.1016/S0140-6736(08)61598-6.
157. Chuang P-T, McMahon AP. Branching morphogenesis of the lung: new molecular insights into an old problem. Trends Cell Biol. 2016;13:86–91. doi:10.1016/S0962-8924(02)00031-4.
158. Andrew DJ, Ewald AJ. Morphogenesis of epithelial tubes: insights into tube formation, elongation, and elaboration. Dev Biol. 2010;341:34–55. doi:10.1016/j.ydbio.2009.09.024.
159. Swarr DT, Morrisey EE. Lung endoderm morphogenesis: gasping for form and function. Annu Rev Cell Dev Biol. 2015;31:553–73. doi:10.1146/annurev-cellbio-100814-125249.
160. Quissell DO, Barzen KA, Redman RS, et al. Development and characterization of SV40 immortalized rat parotid acinar cell lines. Vitr Cell Dev Biol. 1998;34:58–67. doi:10.1007/s11626-998-0054-5.
161. Baker OJ, Schulz DJ, Camden JM, et al. Rat parotid gland cell differentiation in three-dimensional culture. Tissue Eng Part C Methods. 2010;16:1135–44. doi:10.1089/ten.tec.2009.0438.
162. Shirasuna K, Sato M, Miyazaki T. A neoplastic epithelial duct cell line established from an irradiated human salivary gland. Cancer. 1981;48:745–52. doi:10.1002/1097-0142(19810801)48:3<745::AID-CNCR2820480314>3.0.CO;2-7.
163. Royce LS, Kibbey MC, Mertz P, et al. Human neoplastic submandibular intercalated duct cells express an acinar phenotype when cultured on a basement membrane matrix. Differentiation. 1993;52:247–55. doi:10.1111/j.1432-0436.1993.tb00637.x.
164. Hoffman MP, Kibbey MC, Letterio JJ, Kleinman HK. Role of laminin-1 and TGF-beta 3 in acinar differentiation of a human submandibular gland cell line (HSG). J Cell Sci. 1996;109(Pt 8):2013–21.
165. Jung DW, Hecht D, Ho SW, et al. PKC and ERK1/2 regulate amylase promoter activity during differentiation of a salivary gland cell line. J Cell Physiol. 2000;185:215–25. doi:10.1002/1097-4652(200011)185:2<215::AID-JCP6>3.0.CO;2-L.
166. Szlávik V, Vág J, Markó K, et al. Matrigel-induced acinar differentiation is followed by apoptosis in HSG cells. J Cell Biochem. 2008;103:284–95. doi:10.1002/jcb.21404.
167. Capes-Davis A, Theodosopoulos G, Atkin I, et al. Check your cultures! A list of cross-contaminated or misidentified cell lines. Int J Cancer. 2010;127:1–8. doi:10.1002/ijc.25242.
168. Maria OM, Zeitouni A, Gologan O, Tran SD. Matrigel improves functional properties of primary human salivary gland cells. Tissue Eng Part A. 2011;17:1229–38.
169. Szlávik V, Szabó B, Vicsek T, et al. Differentiation of primary human submandibular gland cells cultured on basement membrane extract. Tissue Eng Part A. 2008;14:1915–26. doi:10.1089/ten.tea.2007.0208.
170. Pradhan S, Farach-Carson MC. Mining the extracellular matrix for tissue engineering applications. Regen Med. 2010;5:961–70. doi:10.2217/rme.10.61.
171. Mouw JK, Ou G, Weaver VM. Extracellular matrix assembly: a multiscale deconstruction. Nat Rev Mol Cell Biol. 2014;15:771–85.
172. Giancotti FG, Ruoslahti E. Integrin signaling. Science (80-). 1999;285:1028–33.
173. Durban EM. Mouse submandibular salivary epithelial cell growth and differentiation in long-term culture: influence of the extracellular matrix. In Vitro Cell Dev Biol. 1990;26:33–43.
174. Furue M, Okamoto T, Hayashi H, et al. Effects of hepatocyte growth factor (HGF) and activin A on the morphogenesis of rat submandibular gland-derived epithelial cells in serum-free collagen gel culture. In Vitro Cell Dev Biol Anim. 1999;35:131–5.
175. Joraku A, Sullivan CA, Yoo J, Atala A. In-vitro reconstitution of three-dimensional human salivary gland tissue structures. Differentiation. 2007;75:318–24. doi:10.1111/j.1432-0436.2006.00138.x.
176. Burford-Mason AP, Dardick I, Mackay A. Collagen gel cultures of normal salivary gland: conditions for continued proliferation and maintenance of major cell phenotypes in vitro. Laryngoscope. 1994;104:335–40. doi:10.1288/00005537-199403000-00016.
177. Weisel JW. Fibrinogen and fibrin. Adv Protein Chem. 2005;70:247–99. doi:10.1016/S0065-3233(05)70008-5.
178. Cantara SI, Soscia DA, Sequeira SJ, et al. Selective functionalization of nanofiber scaffolds to regulate salivary gland epithelial cell proliferation and polarity. Biomaterials. 2012;33:8372–82. doi:10.1016/j.biomaterials.2012.08.021.
179. Yin N, Santos TMA, Auer GK, et al. Bacterial cellulose as a substrate for microbial cell culture. Appl Environ Microbiol. 2014;80(6):1926–32. doi:10.1128/AEM.03452-13.
180. Modulevsky DJ, Lefebvre C, Haase K, et al. Apple derived cellulose scaffolds for 3D mammalian cell culture. PLoS One. 2014;9:e97835. doi:10.1371/journal.pone.0097835.
181. Bhattacharya M, Malinen MM, Lauren P, et al. Nanofibrillar cellulose hydrogel promotes three-dimensional liver cell culture. J Control Release. 2012;164:291–8. doi:http://dx.doi.org/10.1016/j.jconrel.2012.06.039
182. Hardingham T. Chemistry and biology of hyaluronan. Chem Biol Hyaluronan. 2004. doi:10.1016/B978-008044382-9/50032-7.
183. Xu X, Jha AK, Harrington DA, et al. Hyaluronic acid-based hydrogels: from a natural polysaccharide to complex networks. Soft Matter. 2012;8:3280–94. doi:10.1039/C2SM06463D.

184. Turley EA, Noble PW, Bourguignon LYW. Signaling properties of hyaluronan receptors. J Biol Chem. 2002;277:4589–92. doi:10.1074/jbc.R100038200.
185. Peach RJ, Hollenbaugh D, Stamenkovic I, Aruffo A. Identification of hyaluronic acid binding sites in the extracellular domain of CD44. J Cell Biol. 1993;122:257–64.
186. Burdick JA, Prestwich GD. Hyaluronic acid hydrogels for biomedical applications. Adv Mater. 2011;23:41–56. doi:10.1002/adma.201003963.
187. Lee JY, Spicer AP. Hyaluronan: a multifunctional, megaDalton, stealth molecule. Curr Opin Cell Biol. 2000;12:581–6.
188. Shubin AD, Felong TJ, Graunke D, et al. Development of poly(ethylene glycol) hydrogels for salivary gland tissue engineering applications. Tissue Eng Part A. 2015;21:1733–51. doi:10.1089/ten.tea.2014.0674.
189. Hosokawa Y, Takahashi Y, Kadoya Y, et al. Significant role of laminin-1 in branching morphogenesis of mouse salivary epithelium cultured in basement membrane matrix. Dev Growth Differ. 1999;41:207–16. doi:10.1046/j.1440-169x.1999.00419.x.
190. Hoffman MP, Nomizu M, Roque E, et al. Laminin-1 and laminin-2 G-domain synthetic peptides bind syndecan-1 and are involved in acinar formation of a human submandibular gland cell line. J Biol Chem. 1998;273:28633–41. doi:10.1074/jbc.273.44.28633.
191. Farach-Carson MC, Brown AJ, Lynam M, et al. A novel peptide sequence in perlecan domain IV supports cell adhesion, spreading and FAK activation. Matrix Biol. 2008;27:150–60. doi:10.1016/j.matbio.2007.09.007.
192. Nakanishi Y, Sugiura F, Kishi JI, Hayakawa T. Collagenase inhibitor stimulates cleft formation during early morphogenesis of mouse salivary gland. Dev Biol. 1986;113:201–6. doi:10.1016/0012-1606(86)90122-3.
193. Fong ELS, Wan X, Yang J, et al. A 3D in vitro model of patient-derived prostate cancer xenograft for controlled interrogation of in vivo tumor-stromal interactions. Biomaterials. 2016;77:164–72. doi:10.1016/j.biomaterials.2015.10.059.

3D Printing Technology in Craniofacial Surgery and Salivary Gland Regeneration

9

Jong Woo Choi, Namkug Kim, and Chang Mo Hwang

Abstract

Patient-specific three-dimensional (3D)-printed phantoms and surgical guides are being utilized more often nowadays to assist diagnosis and treatment planning for surgery, which are tailored to individual's unique needs. 3D printing surgical guides made of temporary materials can be fabricated to fit the surface of the hard or soft tissue organs by 3D modeling of the surgical interface. To date, the value of 3D printing for surgical planning as a guidance tool has been proven in various hard tissue surgical applications, such as craniofacial and maxillofacial surgery, spine surgery, cardiovascular surgery, neurosurgery, pelvic surgery, and visceral surgery. Craniofacial plastic surgery is one of the medical fields that pioneered the use of the 3D printing concept. Rapid prototyping technology was introduced to medicine in the 1990s via CAD-CAM (computer-aided design, computer-aided manufacturing). The medical models or bio-models based on the 3D printing technique represent 1:1 scale reproductions of the human anatomical region of interest that can be obtained via 3D medical imaging. The procedure for the fabrication of medical models comprises multiple steps: (1) acquisition of high-quality volumetric 3D image data of the anatomical structure to be modeled, (2) 3D image processing to extract the region of interest from the surrounding tissues, (3) mathematical surface modeling of the anatomic surfaces, (4) formatting of data for rapid prototyping, (5) model building, and (6) quality assurance of the model and its dimensional accuracy. Furthermore, tissue engineers also experience the advent of a new 3D printing era. The tissue engineering triad comprises cells, scaffolds, and growth factors. Recently, 3D technology has become sufficiently evolved to enable printing of living cells. Although

J.W. Choi, MD, PhD, MMM (✉) • N. Kim, PhD
C.M. Hwang, PhD
Department of Plastic and Reconstructive Surgery, Department of Bioengineering, Ulsan University, College of Medicine, Seoul, South Korea
e-mail: pschoi@amc.seoul.kr; namkugkim@gmail.com; hwang.changmo1@gmail.com

S. Cha (ed.), *Salivary Gland Development and Regeneration*, DOI 10.1007/978-3-319-43513-8_9

many challenging issues remain to be resolved for such complex structures, heart, kidney, and skin regenerations are being investigated using 3D bioprinting technology. A potential candidate for a clinical success resides in the regeneration of the major salivary glands, which consist of various cells encapsulated by a connective tissue membrane.

9.1 Introduction

Three-dimensional (3D) printing is a rapidly developing technology that is applied worldwide in various fields, such as business, fashion, mechanical engineering, and medicine [1, 2]. 3D printing technology has already been used in the mock formation of various products including cellular phones [2], and the extent of its applications has greatly expanded in medicine as this technology has evolved in recent years. A major advantage is that it can generate a unique product in a short period of time, which is suitable for individualized medicine where each patient requires a specific treatment, tailored to a therapeutic approach. As opposed to the products generated by 3D printing, most industrial products are mass produced, and every unit has the same dimensions. As one would expect, in the field of medicine, patients are all different in terms of any shapes and sizes that require surgical attention. The 3D printing technique supports the contemporary aim of implementing personalized medicine by providing a patient-specific product in a short period of time at reasonable prices [3].

Therefore, the clinical applications of the 3D printing technology are expanding more rapidly in recent years. The affordability and convenience of this technology have spurred its adoption in a variety of medical fields. This revolutionary technique may ultimately allow the printing of tissue and organ structures to replace damaged or missing body parts. Although outcomes and efficacy of 3D printing require more scientific research, it is clear that 3D printing technology is unique and has invaluable innovations in medicine. For example, pediatric cardiac surgeons use 3D printing-based tactile models for analyzing and visualizing complex congenital heart diseases. Urologic surgeons simulate surgery of complex renal cell carcinomas in advance of actual surgery using 3D printed tactile prototype models that include the vessels and parenchyma of the kidney. Neurosurgeons utilize similar approaches for neurosurgery of brain cancer. These types of efforts allow surgeons in various specialties to perform advanced analyses of the patient's specific status. In addition, tactile models with a real intraoperative 1:1 scale reference can be very useful for preoperative consultations with patients [3–5].

Craniofacial plastic surgery is one of the medical fields that pioneered the use of the 3D printing concept. Rapid prototype (RP) technology was introduced to medicine in the 1990s via computer-aided design and computer-aided manufacturing (CAD-CAM). The medical models or bio-models based on the 3D printing technique represent 1:1 scale reproductions of the human anatomical region of interest that can be obtained via 3D medical imaging [5]. The procedure for the fabrication of medical models comprises multiple steps: (1) acquisition of high-quality volumetric 3D image data of the anatomical structure to be modeled, (2) 3D image processing to extract the region of interest from the surrounding tissues, (3) mathematical surface modeling of the anatomic surfaces, (4) formatting of data for rapid prototyping, (5) model building, and (6) quality assurance of the model and its dimensional accuracy [3, 6].

For instance, because patients requiring craniofacial surgery tend to have very specific malformations or deformities, mostly in the bone, a 3D printing prototype model can greatly assist with preoperative evaluation and intraoperative procedures. Medical modeling in craniofacial surgery based on 3D printing has mainly been developed over the last 15 years. It can incorporate (1) aiding in the production of surgical implants, (2) improving surgical planning, (3) acting as an orientation aid during surgery, (4) enhancing diagnostic quality, (5) assisting preoperative simulation, (6) obtaining a patient's consent prior to surgery, and (7) preparing a template for resection for surgeons as well as providing an educational tool for medical students and residents [3, 5, 7].

Meanwhile, tissue engineers also experience the advent of a new 3D printing era. The tissue engineering triad comprises cells, scaffolds, and growth factors. Recently, 3D technology has become sufficiently evolved to enable printing of living cells. Although many obstacles need to be overcome, 3D bioprinting provides bioengineers with a new modality such as 3D cell culture on scaffolds that might be superior to conventional cell culture systems. Bioprinting is an emerging technology that is expected to eventually regenerate biological tissues and even solid organs. As a combination of techniques, a nonliving scaffold could be constructed using 3D technology, while bioprinting simultaneously adds a living tissue [1, 8–12]. More specifically, tissue-compatible scaffolds are generated with bioprinting, and living cells are incorporated into them, along with various growth factors, depending on the application. Heart, kidney, and skin regenerations are being investigated using 3D bioprinting technology, although many challenging issues remain to be resolved for such complex structures. The regeneration of the major salivary glands, which consist of various cells encapsulated by connective tissue membrane, certainly requires further investigation and attention for a clinical success of 3D bioprinting. In this book chapter, the current status of 3D printing technology and its clinical applications in craniofacial surgery are reviewed. A potential application of 3D bioprinting for salivary gland regeneration is discussed at the end of the chapter.

9.2 Review of Current 3D Printing in Craniofacial Surgery (Reproduced from Ref. [28])

3D printing technology can be categorized by the techniques, the materials, or the aimed deposition process. The classification based on the techniques includes stereolithography (SL), selective laser sintering (SLS), 3D printing (3D printer-based SLS: 3DP), fused deposition modeling (FDM), direct metal laser sintering (DMLS), laminated object manufacturing (LOM), and electron beam melting (EBM). The materials used for the 3D printing technology include thermoplastic, metal powder, ceramic powder, eutectic metals, alloy metal, photopolymer, paper, foil, plastic film, and titanium alloys. The 3D technology can be classified by the aimed deposition process. PolyJet modeling and 3D plotting technology are based on drop-on-drop deposition. 3D printing is based on drop-on-powder deposition. Fused deposition modeling is based on continuous deposition. The most frequently used representative methods are reviewed and summarized in Table 9.1 [3].

9.2.1 Liquid-Based 3D Printing Technology

9.2.1.1 Stereolithography (SL or SLA)

Stereolithography (SL) has been the most widely used 3D printing technique for craniofacial surgery since it was first applied for grafting a skull defect in 1994 [13]. The SL RP system consists of a bath of photosensitive resin, a model-building platform, and an ultraviolet (UV) laser for curing the resin. A mirror is used to guide the laser focus onto the surface of the resin; the resin becomes cured when exposed to the UV radiation. The mirror is computer controlled and is guided to cure the resin on a slice-by-slice basis. These slice data are fed into the RP machine that directs the exposure path of the UV laser onto the surface of the resin. The layers are cured sequentially and bind together to form a solid object, beginning from the bottom of the model and building upward. Each new layer of resin is wiped across the surface of the previous layer using a wiper blade before being exposed and cured. The model is then removed from the bath and cured for an additional period of time in a UV cabinet. [14].

Generally, SL is considered to provide the greatest accuracy and best surface finish of any

Table 9.1 A comparison of current 3D printing technologies

	3D printing technology	Materials	Aimed deposition process
Liquid base	SL (Stereolithography)	Photopolymer	
	Polyjet or Multijet Printing	ABS, Acryl	Drop-on-drop deposition
Powder base	SLS (Selective Laser Sintering)	Thermoplastics	
		Metal powder	
	3DP (3D printing)	Plastic powder	Drop-on-powder deposition
	DMLS (Direct Metal Laser Sintering)	Alloy metal	
		Ceramic powder	
	EBM (Electron beam melting)	Titanium alloys	
Solid base	FDM (Fused Deposition Modeling)	Thermoplastics	Continuous deposition
		Eutectic metals	
		ABS	
	LOM (Laminated object manufacturing)	Paper	
		Foil	
		Plastic film	

Reprinted from ref [18]. ABS, acrylonitrile butadiene styrene

RP technology. The model material is robust, slightly brittle, and relatively light [15]. SL accuracy is 1.2 mm (range, 0–4.8 mm) for skull base measures, 1.6 mm (range, 0–5.8 mm) for midface measures, 1.9 mm (range, 0–7.9 mm) for maxilla measures, and 1.5 mm (range, 0–5.7 mm) for orbital measures. The mean differences in defect dimensions are 1.9 mm (range, 0.1–5.7 mm) for unilateral maxillectomy, 0.8 mm (range, 0.2–1.5 mm) for bilateral maxillectomy, and 2.5 mm (range, 0.2–7.0 mm) for orbitomaxillectomy defects [16]. Midface SL models may be more prone to error than those of other craniofacial regions because of the presence of thin walls and small projections. Choi et al. [29] found that the absolute mean deviation between an original dry skull and an SL RP model over 16 linear measurements was 0.62 ± 0.5 mm (0.56 ± 0.39 %) [15, 17]. The accuracy of computed tomography (CT) and SL models was compared. The accuracy for SL models expressed as the arithmetic mean of the relative deviations ranged from 0.8 to 5.4 %, with an overall mean deviation of 2.2 %. The mean deviations of the investigated anatomical structures ranged from 0.8 to 3.2 mm. An overall mean deviation (comprising all structures) of 2.5 mm was found.

9.2.1.2 PolyJet Modeling

PolyJet modeling is performed by jetting state-of-the-art photopolymer materials in ultrathin layers (16 μm) onto a build tray layer by layer until the model is completed. Each photopolymer layer is cured by UV light immediately after it is jetted, producing fully cured models that can be handled and used immediately without post-curing. The gel-like support material used, which is specially designed to support complicated geometries, is easily removed by hand and water jetting [14]. At present, this technique is too time-consuming and expensive to be used in craniofacial surgery clinical applications. Ibrahim et al. reported a dimensional error of 2.14 % in reproducing a dry mandible when using this technique [18].

9.2.2 Powder-Based 3D Printing Technology

9.2.2.1 Selective Laser Sintering (SLS)

The selective laser sintering (SLS) technique uses a CO_2 laser beam to selectively fabricate models in consecutive layers. First, the laser beam scans over a thin layer of powder previously deposited on the build tray and leveled with a roller. The laser heats the powder particles, fusing them to form a solid layer, and then moves along the x- and y-axes to design the structures according to the CAD data. After the first layer fuses, the build tray moves downward, and a new layer of powder is deposited and sintered, and the process is repeated until the object is completed. The prototype surface is finished by sandblasting [14]. The SLS prototype is opaque, and its surface is abrasive and porous. Prototype fabrication time is 15 h. The accuracy of the SLS model is relatively high, with maximum standard errors of 0.1–0.6 mm. This accuracy depends on the thickness of the CT scans used, which should be as thin as possible (1–2 mm is a good compromise for a skull study). Because of the high cost of the materials, several parts are fabricated simultaneously. The long fabrication time for the SLS technique (16 h) is close to the time required for fabrication with the SL system [19].

9.2.2.2 3D Printer-Based SLS (3D Printing)

The 3D printing system uses a print head to selectively disperse a binder onto powder layers. This technology has a lower cost than similar techniques. First, a thin layer of powder is spread over a tray using a roller similar to that used in the SLS system. The print head scans the powder tray and delivers a continuous jet of a solution that binds the powder particles as it touches them. No support structures are required while the prototype is being fabricated because the surrounding powder supports the unconnected parts. When the process is complete, the surrounding powder is aspirated. In the finishing process, the prototype surfaces are infiltrated with a cyanoacrylate-based material to harden the structure [19]. The printing technique enables the formation of complex geometrical structures, such as hanging partitions inside the cavities, without artificial support structures [14].

After the CT scan, the rendering of the DICOM data and transformation into STL data files take a maximum of 30 min, and the printing and infiltration process takes approximately 4–6 h. Simpler models can be purchased for as little as $300–$400 [19]. The 3D printers used in this process are relatively inexpensive ($2500–$3000), have fast build times (4 h for a full skull), and are easy to maintain. Additionally, 3D printers are cost-effective, associated with low waste, and accurate (± 0.1 mm in the Z plane, ± 0.2 mm in the X and Y planes), and they can make hard, soft, or flexible models. These printers can also be used to identify different types of body tissue depending on the predefined threshold setting selected. Silva et al. reported a mean dimensional error of 2.67 % in prototypes produced using 3D printing technologies in comparison with a dry human skull [19].

9.2.3 Solid-Based 3D Printing Technology

9.2.3.1 Fused Deposition Modeling

Fused deposition modeling (FDM) uses a similar principle to SL in that it builds models on a layer-by-layer basis. The main difference is that the layers are deposited as a thermoplastic that is extruded from a fine nozzle. A commonly used material for this procedure is acrylonitrile butadiene styrene (ABS). The 3D model is constructed by extruding the heated thermoplastic material onto a foam surface along a path indicated by the model data. Once a layer has been deposited, the nozzle is raised between 0.278 and 0.356 mm, and the next layer is deposited on top of the previous layer. This process is repeated until the model is completed [14]. As with SL, support structures are required for FDM models because time is needed for the thermoplastic to harden and the layers to bond together [20].

9.3 Patient-Specific Modeling and Its Clinical Application Using 3D Printing Technology

9.3.1 Patient-Specific Modeling from Medical Images and Computer-Aided Design

As depicted in Fig. 9.1, after patient scanning with CT and/or MRI, the DICOM data can be transferred and processed into STL data files or other 3D file formats by using segmentation, surface extraction, and 3D model post-processing. Less than a 1-mm CT slice thickness and voxel with iso-

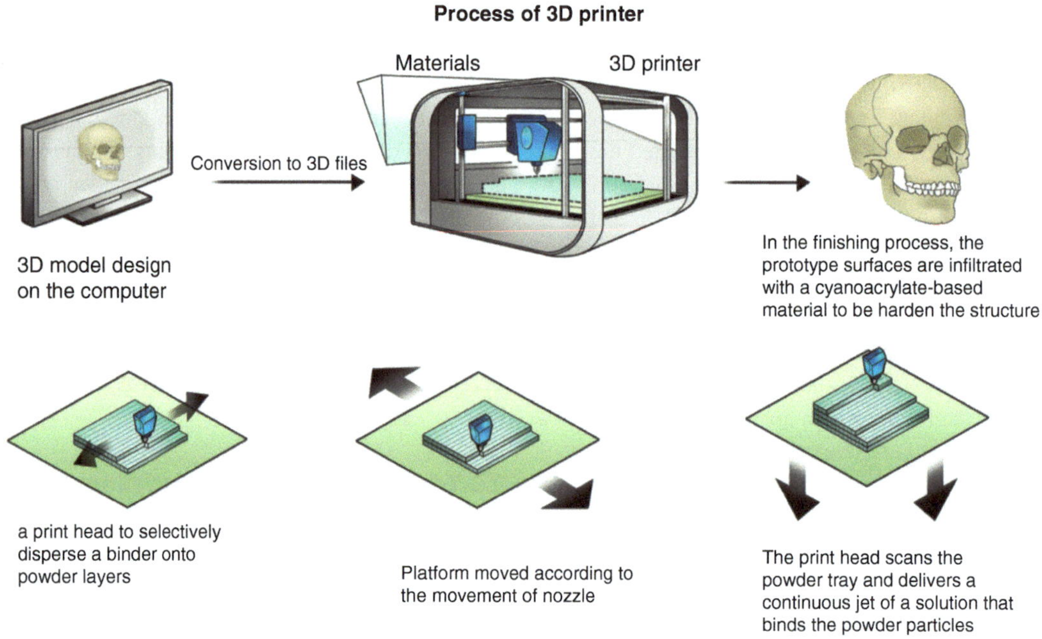

Fig. 9.1 The overall process of 3D printing in craniofacial surgery. After the patient is scanned via CT, DICOM files should be exported. Less than a 1-mm CT slice thickness is recommended. DICOM data are imported and converted to stereolithography (STL) files. Rendering of CT scan DICOM data into STL data files takes about 30 min. Converted 3D files are uploaded into the 3D printer. Rapid prototyping (RP) uses layer-by-layer stereolithographic accumulation. The RP model is then fabricated on plaster via jetting of a material that consists of plaster (<90 %), vinyl polymer (<20 %), and carbohydrate (<10 %). The printing and infiltration process takes about 4–6 h. Finally, unsintered sections are removed (Reprinted from Ref. [18])

cubic spacing are recommended. The time required mainly rests on the clinical application. In particular, segmentation is a critical procedure for improving the overall accuracy and needs considerable time. No satisfactory fully automated medical image segmentation algorithms have been established. Therefore, manual or semiautomated segmentation algorithms have generally been used, which have enhanced the importance of operator experience. After segmentation, a surface model should be produced by a marching cube [21, 22] or other 3D contour extraction algorithms [23]. For medical visualization, these kinds of shaded surface display techniques are well established. However, this 3D model by itself is not good enough for 3DP, due to, for example, too many mesh units and incomplete topological soundness. Therefore, topological correction [24], decimation [25], Laplacian smoothing [26], and local smoothing [27] are required to make a 3D model for 3DP. In addition, virtual simulation, including determination of the entry point and direction of the screw and surgical line, is accomplished for patient-specific surgical planning. Based on this planning, surgical guides are designed by computer-aided design (CAD) software. After the generation of a 3D model, the most suitable 3D printer for their applications is selected among various kinds of 3DP techniques. The 3D model file is uploaded into the 3D printer. The 3D printer uses layer-by-layer STL accumulation to fabricate the 3D physical model.

9.3.2 Applications for Personalized Treatment

9.3.2.1 Surgical Planning and Guidance Tools

Patient-specific 3D printed phantoms and surgical guides are being used more often to aid diagnosis and treatment planning for surgery, which allow individual customization. 3D printing surgical guides made of temporary materials can be fabricated to fit the surface of the hard or soft tissue organs by 3D modeling of the surgical interface. To date, the value of 3D printing for surgical planning as a guidance tool has been proven in various hard tissue surgical applications, such as craniofacial and maxillofacial surgery [28–33], spine surgery [34], cardiovascular surgery [35, 36], neurosurgery [37, 38], pelvic surgery [39, 40], and visceral surgery [41].

Recent advances in 3D printable materials have increased the level of realism of the 3D phantoms used for surgical planning. Improved diversity due to better transparency, color, and softness facilitates better understanding of complex 3D anatomical structures and guidance functions for soft tissues [42]. Yang et al. [43] used a full-colored and flexible 3D printed phantom as a preplanning simulator for extended septal myectomy. From the cardiac CT data, a myocardial 3D model was made by in-house software (A-view Cardiac; Asan Medical Center, Seoul, Republic of Korea). Using a 3D printer (Connex3 Objet500; Stratasys Corporation, Rehovot, Israel), the left ventricular (LV) myocardium, papillary muscle, and intraventricular muscle band (including the accessory papillary muscle) were fabricated with differently colored materials, whose flexibility could be controlled by adding a rubberlike and transparent material (Fig. 9.1). The 3D printed phantom provided invaluable information on the LV geometry. It is known that the softest 3D printable materials cannot be directly used as surgical simulators because they are still too hard for scalpel incision and suturing. Therefore, additional post-processing using gelatin or silicone molding techniques or a novel 3D printing system that can directly jet a variety of silicone materials needs to be developed.

9.3.2.2 Implantable Devices

3D printing techniques are also used in implant design to make patient-specific prosthetics, outside the standard range of ready-made commercial implants (Fig. 9.2). In addition, this approach has improved surgical performance by enabling the creation of patient-specific anatomy-based implants. For hard tissue structures, metal implants have, in particular, been successfully used in various applications [44, 45], which were mostly FDA cleared, such as mandible [33] and dental [46] restoration and hip [47], femoral [48], and hemi-knee joint reconstruction [44, 45]. In

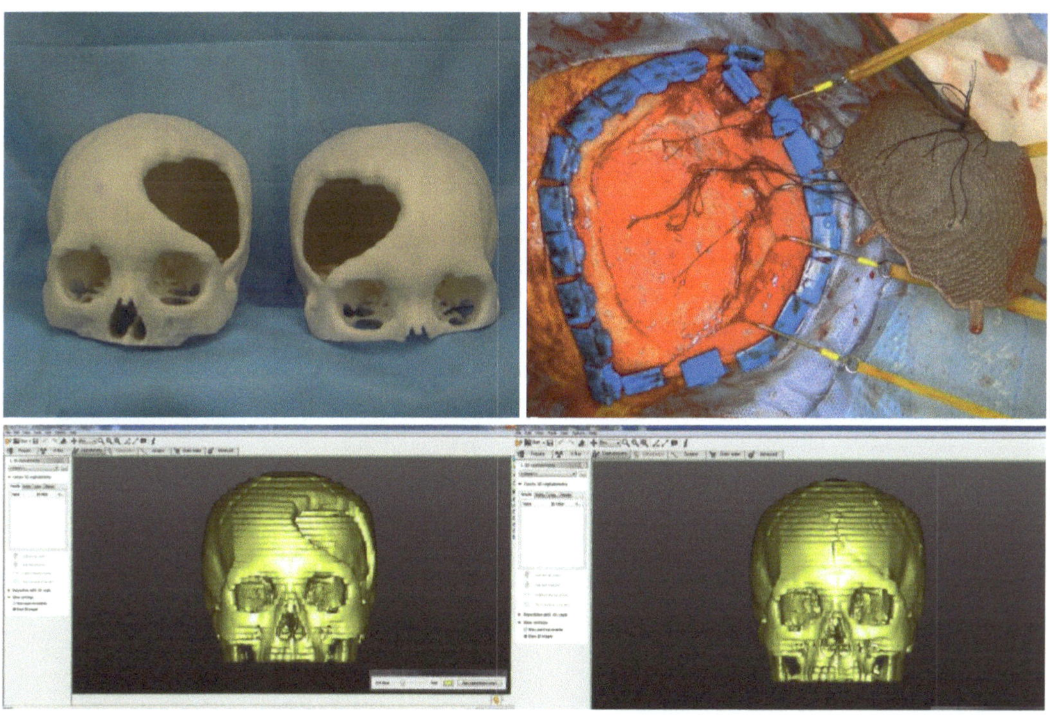

Fig. 9.2 Large cranial defect reconstructed with 3D printed titanium implant. *Top panels*: The contralateral normal cranium was mirrored and the 3D printed titanium implant was inserted for the correction of the calvarial bone defect (Modified from Ref. [18]). *Bottom panel*: Computer-simulated skull defect images before (*left*) and after (right) titanium implant was inserted

addition, the biocompatible ceramic hydroxyapatite [49] and the biodegradable polymer polycaprolactone [50] have been used in 3D printing-based applications to substitute hard tissues with customized implants.

Beyond the hard tissue applications, customized implants created using 3D printing have recently been used in the interventional field. Amerini et al. [51] revealed the feasibility of a personalized interventional treatment for tricuspid regurgitation using a braided stent in an animal study. From the cardiac CT data, the 3D reconstructed model of the right-sided cardiac cavities of a pig was obtained (OsiriX® Imaging Software; Pixmeo, Switzerland). A solid Alumide® mold was manufactured using a 3D printing system, and then a personalized compressible nitinol stent was subsequently produced and fitted onto the 3D printing mold (Fig. 9.3). This customized stent was almost completely fitted onto the right atrium,

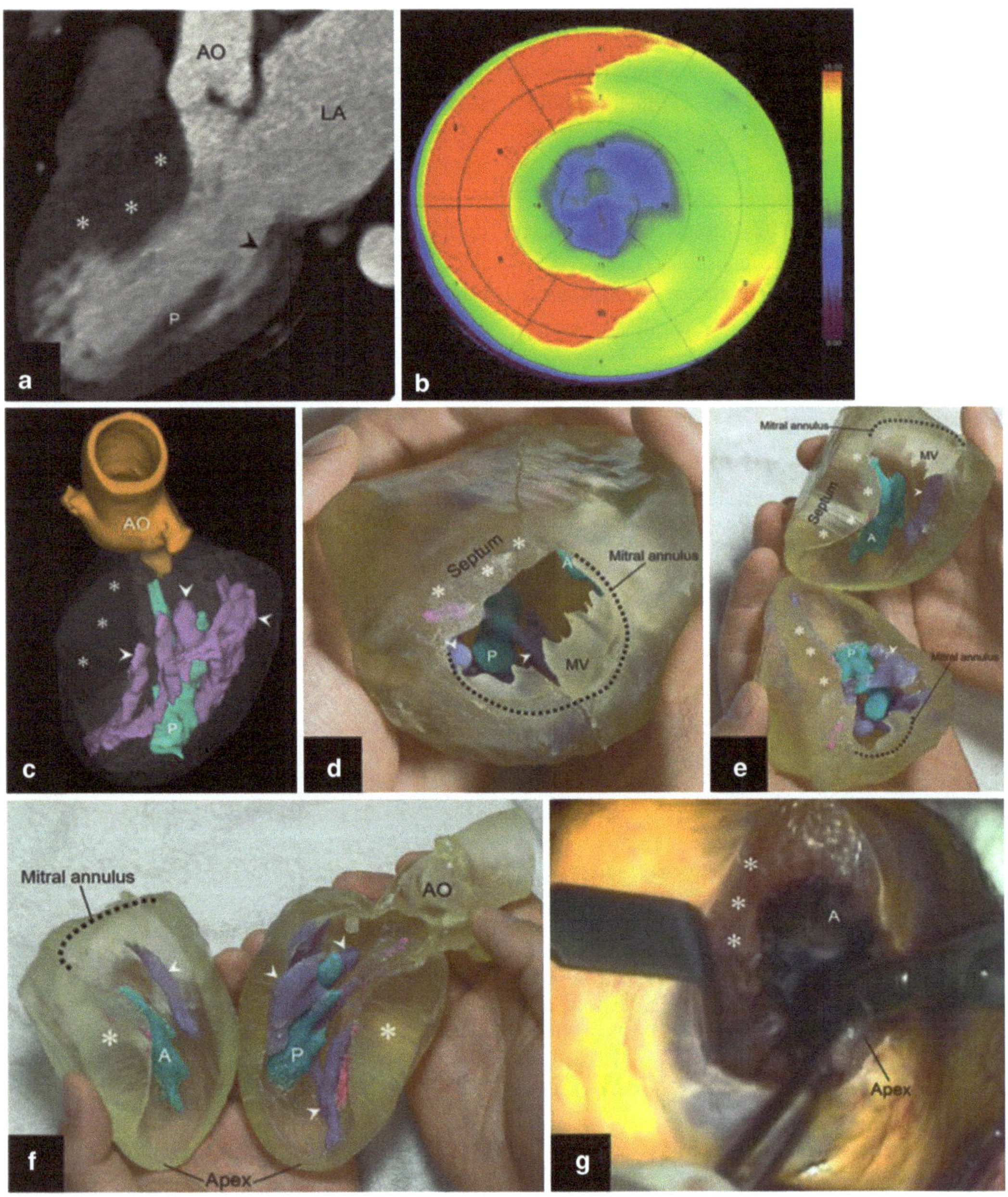

and an additional tubular stent component containing a tissue valve prosthesis was established. In the feasibility study performed in animals, they found that the 3D printing-based stent could stabilize the biological valve prostheses by force transmission from the annulus to the atrial wall and the adjacent vena cava.

In this book chapter, only clinical applications with previously developed 3D printing technologies were discussed. However, other approaches for personalized implants have been proposed, including bioprinting of tissues and organs [52–54] and the organ-on-a-chip technique [55, 56].

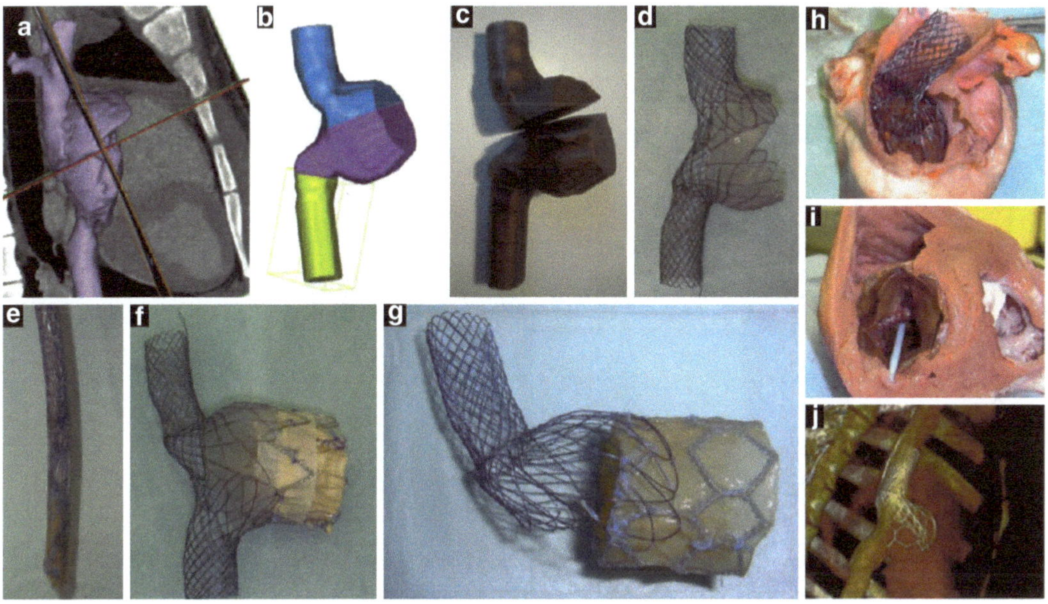

Fig. 9.4 In silico tridimensional reconstruction of the right-sided cardiac cavities of a female pig. (**a**) CT-based primary 3D reconstruction. (**b**) 3D reconstructed model of main structural parts. (**c**) 3D printed phantom mold of the main structural parts with alumide material. (**d**) A personalized stent with nitinol material. (**e**) An equipped state of the developed stent in an introducer. (**f**, **g**) Two different types of the prototype equipped with a self-expanding bioprosthetic valve. (**h**, **j**) Study results [51] showing implantation of the developed stent. Postmortem autopsy (h, i) and CT fluoroscopy (j) both revealed accurate positioning of the valve prostheses (Reproduced from Ref. [51])

Fig. 9.3 A cardiac three-chamber CT image and 3D printing of the heart. (**a**) CT imaging demonstrating a hypertrophied interventricular septum (*asterisks*), posterior papillary muscle (*P*), and intraventricular muscle band or accessory papillary muscles (*arrowhead*). (**b**) A bull's-eye map generated by using the end-diastolic phase of the CT imaging shows the extent of the hypertrophied myocardium (*red area*, >15 mm in thickness). (**c**) 3D reconstructed model. (**d–f**) 3D printed phantom of the myocardium showing the geometric relationship among the hypertrophied septum (*asterisks*), papillary muscle (*A* anterior, *P* posterior), and intraventricular muscle band (*asterisks*). (**g**) Intraoperative photography via the apical approach shows the limited visual field of the LV cavity. The base of the anterior papillary muscle is exposed after excision of the muscle band (not shown) near the anterior papillary muscle. *LV* left ventricle (Reprinted from Ref. [43])

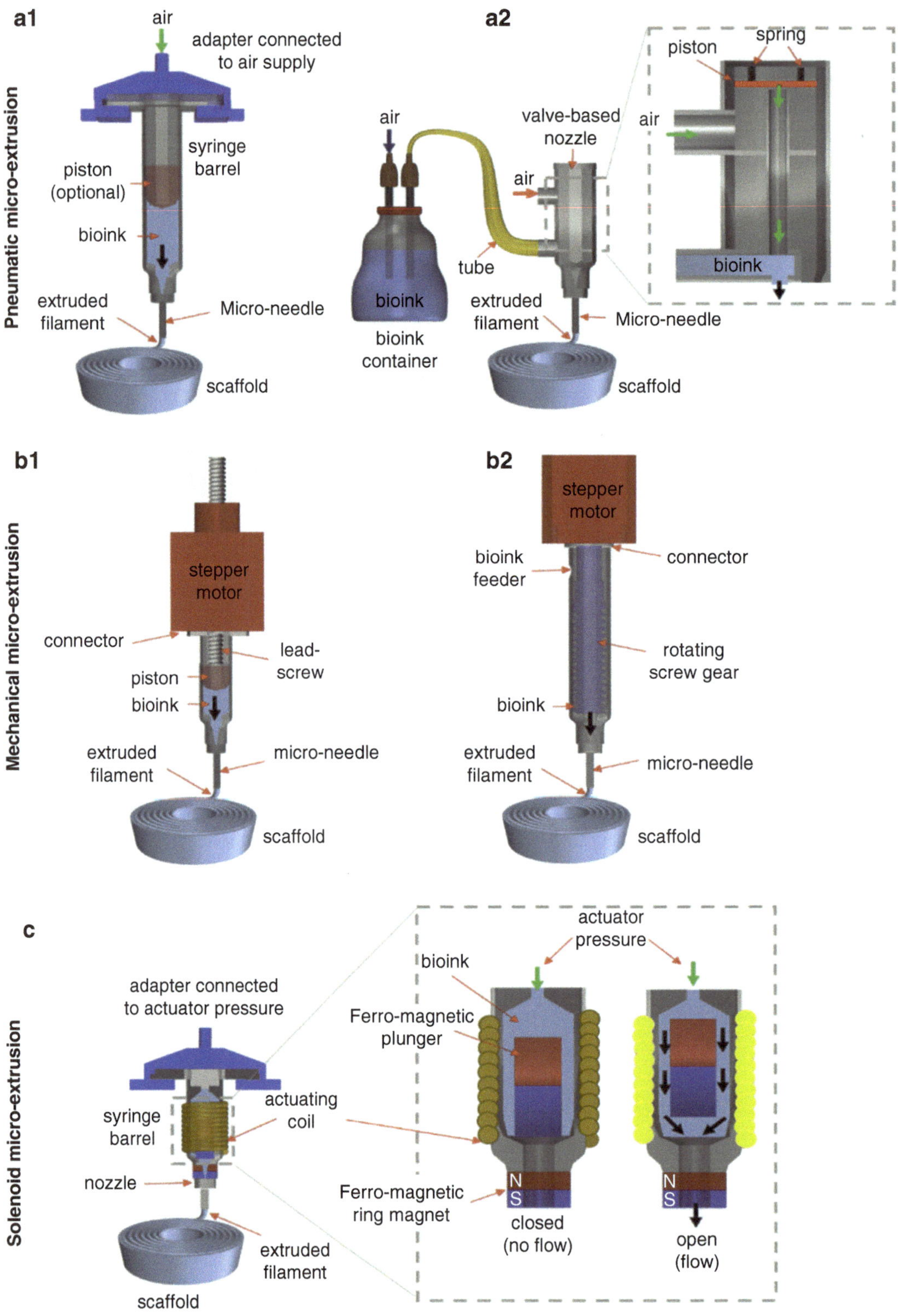

Fig. 9.5 Extrusion-based bioprinting systems: (**a**) pneumatic micro-extrusion including valve-free (*A1*) and valve based (*A2*) and (**b**) motor-driven micro-extrusion including piston (*B1*) and screw-driven (*B2*) and (**c**) solenoid micro-extrusion (Reprinted from Ref. [71])

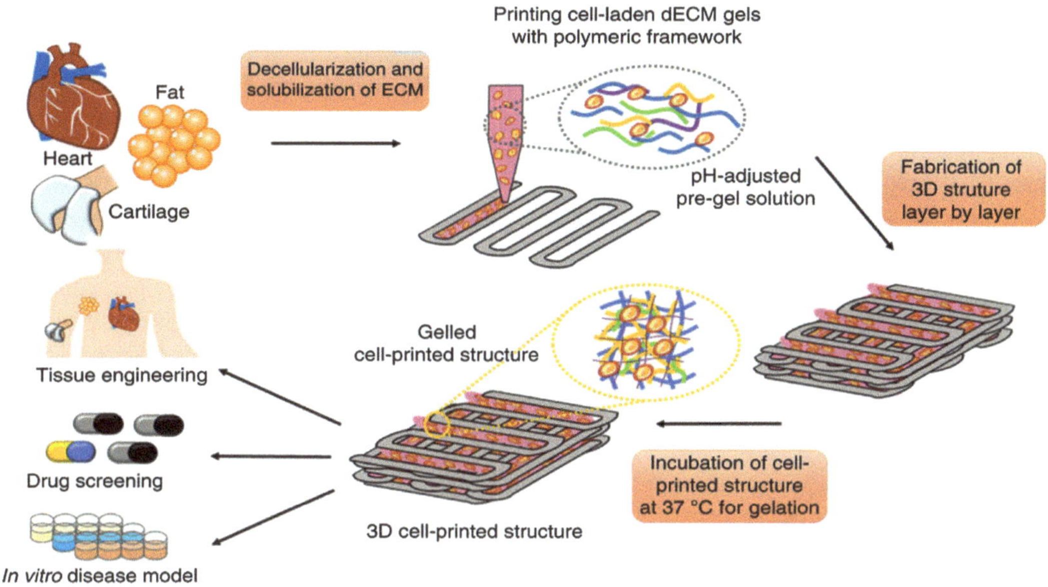

Fig. 9.6 Extrusion bioprinting with tissue originated bioinks. Biodegradable synthetic polymer scaffold is coextruded side by side to hydrogel bioink to maintain 3D architecture of printed objects (Reprinted from Ref. [70])

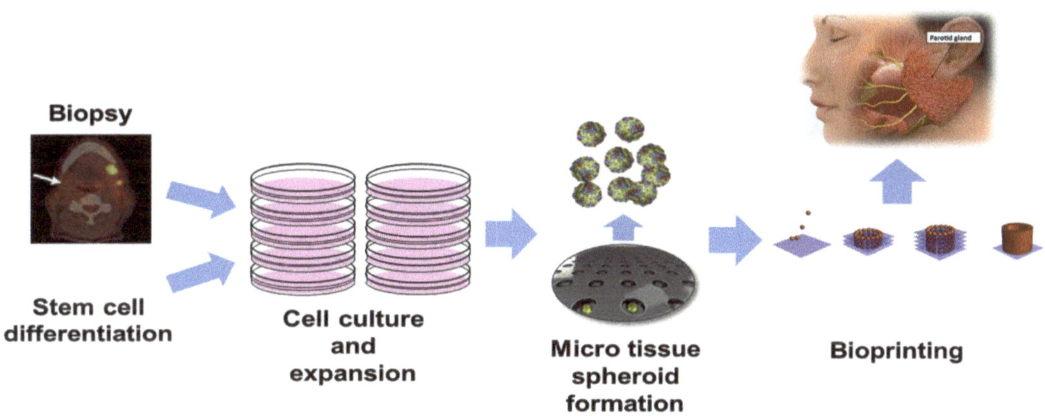

Fig. 9.7 Salivary gland regeneration with microsphere 3D bioprinting (Reproduced from Refs. [82, 88])

9.4 3D Bioprinting and Salivary Gland Regeneration

9.4.1 3D Bioprinting Considerations

An increasing number of publications for 3D bioprinting report significant progress and successes in vitro and in vivo. The three major components of tissue engineering include cells, scaffolds, and biological factors that facilitate tissue growth and organization. Similarly, key elements of bioprinting consist of cells, bioprinter, bioink, and the bioreactor system. The choice of cells for tissue reconstruction depends on the types of cells in the target tissues and organs. For example, vascular endothelial cells and smooth muscle cells would be appropriate for blood vessel printing and fibroblasts for connective tissues. Stem cells are frequently considered as a potential source of cells as well.

The selection of cell types has been widely investigated in the tissue engineering literature [57–61]. Therefore, it will not be explained further here. Bioreactors can be employed for the maturation of printed tissue constructs into functional tissue units and organs [62–64]. Generally, bioprinted three-dimensional tissue constructs are formed layer by layer by printing bioinks that contain living cells. Current 3D bioprinting research mainly focuses on the printing device and material compositions. The two most widely employed bioprinting mechanisms would be the inkjet printing and extrusion printing [65].

9.4.1.1 Inkjet Printing

Inkjet printing has a mechanism which is very similar to conventional office inkjet printers, with serial deposition of cell-containing bioink droplets. Piezoelectric actuators, heat-assisted bubble jet actuators, and pneumatic pressurization with solenoid valves are examples of inkjet printing techniques used to generate droplets [66–68]. The electronic control system of inkjet bioprinters enables relatively precise cell positioning, which can be used for drug testing or small-scale tissue unit fabrication.

9.4.1.2 Extrusion Printing

Extrusion-based printing is the most widely used bioprinting system [57, 69, 70]. Cell-containing material (bioink) is extruded from a reservoir to the printing bed through printer nozzles as shown in Fig. 9.5. The driving mechanism of extrusion can be pneumatic pressure or motor-driven syringe plunger movement [70, 71].

9.4.1.3 Bioinks

Cell behaviors including adhesion, migration, proliferation, differentiation, and tissue formation are influenced by the extracellular microenvironments, both in vivo and in vitro. After printing, cells are encapsulated in the bioink, and cell behavior is mainly affected by the biophysicochemical properties of the bioink, such as stiffness, molecular structure, cytokines or growth factors, degradability, and permeability [71–73].

The first consideration of materials as a bioink is printability. An appropriate shape holding mechanism is necessary to maintain 3D configuration of printed objects. The transition of bioinks from liquid to solid (semisolid) should be shorter than significant shape change. The stability of bioprinted constructs mainly depends on the viscosity of the bioinks after printing.

Liquid phase bioinks out of the printer nozzle are subject to surface tension and gravitational force. These external forces affect shape change of printed bioinks until possessing high enough viscosity.

From the viscosity point of view, materials with short cross-linking time can be a first consideration as bioink candidates. Bioink materials modified to have a short gelling time are widely used in 3D bioprinting. Hydrogels with short cross-linking time are widely employed as bioinks because these materials have dimensional stability in a relatively short time after printing [65, 66, 69, 74, 75].

Another approach for high viscosity is employing thixotropic materials to improve the stability of printed 3D constructs during cross-linking. Thixotropic materials are usually semisolid and have a shear thinning property (thixotropic means "shear thinning"). During bioink printing through the printer nozzle, shear forces induce a lowering of the viscosity, and bioinks have low flow resistance, with minimal harmful effect to the suspended cells. After exiting the nozzle, the thixotropic materials regain their high viscosity, and shape changes are minimized [76]. This prevents the collapse of printed 3D constructs.

One additional advantage of thixotropic bioink is the absence of cell sedimentation in the reservoir during the printing process. As the specific gravity of a cell is slightly higher than water, suspended cells tend to localize on the bottom of a reservoir. This effect is significant when cells are suspended in a low viscosity liquid. In thixotropic bioinks, suspended cells may show no or negligible displacement. As printing time is proportionally increased with the volume of an object, inhomogeneous cell distribution would be a significant defect in human-sized organs made with non-thixotropic aqueous bioinks. While extruding with thixotropic bioinks, care must be taken to keep the proper shear stress range to avoid lowering suspended cell viability. In side-by-side polymer printing the nonporous structure of each layer can be employed as a supporting structure to improve dimensional stability of constructs during 3D bioprinting as shown in Fig. 9.6 [57, 70, 77, 78].

Hydrogels can provide cells with a minimum damage environment during the bioprinting process. Hydrogels are widely used as a bioink material with a cell compatible pH and appropriate osmolarity. Examples of biomaterials with natural origins are alginate, fibrin, gelatin, hyaluronic acid, and collagen, and synthetic biomaterials are polyethylene glycol and Pluronic® F-127 [70, 72–74, 78]. Mixtures of these materials are also used with optimized printability, low cell damage, and higher 3D printed construct stability.

The cross-linking mechanism depends on hydrogels' intrinsic characteristics. Alginate has ionically cross-linking, and simple contact of alginate solution with divalent cationic solutions, such as calcium, barium, and strontium, can generate cross-linked hydrogel. Due to its low cost and simple cross-linking process, alginate is often employed as an initial test material for various bioprinters. Collagen and decellularized extracellular matrix (dECM) have pH- and temperature-dependent cross-linking manner [57, 70, 74, 77]. Under the physiologic pH condition and temperature (pH 7.4 and 37 °C, respectively), these materials cross-link to form stable hydrogel matrix. Further, with high cytocompatibility, cells in collagen and dECM show high tissue formation superior to alginate. However, relatively long cross-linking time (~30 min under 37 °C) hampers widespread use of these materials as bioink [57, 75]. Fibrin has enzyme-activated cross-linking mechanism. By mixing fibrinogen solution with thrombin solution, a stable fibrin hydrogel forms. Fibrinogen is a blood coagulation protein and has high cytocompatibility but still has relatively longer cross-linking time (0.5~10 min) than alginate (0.5~5 s).

Photo-cross-linking polymers are also being widely investigated as bioinks. Hydrogel precursors, including methacrylated gelatin (GelMA), star poly(ethylene glycol-co-lactide)-acrylate (SPELA), poly(ethylene glycol) dimethacrylate (PEGDMA), and polyethylene glycol) diacrylate (PEGDA), can be cross-linked using UV light [73, 79, 80]. A brief summary of bioinks currently used are listed in Table 9.2. Bioink materials that support cell viability and proliferation and have short cross-linking time are still needed to be developed for employing 3D bioprinting process for tissue regeneration. Bioprinters should have appropriate design compatible to bioink's cross-linking mechanism. Dual or multiple mixing nozzle configuration is required for mixing precursor solutions. Cooling or heating temperature control should be considered for temperature-induced cross-linking materials [80, 81].

9.4.2 Salivary Gland Regeneration by 3D Bioprinting

The ultimate goal of 3D bioprinting is to provide vascularized functional living organs, which can be applied to the replacement of missing or disabled tissues and organs. Observations and lessons from developmental biology can provide fundamental and practical ideas for tissue engineering approaches. Specific tissues or organs at different stages of development will have varying structural requirements. The essential morphogenetic steps and events of organogenesis during developmental stage can provide insights for salivary gland regeneration through 3D bioprinting [82].

Salivary glands consist of saliva-secreting acinar cells and various other types of cells. Tissue engineering of salivary glands was tried with several different approaches with hydrogel material for tissue regeneration [83–85]. Tissue spheroids, which have been used as an in vitro 3D model system in biomedical and tumor research for several decades, may be a useful candidate in salivary gland regeneration with 3D bioprinting technology (Fig. 9.7) [77, 82].

Table 9.2 Hydrogels used in extrusion-based bioprinting

Hydrogel type	Bioink	Cross-linking mechanism in extrusion bioprinting	Solidification reversibility	Extrusion bioprinting system	Advantages	Disadvantages
Alginate	Aggregates, proteins, encapsulated cells (skeletal myoblasts, BMSC, SMC, MSC, ASC, CPC, chondrocytes, cardiomyocytes)	Ionic	–	Pneumatic micro-extrusion and bioplotter	Biocompatibility, good extrudability and bioprintability, fast gelation, good stability and integrality of printed construct, medium elasticity, low cost, nonimmunogenic	Low cell adhesion and spreading without modification of hydrogel
Collagen type I	Encapsulated cells (bovine aortic endothelial cells, keratinocytes, fibroblasts, rat neural cells, MSC, AFS)	pH mediated or thermal	–	Pneumatic micro-extrusion	Cell adherent, promote proliferation, signal transducer, good extrusion and bioprinting abilities, nonimmunogenic	Poor mechanical properties, slow gelation, unstable
Gelatin	Encapsulated cells (HepG2, hepatocytes, fibroblasts, SMC)	Thermal	+	Mechanical and pneumatic micro-extrusion	Cell adherent, biocompatible, nonimmunogenic	Unstable, fragile, weak mechanical properties at physiological temperature and low abilities to extrude and print without modification
PEG	Encapsulated cells (bone marrow stem cells or porcine aortic valve interstitial cells)	Ionic, physical, or covalent agents	–	Pneumatic micro-extrusion	Support cell viability, biocompatible, nonimmunogenic, widely used in tissue engineering when modified	Low proliferation rate, low cell adhesion, weak mechanical properties and stability without modification
Fibrin	Acellular scaffolds or encapsulated cells (AFS, HUVEC)	Enzymatic	–	Pneumatic micro-extrusion	Promote angiogenesis (causes inflammatory response), fast gelation, good integrality, medium elasticity	Difficult to control geometry, low mechanical properties, limited EBB printability

Matrigel	Encapsulated cells (HepG2, BMSCs, gMSC, gEPC)	Thermal	–	Pneumatic micro-extrusion	Promote cell differentiation and vascularization of construct, support cell viability, good bioprintability, highly suitable particularly for cardiac tissue engineering	Slow gelation, which affects mechanical stability, requires cooling system for extrusion bioprinting, expensive
Agarose	Encapsulated cells (BMSCs osteosarcoma cells, MSC)	Thermal	+	Pneumatic and mechanical micro-extrusion	High mechanical properties, stable, resistant for protein adsorption, low cost, good integrality, nonimmunogenic	Low cell adhesion, fragile, require heating system for extrusion bioprinting
Chitosan	Acellular scaffolds, encapsulated cells (cartilage progenitor cells, MSC, CPC)	Ionic or covalent agents	–	Pneumatic micro-extrusion	Antibacterial and antifungal, medium printability, nonimmunogenic	Weak mechanical and stability properties without modification, slow gelation rate
Pluronic® F-127	Encapsulated cells (human primary fibroblasts, BMSC, HepG2)	Thermal	+	Pneumatic and mechanical micro-extrusion	High printability, good bioprintability, nonimmunogenic	Poor mechanical and structural properties, slow gelation, rapid degradation, require heating system for extrusion bioprinting
Hyaluronic acid	Encapsulated cells (chondrocytes, HepG2, C3A, fibroblasts)	Ionic, covalent agents	–	Pneumatic and mechanical micro-extrusion	Promote proliferation and angiogenesis, fast gelation, good bioprintability, nonimmunogenic	Rapid degradation, poor mechanical properties and low stability without modification
Methylcellulose	Encapsulated chondrocytes	Thermal, pH mediated	+	Mechanical micro-extrusion	High printability, nonimmunogenic	Low bioprintability, sensitive on common cell culture media, unstable

Reprinted from Ref. [71]

Bioprinting, or robotic additive biomanufacturing, could be implemented by a precise layer-by-layer placement of self-assembled tissue spheroids in advanced hydrogels. The rapid process of tissue spheroids to self-assemble and to form mature tissue in a relatively short time scale may provide the versatility needed for successful 3D bioprinting. Advancement of the tissue spheroids-based approach demands the synthesis of sophisticated soft biomaterials and extracellular matrices, such as bio-processible and biomimetic stimuli-sensitive functional hydrogels as bioink materials [71].

Salivary gland regeneration is also possible using 3D bioprinting with cells and hydrogels. Cells in the duct close to the acini are believed to provide all the cell types required for the formation of acini and ducts. In vitro cultured salivary cells could be assembled into three-dimensional acinar and ductal structures in the presence of collagen and Matrigel® [86]. Bioprinting of three-dimensional salivary gland structures may be guided by present experience with 3D bioprinting of vascular branch formation [72, 87]. Advancement in 3D bioprinting technology, in combination with a fundamental understanding of the molecular mechanisms of development, provides a novel strategy for salivary gland regeneration.

Conclusions

3D printing technology enables more effective patient consultations, increases diagnostic quality, improves surgical planning, acts as an orientation aid during surgery, and provides a template for surgical resection. In addition, as bioprinting technology further evolves, tissues or organs might one day be made with patient-specific shapes and dimensions, thus substantializing the goal of individualized medicine.

References

1. Müller A, Krishnan KG, Uhl E, Mast G. The application of rapid prototyping techniques in cranial reconstruction and preoperative planning in neurosurgery. J Craniofac Surg. 2003;14(6):899–914.
2. Poukens J, Haex J, Riediger D. The use of rapid prototyping in the preoperative planning of distraction osteogenesis of the cranio-maxillofacial skeleton. Comput Aided Surg. 2003;8(3):146–54.
3. Wagner JD, Baack B, Brown GA, Kelly J. Rapid 3-dimensional prototyping for surgical repair of maxillofacial fractures: a technical note. J Oral Maxillofac Surg. 2004;62(7):898–901.
4. Faber J, Berto PM, Quaresma M. Rapid prototyping as a tool for diagnosis and treatment planning for maxillary canine impaction. Am J Orthod Dentofacial Orthop. 2006;129(4):583–9.
5. Mavili ME, Canter HI, Saglam-Aydinatay B, Kamaci S, Kocadereli I. Use of three-dimensional medical modeling methods for precise planning of orthognathic surgery. J Craniofac Surg. 2007;18(4):740–7.
6. D'Urso PS, Effeney DJ, Earwaker WJ, Barker TM, Redmond MJ, Thompson RG, et al. Custom cranioplasty using stereolithography and acrylic. Br J Plast Surg. 2000;53(3):200–4.
7. Paiva WS, Amorim R, Bezerra DAF, Masini M. Aplication of the stereolithography technique in complex spine surgery. Arq Neuropsiquiatr. 2007;65(2B):443–5.
8. Armillotta A, Bonhoeffer P, Dubini G, Ferragina S, Migliavacca F, Sala G, et al. Use of rapid prototyping models in the planning of percutaneous pulmonary valved stent implantation. Proceedings of the Institution of Mechanical Engineers, Part H. J Eng Med. 2007;221(4):407–16.
9. Kim MS, Hansgen AR, Wink O, Quaife RA, Carroll JD. Rapid prototyping a new tool in understanding and treating structural heart disease. Circulation. 2008;117(18):2388–94.
10. Wurm G, Tomancok B, Pogady P, Holl K, Trenkler J. Cerebrovascular stereolithographic biomodeling for aneurysm surgery: technical note. J Neurosurg. 2004;100(1):139–45.
11. Giesel FL, Hart AR, Hahn HK, Wignall E, Rengier F, Talanow R, et al. 3D reconstructions of the cerebral ventricles and volume quantification in children with brain malformations. Acad Radiol. 2009;16(5):610–7.
12. Guarino J, Tennyson S, McCain G, Bond L, Shea K, King H. Rapid prototyping technology for surgeries of the pediatric spine and pelvis: benefits analysis. J Pediatr Orthop. 2007;27(8):955–60.
13. Hurson C, Tansey A, O'Donnchadha B, Nicholson P, Rice J, McElwain J. Rapid prototyping in the assessment, classification and preoperative planning of acetabular fractures. Injury. 2007;38(10):1158–62.
14. Hiramatsu H, Yamaguchi H, Nimi S, Ono H. Rapid prototyping of the larynx for laryngeal frame work surgery. Nihon Jibiinkoka Gakkai Kaiho. 2004; 107(10):949–55.
15. Winder J, Bibb R. Medical rapid prototyping technologies: state of the art and current limitations for application in oral and maxillofacial surgery. J Oral Maxillofacial Surgery: official journal of the American Association of Oral and Maxillofacial Surgeons. 2005;63(7):1006–15.

16. Fullerton JN, Frodsham GC, Day RM. 3D printing for the many, not the few. Nat Biotechnol. 2014;32(11): 1086–7.
17. Hoy MB. 3D printing: making things at the library. Med Ref Serv Q. 2013;32(1):94–9.
18. Choi JW, Kim N. Clinical application of three-dimensional printing technology in craniofacial plastic surgery. Arch Plast Surg. 2015;42(3):267–77.
19. Ventola CL. Medical Applications for 3D Printing: Current and Projected Uses. P T: a peer-reviewed journal for formulary management. 2014;39(10): 704–11.
20. Olszewski R, Tranduy K, Reychler H. Innovative procedure for computer-assisted genioplasty: three-dimensional cephalometry, rapid-prototyping model and surgical splint. Int J Oral Maxillofac Surg. 2010;39(7):721–4.
21. Winder J, McRitchie I, McKnight W, Cooke S. Virtual surgical planning and CAD/CAM in the treatment of cranial defects. Stud Health Technol Inform. 2005;111:599–601.
22. Dankowski R, Baszko A, Sutherland M, Firek L, Kalmucki P, Wroblewska K, et al. 3D heart model printing for preparation of percutaneous structural interventions: description of the technology and case report. Kardiol Pol. 2014;72(6):546–51.
23. Rengier F, Mehndiratta A, von Tengg-Kobligk H, Zechmann CM, Unterhinninghofen R, Kauczor HU, et al. 3D printing based on imaging data: review of medical applications. Int J Comput Assist Radiol Surg. 2010;5(4):335–41.
24. McGowan J. 3D printing technology speeds development. Health Estate. 2013;67(9):100–2.
25. Lee H, Fang NX. Micro 3D printing using a digital projector and its application in the study of soft materials mechanics. J Vis Exp: JoVE. 2012;69: e4457.
26. Mironov V, Boland T, Trusk T, Forgacs G, Markwald RR. Organ printing: computer-aided jet-based 3D tissue engineering. Trends Biotechnol. 2003;21(4): 157–61.
27. Mankovich NJ, Samson D, Pratt W, Lew D, Beumer 3rd J. Surgical planning using three-dimensional imaging and computer modeling. Otolaryngol Clin North Am. 1994;27(5):875–89.
28. Olszewski Raphael RH. Clinical applications of rapid prototyping models in cranio-maxillofacial surgery. In: Hoque M, editor. Advanced applications of rapid prototyping technology in modern engineering. InTech; 2011. p. 173–206.
29. Choi JY, Choi JH, Kim NK, Kim Y, Lee JK, Kim MK, et al. Analysis of errors in medical rapid prototyping models. Int J Oral Maxillofac Surg. 2002;31(1):23–32.
30. Chang PS, Parker TH, Patrick Jr CW, Miller MJ. The accuracy of stereolithography in planning craniofacial bone replacement. J Craniofac Surg. 2003;14(2): 164–70.
31. Schicho K, Figl M, Seemann R, Ewers R, Lambrecht JT, Wagner A, et al. Accuracy of treatment planning based on stereolithography in computer assisted surgery. Med Phys. 2006;33(9):3408–17.
32. Ibrahim D, Broilo TL, Heitz C, de Oliveira MG, de Oliveira HW, Nobre SM, et al. Dimensional error of selective laser sintering, three-dimensional printing and PolyJet models in the reproduction of mandibular anatomy. J Cranio Maxillo Facial Surg: official publication of the European Association for Cranio-Maxillo-Facial Surgery. 2009;37(3):167–73.
33. Silva DN, de Oliveira MG, Meurer E, Meurer MI, da Silva JVL, Santa-Barbara A. Dimensional error in selective laser sintering and 3D-printing of models for craniomaxillary anatomy reconstruction. J Cranio Maxillo Facial Surg: official publication of the European Association for Cranio-Maxillo-Facial Surgery. 2008;36(8):443–9.
34. Ohtani T, Kusumoto N, Wakabayashi K, Yamada S, Nakamura T, Kumazawa Y, et al. Application of haptic device to implant dentistry--accuracy verification of drilling into a pig bone. Dent Mater J. 2009;28(1): 75–81.
35. WE L, HE C, editors. Marching cubes: a high resolution 3D surface construction algorithm. ACM siggraph computer graphics. New York: ACM; 1987.
36. Tiede U, Hoehne KH, Bomans M, Pommert A, Riemer M, Wiebecke G. Investigation of medical 3D-rendering algorithms. IEEE Computer Graphics and Applications. 1990;10(2):41–53.
37. Yushkevich PA, Piven J, Hazlett HC, Smith RG, Ho S, Gee JC, et al. User-guided 3D active contour segmentation of anatomical structures: significantly improved efficiency and reliability. Neuroimage. 2006;31(3):1116–28.
38. Shattuck DW, Leahy RM. BrainSuite: an automated cortical surface identification tool. Med Image Anal. 2002;6(2):129–42.
39. Schroeder WJ, Zarge JA, Lorensen WE, editors. Decimation of triangle meshes. ACM Siggraph Computer Graphics. New York: ACM; 1992.
40. Field DA. Laplacian smoothing and Delaunay triangulations. Comm Appl Numer Methods. 1988;4(6): 709–12.
41. Hinton E, Campbell J. Local and global smoothing of discontinuous finite element functions using a least squares method. Int J Numer Methods Eng. 1974;8(3): 461–80.
42. Mahmood F, Owais K, Taylor C, Montealegre-Gallegos M, Manning W, Matyal R, et al. Three-dimensional printing of mitral valve using echocardiographic data. J Am Coll Cardiol Img. 2015;8(2):227–9.
43. Yang DH, Kang J-W, Kim N, Song J-K, Lee J-W, Lim T-H. Myocardial 3-dimensional printing for septal myectomy guidance in a patient with obstructive hypertrophic cardiomyopathy. Circulation. 2015; 132(4):300–1.
44. Jiankang H, Dichen L, Bingheng L, Zhen W, Tao Z. Custom fabrication of composite tibial hemi-knee joint combining CAD/CAE/CAM techniques. Proceedings of the Institution of Mechanical Engineers, Part H. J Eng Med. 2006;220(8):823–30.

45. Wang Z, Teng Y, Li D. Fabrication of custom-made artificial semi-knee joint based on rapid prototyping technique: computer-assisted design and manufacturing. Zhongguo xiu fu chong jian wai ke za zhi= Zhongguo xiufu chongjian waike zazhi= Chinese journal of reparative and reconstructive surgery. 2004;18(5):347–51.
46. Lee M-Y, Chang C-C, Ku Y. New layer-based imaging and rapid prototyping techniques for computer-aided design and manufacture of custom dental restoration. J Med Eng Technol. 2008;32(1):83–90.
47. Dai K-R, Yan M-N, Zhu Z-A, Sun Y-H. Computer-aided custom-made hemipelvic prosthesis used in extensive pelvic lesions. J Arthroplasty. 2007;22(7): 981–6.
48. Harrysson OL, Hosni YA, Nayfeh JF. Custom-designed orthopedic implants evaluated using finite element analysis of patient-specific computed tomography data: femoral-component case study. BMC Musculoskelet Disord. 2007;8(1):91.
49. Stevens B, Yang Y, Mohandas A, Stucker B, Nguyen KT. A review of materials, fabrication methods, and strategies used to enhance bone regeneration in engineered bone tissues. J Biomed Mater Res B Appl Biomater. 2008;85(2):573–82.
50. Peltola SM, Melchels FP, Grijpma DW, Kellomäki M. A review of rapid prototyping techniques for tissue engineering purposes. Ann Med. 2008;40(4):268–80.
51. Amerini A, Hatam N, Malasa M, Pott D, Tewarie L, Isfort P, et al. A personalized approach to interventional treatment of tricuspid regurgitation: experiences from an acute animal study. Interact Cardiovasc Thorac Surg. 2014;19(3):414–8.
52. Griffith LG, Naughton G. Tissue engineering--current challenges and expanding opportunities. Science (New York, NY). 2002;295(5557):1009–14.
53. Melchels FP, Domingos MA, Klein TJ, Malda J, Bartolo PJ, Hutmacher DW. Additive manufacturing of tissues and organs. Prog Polym Sci. 2012;37(8):1079–104.
54. Murphy SV, Atala A. 3D bioprinting of tissues and organs. Nat Biotechnol. 2014;32(8):773–85.
55. Huh D, Matthews BD, Mammoto A, Montoya-Zavala M, Hsin HY, Ingber DE. Reconstituting organ-level lung functions on a chip. Science (New York, NY). 2010;328(5986):1662–8.
56. Huh D, Hamilton GA, Ingber DE. From 3D cell culture to organs-on-chips. Trends Cell Biol. 2011;21(12):745–54.
57. Shim JH, Jang KM, Hahn SK, Park JY, Jung H, Oh K, et al. Three-dimensional bioprinting of multilayered constructs containing human mesenchymal stromal cells for osteochondral tissue regeneration in the rabbit knee joint. Biofabrication. 2016;8(1):014102.
58. Lewis PL, Shah RN. 3D printing for liver tissue engineering: current Approaches and future challenges. Curr Transpl Reports. 2016;3(1):100–8.
59. Maeshima A, Takahashi S, Nakasatomi M, Nojima Y. Diverse cell populations involved in regeneration of renal tubular epithelium following acute kidney injury. Stem Cells Int. 2015;2015:964849.
60. Moon KH, Ko IK, Yoo JJ, Atala A. Kidney diseases and tissue engineering. Methods. 2016;99:112–9.
61. Ohashi K, Okano T. Functional tissue engineering of the liver and islets. Anat Rec. 2014;297(1):73–82.
62. Scarritt ME, Pashos NC, Bunnell BA. A review of cellularization strategies for tissue engineering of whole organs. Front Bioeng Biotechnol. 2015;3:43.
63. Price AP, Godin LM, Domek A, Cotter T, D'Cunha J, Taylor DA, et al. Automated decellularization of intact, human-sized lungs for tissue engineering. Tissue Eng Part C Methods. 2015;21(1):94–103.
64. Bueno EM, Diaz-Siso JR, Sisk GC, Chandawarkar A, Kiwanuka H, Lamparello B, et al. Vascularized composite allotransplantation and tissue engineering. J Craniofac Surg. 2013;24(1):256–63.
65. Sears NA, Dhavalikar PS, Seshadri D, Cosgriff-Hernandez E. A review of 3D printing of tissue engineering constructs. Tissue Eng Part B Rev. 2016; 22(4):298–310.
66. Rodriguez-Devora JI, Zhang B, Reyna D, Shi ZD, Xu T. High throughput miniature drug-screening platform using bioprinting technology. Biofabrication. 2012;4(3):035001.
67. Hendriks J, Willem Visser C, Henke S, Leijten J, Saris DB, Sun C, et al. Optimizing cell viability in droplet-based cell deposition. Sci Rep. 2015;5:11304.
68. Gao G, Cui X. Three-dimensional bioprinting in tissue engineering and regenerative medicine. Biotechnol Lett. 2016;38(2):203–11.
69. Wu Z, Su X, Xu Y, Kong B, Sun W, Mi S. Bioprinting three-dimensional cell-laden tissue constructs with controllable degradation. Sci Rep. 2016;6:24474.
70. Pati F, Jang J, Ha DH, Won Kim S, Rhie JW, Shim JH, et al. Printing three-dimensional tissue analogues with decellularized extracellular matrix bioink. Nat Commun. 2014;5:3935.
71. Ozbolat IT, Hospodiuk M. Current advances and future perspectives in extrusion-based bioprinting. Biomaterials. 2016;76:321–43.
72. Duan B. State-of-the-art review of 3D bioprinting for cardiovascular tissue engineering. Ann Biomed Eng. 2016, in press.
73. Skardal A, Devarasetty M, Kang HW, Mead I, Bishop C, Shupe T, et al. A hydrogel bioink toolkit for mimicking native tissue biochemical and mechanical properties in bioprinted tissue constructs. Acta Biomater. 2015;25:24–34.
74. Stanton MM, Samitier J, Sanchez S. Bioprinting of 3D hydrogels. Lab Chip. 2015;15(15):3111–5.
75. Chua CK, Yeong WY. Materials for bioprinting. In: Bioprinting: principles and applications. Singapore: World Scientific Publishing Co. Pte. Ltd.; 2014. p. 117–64.
76. Ozbolat IT, Yu Y. Bioprinting toward organ fabrication: challenges and future trends. IEEE Trans Biomed Eng. 2013;60(3):691–9.
77. Murphy SV, Atala A. 3D bioprinting of tissues and organs. Nat Biotechnol. 2014;32(8):773–85.
78. Kang H-W, Lee SJ, Ko IK, Kengla C, Yoo JJ, Atala A. A 3D bioprinting system to produce human-scale

tissue constructs with structural integrity. Nat Biotechnol. 2016;34(3):312–9.
79. Luiz EB, Juliana CC, Vijayan M, Ana LC, Nupura SB, Wesleyan AA, et al. Direct-write bioprinting of cell-laden methacrylated gelatin hydrogels. Biofabrication. 2014;6(2):024105.
80. Skardal A, Zhang J, Prestwich GD. Bioprinting vessel-like constructs using hyaluronan hydrogels crosslinked with tetrahedral polyethylene glycol tetracrylates. Biomaterials. 2010;31(24):6173–81.
81. Kolesky DB, Homan KA, Skylar-Scott MA, Lewis JA. Three-dimensional bioprinting of thick vascularized tissues. Proc Natl Acad Sci U S A. 2016;113(12):3179–84.
82. Holmberg KV, Hoffman MP. Anatomy, biogenesis and regeneration of salivary glands. Monogr Oral Sci. 2014;24:1–13.
83. Jakab K, Norotte C, Marga F, Murphy K, Vunjak-Novakovic G, Forgacs G. Tissue engineering by self-assembly and bio-printing of living cells. Biofabrication. 2010;2(2):022001.
84. Chan YH, Huang TW, Young TH, Lou PJ. Human salivary gland acinar cells spontaneously form three-dimensional structures and change the protein expression patterns. J Cell Physiol. 2011;226(11):3076–85.
85. Joraku A, Sullivan CA, Yoo J, Atala A. In-vitro reconstitution of three-dimensional human salivary gland tissue structures. Differentiation. 2007;75(4):318–24.
86. Maria OM, Maria O, Liu Y, Komarova SV, Tran SD. Matrigel improves functional properties of human submandibular salivary gland cell line. Int J Biochem Cell Biol. 2011;43(4):622–31.
87. Mandrycky C, Wang Z, Kim K, Kim DH. 3D bioprinting for engineering complex tissues. Biotechnol Adv. 2016;34(4):422–34.
88. Jeong GS, Song JH, Kang AR, Jun Y, Kim JH, Chang JY, et al. Surface tension-mediated, concave-microwell arrays for large-scale, simultaneous production of homogeneously sized embryoid bodies. Adv Healthc Mater. 2013;2(1):119–25.

10 Functional Salivary Gland Regeneration by Organ Replacement Therapy

Miho Ogawa and Takashi Tsuji

Abstract

The salivary glands are exocrine organs that secrete saliva to maintain oral health and homeostasis. Dysfunctional salivary glands exhibit symptoms of dry mouth, including dental caries and dysfunction in speech and swallowing. Current clinical therapies for dry mouth disease include artificial saliva substitutes or parasympathetic stimulants, but these are transient and palliative approaches. To achieve the functional recovery of dysfunctional salivary glands, salivary gland tissue stem cells are thought to be candidate cell sources for salivary gland tissue repair therapies. In addition, whole salivary gland replacement therapy is expected to be a novel therapy resulting in the regeneration of fully functional salivary glands. The salivary glands arise from their organ germs, which are induced by epithelial-mesenchymal interactions. Recently, we developed a novel bioengineering method, i.e., the organ germ method, which can regenerate the ectodermal organs, including the teeth, hair, lacrimal glands, and salivary glands. The bioengineered salivary glands successfully secrete saliva into the oral cavity and can also improve the symptoms of dry mouth, such as bacterial infection and swallowing dysfunction. In this review, we summarize recent findings and bioengineering methods for salivary gland regeneration therapy.

10.1 Introduction

Exocrine glands, such as the sweat glands, lacrimal glands, and salivary glands, produce secretory fluids such as sweat, tears, and saliva. These secretory fluids have important roles in maintaining health and homeostasis. For example, saliva is secreted into the oral cavity and functions during chewing, digestion, cleaning, and swallowing [1, 2]. The salivary glands arise from the salivary

M. Ogawa, PhD • T. Tsuji, PhD (✉)
Organ Technologies Inc., Tokyo 101-0048, Japan

RIKEN Center for Developmental Biology,
Kobe, Hyogo 650-0047, Japan
e-mail: miho_ogawa@organ-technol.co.jp;
t-tsuji@cdb.riken.jp

S. Cha (ed.), *Salivary Gland Development and Regeneration*, DOI 10.1007/978-3-319-43513-8_10

gland organ germ, which is generated by the interaction of epithelial-mesenchymal stem cells during embryonic development [3, 4]. Salivary glands consist of three major glands, including the submandibular gland (SMG), sublingual gland (SLG), and parotid gland (PG), and many minor glands. Overall, 95 % of the saliva secreted per day is secreted by the SMG, SLG, and PG, and 5 % is secreted by the minor salivary glands. The SMG and PG secrete serous saliva that contributes mainly to the digestion of food. The SLG secretes mucous saliva, which protects the oral cavity from drying. Therefore, salivary gland dysfunction induces xerostomia and has an adverse effect on bodily health [5, 6].

Xerostomia induces some clinical problems, including dental decay, bacterial infection, mastication and swallowing dysfunction, and a general reduction in quality of life [5–7]. Xerostomia develops due to autoimmune diseases, such as Sjögren's syndrome, aging, and radiation therapy for head and neck cancer. Current therapies for xerostomia rely on the use of artificial saliva substitutes or parasympathetic stimulants to promote saliva secretion and to prevent dry mouth [8, 9]. However, these therapies only provide temporary effects and cannot result in the recovery of salivary gland dysfunction, which is why the development of novel therapies for restoration of salivary gland function is necessary [10].

Regenerative therapies utilizing stem cell transplantation have been conducted in various organs, including salivary glands [11, 12]. In addition, salivary gland regeneration therapy involving gene modification and tissue engineering may eventually be used to restore damaged tissue and recover the flow of saliva [13]. Similar organ replacement therapy approaches for ectodermal organs such as the teeth and hair follicles, which can be achieved by transplantation of bioengineered organ germs that have been reconstituted using organ germ methods, have been reported [14–16]. Recently, we induced the regeneration of salivary glands and lacrimal glands using this method [17, 18]. In this book chapter, we discuss the novel findings and bioengineering methods used in salivary gland regeneration and the feasibility of these methods for future organ replacement regenerative therapy.

10.2 Development of Salivary Glands During Embryogenesis

The salivary glands are generated from the organ germ, which is produced by epithelial and mesenchymal stem cell interactions during early embryonic development. The SMG, SLG, and PG are generated through similar morphogenetic events but differ in the timing and position at which generation begins [2–4, 19–22]. The development of the SMG is produced by the invagination of the oral epithelium into the mesenchymal region derived from the base of the tongue on embryonic day (ED) 11 (prebud) (Fig. 10.1). The invaginated epithelial tissue proliferates to form an epithelial stalk and a terminal bud at the tip (initial bud). The epithelial stalk differentiates into the ducts, which are called the intercalated, striated, and excretory ducts depending on their position relative to the side of the opening. The terminal bud forms the branched structure by forming a cleft and by repeating the elongation and branching process during ED 12.5–13.5 (pseudoglandular) [23–25]. From ED 15.0, the terminal bulbs differentiate into the acinar cells and begin the synthesis of secretory proteins, which differ depending on the type of salivary gland [26]. The SMG and PG secrete serous saliva, which contains a large amount of digestive enzymes such as α-amylase, which degrades starches and aids in digestion. The SLG secretes mucous saliva, which contains rich mucin protein to protect the mouth against dryness. The epithelial cells also differentiate into myoepithelial cells, and adult epithelial tissue stem cells are maintained in the excretory duct to contribute to the repair of injured tissue [27–29].

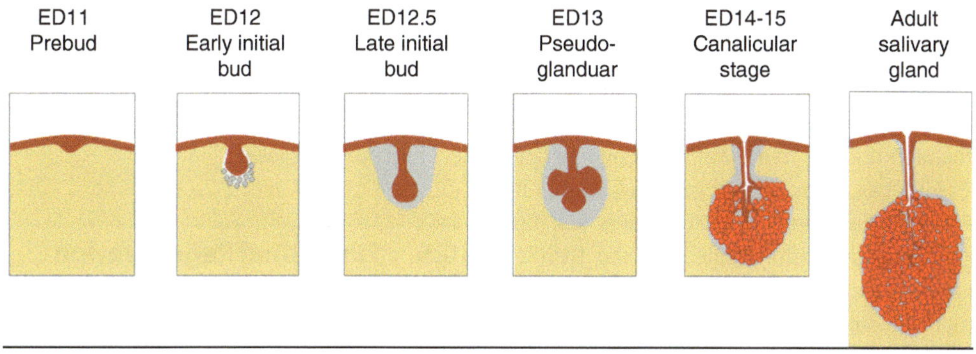

Fig. 10.1 Schematic representation of the developmental stages of the salivary gland. The salivary glands, including SMG, SLG, and PG, are produced from organ germs induced by the interaction of the epithelial tissue and the mesenchymal tissue. The epithelial tissue invaginates into the mesenchymal tissue and forms a certain morphology according to the development of each organ. The salivary gland epithelial tissue is formed by the epithelial stalk and terminal bulb, which form the duct and acinar cells. The acinar cells mature and begin to synthesize and secrete secretory proteins

10.3 Diseases and Treatments of Salivary Glands

The various types of salivary gland diseases include salivary gland-specific diseases, such as salivary tumors, obstructive disorders, and infections, as well as the symptoms of systemic diseases, such as Sjögren's syndrome (SS), lymphoma, and metabolic diseases [2]. Salivary dysfunction resulting from the atrophy of acinar cells and saliva reduction leads to xerostomia (dry mouth syndrome). In Europe, approximately 20 % of the population is thought to suffer from dry mouth syndrome, and this disease has been estimated to occur in approximately 800 million people in Japan [30]. The treatment of head and neck cancer, including salivary tumors, has been performed using radiation therapy. However, as the salivary glands are more sensitive to radiation, this treatment can cause atrophy of the acinar cells. Another condition that affects the salivary glands is SS, an autoimmune disease that occurs frequently in middle-aged and elderly women. It also affects the salivary glands as well as other glands such as the lacrimal glands, resulting in dry eyes. The annual number of SS patients has been reported to be approximately 15,000–20,000 [30]. Of all SS patients, approximately 70 % are positive for the SS antibody SSA (anti-Ro), and 40 % are positive for the SS antibody SSB (anti-La) [31–33]. However, these antibodies are not common to all patients, and the details of the pathogenic mechanism are not clear. Current therapies for dry mouth syndrome include symptomatic treatments such as the administration of artificial saliva and sialogogues to enhance moisture retention in the oral cavity [9]. In addition, parasympathomimetic drugs such as pilocarpine and cevimeline have been used to stimulate the muscarinic M3 receptor and induce salivary flow [32].

10.4 Salivary Gland Regeneration Using Stem Cells and Gene Therapy

10.4.1 Tissue Regeneration Using Adult Tissue Stem Cells

Transplantation of adult tissue stem cells has become a recognized method for regenerative therapy to restore damaged tissues and organs in diverse diseases [11, 34]. Regarding salivary gland regeneration, tissue stem/progenitor cell studies have reported that tissue stem cells have the capacity for tissue repair. The c-kit- and sca-1-positive tissue stem cells are localized to the intercalated duct of adult salivary glands, where these cells can induce the acinar and duct cells [35–38]. Furthermore, these stem cells are pluripotent and can differentiate into the liver or pancreas' tissues [39, 40]. The c-kit-positive salivary gland stem cells can be cultured while maintaining the tissue repair capacity in vitro [12, 41–43]. A transplant of these cells can recover the decreased salivation resulting from irradiation-induced atrophy of the acinar cells. In addition, it has been reported that the bone marrow-derived mesenchymal stem cells have the ability to promote the regenerative capacity of the salivary gland stem cells that remain in the damaged salivary glands after irradiation [44]. Stem cell transplantation is expected to serve as an effective means of achieving salivary gland regeneration.

10.4.2 Gene Therapy for Salivary Gland Regeneration

Because the salivary glands are located close to the body surface, regeneration of damaged salivary glands via gene therapy has been studied. The salivary glands open via the duct into the oral cavity, and thus methods of direct gene transfection into the salivary glands via the duct have been reported. After the transfection of the water channel aquaporin-1 (AQP1) gene using adenovirus or adeno-associated virus, the saliva secretion of irradiated salivary glands was significantly recovered [45, 46]. However, salivary glands are known to function as exocrine glands that secrete saliva in the oral cavity and as endocrine glands that secrete substances into the bloodstream. Gene therapy using the salivary glands has also been performed as a treatment for other diseases, including SS and other genetic diseases [47, 48]. It has been reported that some materials, such as IL-17 receptor antibodies, growth hormones, parathyroid hormones, and erythropoietin, can be expressed in adult salivary glands via gene transfer and circulated throughout the body by the bloodstream [49–53]. Stem cell transplantation therapy and gene therapy are expected to be a new treatment strategy for salivary gland disorders and other diseases.

10.5 Functional Regeneration of a Bioengineered Salivary Gland

The current research for regenerating three-dimensional organs mimics organogenesis in the developing embryo. In the salivary gland regeneration field, epithelial cell aggregates are used to elucidate the mechanism of regeneration and branching morphogenesis in vitro [54]. In addition, the aggregate mix of epithelial and mesenchymal stem cells has been reported to increase the number of branches and rate of branch formation [54].

Recently, we demonstrated the possibility of full functional regeneration of the ectodermal organs, including the teeth, hair follicles, lacrimal glands, and salivary glands, using "organ germ methods" that involved epithelial and mesenchymal stem cell manipulation techniques to induce the formation of an organ germ (Fig. 10.2a) [14–18]. Using this method, it is possible to control the size, number, morphology, and invagination direction of the regenerated organ [16, 55]. For successful salivary gland replacement therapy, it is important that the invagination direction is controlled in such a way that the invaginated tissue connects to the ducts to secrete saliva into the oral cavity.

10.5.1 Development of a Bioengineered Salivary Gland

To reconstruct the bioengineered salivary gland, the germs including the SMG, SLG, and PG were isolated from mice at embryonic day (ED) 13.5–14.5. The bioengineered SMG germ showed epithelial-mesenchymal interactions and epithelial bud formation in organ culture (Fig. 10.2b) [17].

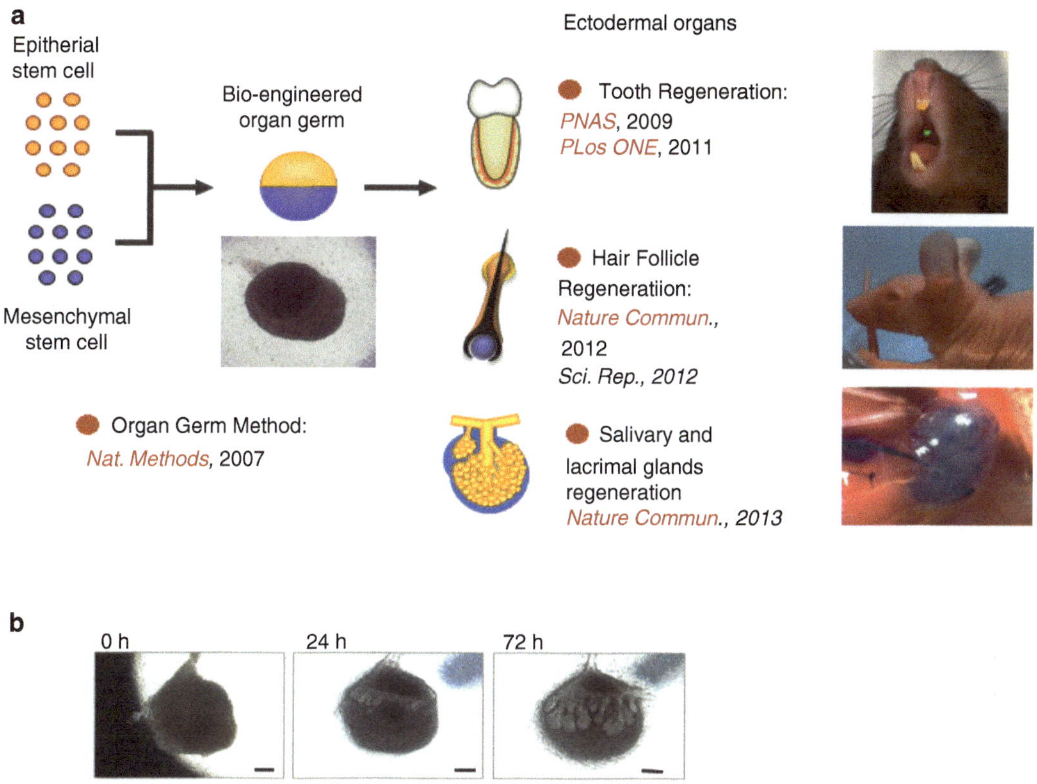

Fig. 10.2 Regeneration of salivary gland germs using organ germ methods. (**a**) Ectodermal organs including the teeth, hair follicle, and secretory glands can be regenerated in vivo by transplanting bioengineered organ germs that are reconstituted by organ germ methods. (**b**) Phase-contrast images of the bioengineered submandibular gland germ at 0, 24, and 72 h of in vitro culture. The bioengineered submandibular gland showed the interaction between the epithelial and mesenchymal cells (24 h) and invagination of the epithelial tissue (72 h)

The regenerated SLG and PG germs also showed patterns that were similar to that of the SMG germ and structurally correct based on the natural salivary gland germ. The correct transplantation of the bioengineered salivary gland is important to achieving the secretion of saliva into the oral cavity. A bioengineered salivary gland germ was engrafted into the PG duct of the model mice with salivary gland defects using an intraepithelial tissue-connecting plastic method. In these mice, the SMG, SLG, and PG were excised. After 30 days, the growth of the bioengineered salivary gland and its connection to the PG duct was successfully achieved (Fig. 10.3a) [17]. The bioengineered salivary gland structures, including the localization of myoepithelial cells, the water channel aquaporin 5 (AQP5), and neuronal connections, were similar to those of a natural tissue (Fig. 10.3b) [17].

10.5.2 Secretion of a Bioengineered Saliva

About 1–1.5 L of saliva is secreted per day from the salivary glands. This secretion is induced by eating, heat, and painful stimulation to the oral cavity. These stimulations are transmitted via the afferent and efferent neural networks from the oral cavity to the salivary glands (Fig. 10.4a) [56–60]. Moreover, because secreted saliva also plays an important role in taste perception, the hyposecretion of saliva has been known to cause taste disorders [61–63]. In the medical field, the secretion of saliva from the salivary gland has been analyzed using five tastes, including sour (citrate), bitter (quinine hydrochloride), salty (NaCl), sweet (sucrose), and umami (glutamate) [63, 64]. Compared to the control substance, citrate

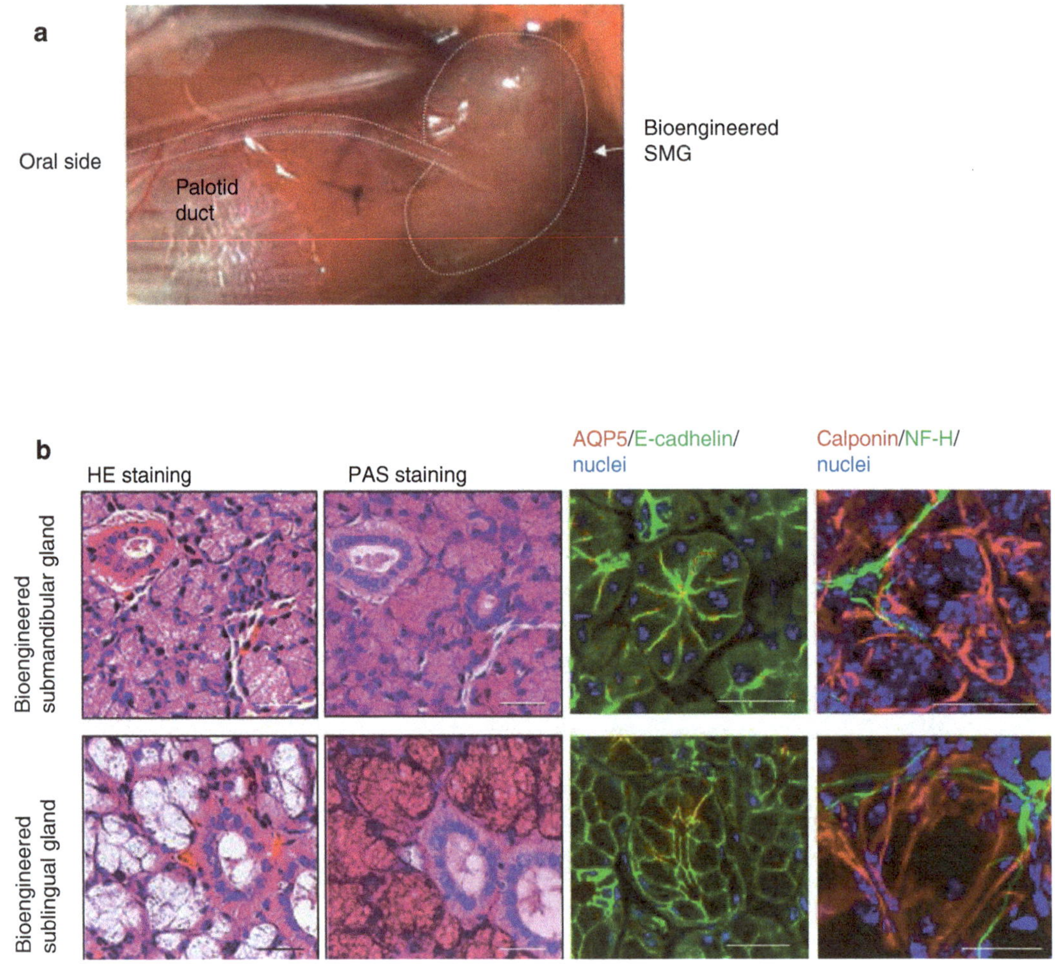

Fig. 10.3 In vivo transplantation of a bioengineered salivary gland. (**a**) Photographs of the bioengineered submandibular gland at day 30 after transplantation (*left*). The bioengineered submandibular gland duct connected with natural PG duct (*right*). (**b**) Histological analysis of the bioengineered SMG (*upper columns*) and the SLG (*lower columns*). Images of HE staining (*left*) and periodic acid and Schiff (*PAS*) staining (*second from the left*). The bioengineered SLG showed a strongly positive PAS staining. Immunohistochemical images of calponin (*red*), E-cadherin (*green, third from the left*), and NF-H (*green, right*) are shown (Modified from Ref. Ogawa et al. [17])

stimulation induced significant quantities of saliva secretion from both the natural and bioengineered salivary gland (Fig. 10.4b) [17]. Saliva secretion was induced in response to all tastes in addition to the sour stimulus. The secretion amount depended on the type of stimulus and exhibited the following order: sour > bitter > umami > salty = sweet [64]. In addition, the secreted bioengineered saliva contained the amylase protein, which has starch-degrading activity [17]. Salivation was measured about 1–3 months after transplantation and followed up for 6 months. These findings demonstrate that saliva secretion by the bioengineered salivary gland may be controlled through the afferent-efferent neural network.

10.5.3 Functional Restoration of Swallowing Dysfunction Using a Bioengineered Salivary Gland

Oral health and homeostasis are maintained by saliva and saliva proteins, such as amylase,

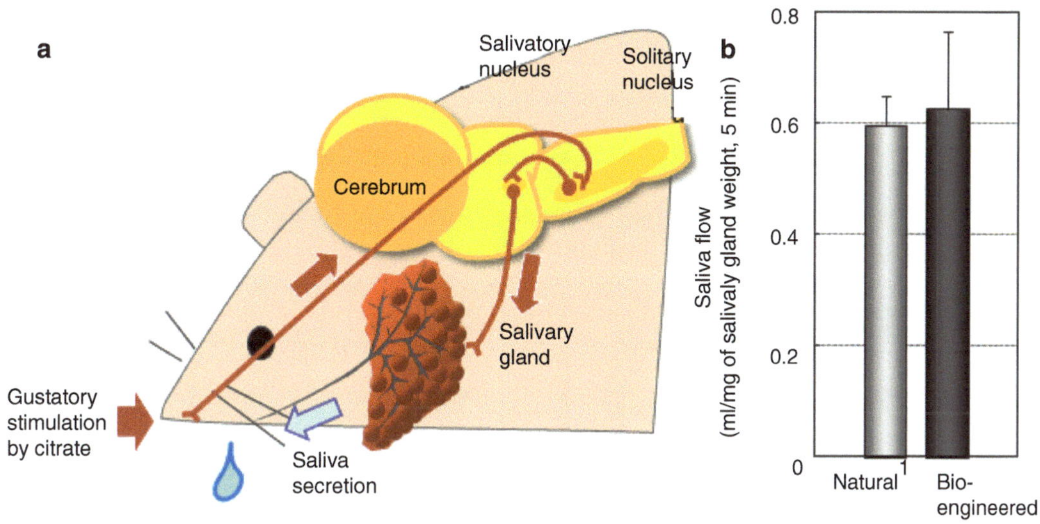

Fig. 10.4 Saliva secretion induced by gustatory stimulation. (**a**) Schematic representation of saliva secretion induced by gustatory stimulation via the central nervous system. (**b**) Assessment of the amount of saliva secretion from natural SMG (*light bar*) and bioengineered SMG (*dark bar*) after gustatory stimulation by citrate. The amount of secreted saliva exhibited no significant difference (**b** reprinted from Ref. Ogawa et al. [17])

lysozyme, IgA, lactoferrin, myeloperoxidase, NGF, EGF, and parotin. Therefore, the hyposecretion of saliva causes various problems, including dental caries, bacterial infection, sleep disorders, and swallowing dysfunction [65, 66]. The bioengineered salivary gland-engrafted mouse had fewer bacteria compared with the salivary gland defect model mouse. Among the salivary gland functions, the swallowing function is critical for reducing the risk of aspiration. The saliva promotes the formation of a bolus of food or water and triggers the swallowing reflex. Therefore, salivary gland dysfunction can cause chronic lung disease and can affect the survival, quality of life, and overall health of an individual [67]. In the salivary gland defect model mouse, the body weight was abnormally decreased, and all mice died within 5 days, despite having free access to food and water (Fig. 10.5) [17]. Dry mouth patients often drink high-viscosity water because they cannot swallow water. Similarly, the salivary gland defect model mouse exhibited a recovery of body weight and an increased survival rate when drinking high-viscosity water; this result in the model mouse raised the possibility that dysphagia may occur. In contrast, all of the bioengineered salivary gland-engrafted mice survived, and their body weight increased within 4 days after transplantation [17]. These results indicated that the bioengineered salivary gland can improve the swallowing function associated with the maintenance of oral health.

10.6 Future Directions of Salivary Gland Regeneration

Organ regenerative technology has advanced significantly. To achieve future clinical applications of salivary gland replacement therapy, it is important to identify suitable cell sources. The ideal cell source is the patient's own cells because there is no immunological rejection. Recent stem cell biology studies have revealed the presence of adult tissue stem cells in the salivary gland. These adult tissue-derived stem cells, which include c-kit- and sca-1-positive cells, can repair the acinar cells injured by radiation and can partially recover the total amount of secreted saliva [12, 41–44]. The salivary gland of adult stem cells would be valuable cell sources for achieving salivary gland tissue regeneration via stem cell transplantation therapy. In contrast, pluripotent stem cells and induced pluripotent stem cells are also potential cell sources

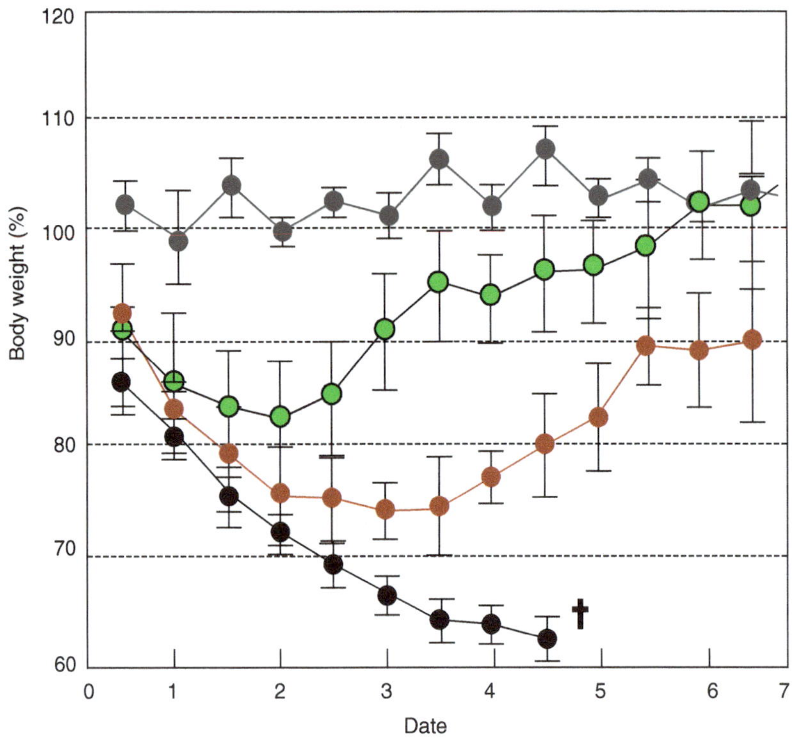

Fig. 10.5 Analysis of body weight. Measurement of body weight (*left graphs*) every 0.5 days after transplantation in normal mice (*gray dots*); salivary gland defect model mice (*black dots*), salivary gland-engrafted mice (*red dots*), and salivary gland defect model mice were given high-viscosity water (*green dots*). All salivary gland defect model mice died within 5 days (**†**) after removing all of the major salivary glands. Salivary gland-engrafted mice recovered the body weight (Reprinted from Ref. Ogawa et al. [17])

for salivary gland regeneration because these cells can differentiate into all types of cells, including endodermal, ectodermal, and mesodermal cells [68–70]. It has been reported that some organs, such as the optic cup and pituitary gland, can be derived from ES cells or iPS cells. In the future, the method of salivary gland regeneration using iPS cells is expected to be established [71–73].

Another important direction for future salivary gland regeneration therapy is to establish the mechanisms by which autoimmune diseases such as SS cause xerostomia [31–33]. In autoimmune diseases, salivary gland damage, such as the atrophy of acinar cells, is caused by self-antigens. Therefore, even if the bioengineered salivary gland is transplanted and can temporarily recover the saliva secretion, there is the possibility that the bioengineered acinar cells will again undergo atrophy. To achieve future clinical applications of salivary gland replacement therapy, genetic modifications of patient-derived stem cells will be necessary to decrease the expression of autoantigens.

Current whole organ regenerative therapy has the potential as a future therapeutic technology for several diseases. Salivary gland regenerative therapy is regarded as a model for future secretory organ replacement therapies that will substantially contribute to achieving an understanding related to organ regeneration technology.

Acknowledgments This work was partially supported by a Grant-in-Aid for KIBAN (A) from the Ministry of Education, Culture, Sports, Science and Technology (no. 25242041). This work was also partially supported by Organ Technologies, Inc.

Conflict of Interest This work was partially funded by Organ Technologies Inc. M.O. is a researcher and T.T. is a director at Organ Technologies Inc. This work was performed under an Invention Agreement between Tokyo University of Science, RIKEN and Organ Technologies Inc.

References

1. Edgar M, Dawes C, Mullane OD. Saliva and oral health. 3rd ed. London: British Dental Association; 2004.
2. Tucker AS, Miletich I. Salivary glands; development, adaptations, and disease. London: Karger; 2010.
3. Avery JK. Oral development and histology. New York: Thieme Press; 2002. p. 292–330.
4. Tucker AS. Salivary gland development. Semin Cell Dev Biol. 2007;18:237–44.
5. Vissink A, Burlage FR, Spijkervet FK, Jansma J, Coppes RP. Prevention and treatment of the consequences of head and neck radiotherapy. Crit Rev Oral Biol Med. 2003;14:213–25.
6. Vissink A, Jansma J, Spijkervet FK, Burlage FR, Coppes RP. Oral sequelae of head and neck radiotherapy. Crit Rev Oral Biol Med. 2003;14:199–212.
7. Atkinson JC, Grisius M, Massey W. Salivary hypofunction and xerostomia: diagnosis and treatment. Dent Clin N Am. 2005;49:309–26.
8. Ship JA, Pillemer SR, Baum BJ. Xerostomia and the geriatric patient. J Am Geriatr Soc. 2002;50:535–43.
9. Fox PC. Salivary enhancement therapies. Caries Res. 2004;38:241–6.
10. Kagami H, Wang S, Hai B. Restoring the function of salivary glands. Oral Dis. 2008;14:15–24.
11. Körbling M, Estrov Z. Adult stem cells for tissue repair – a new therapeutic concept? N Engl J Med. 2003;349:570–82.
12. Lombaert IM, Brunsting JF, Wierenga PK, Faber H, Stokman MA, Kok T, Visser WH, Kampinga HH, de Haan G, Coppes RP. Rescue of salivary gland function after stem cell transplantation in irradiated glands. PLoS ONE. 2008;3:e2063.
13. O'Connell AC, Baccaglini L, Fox PC, O'Connell BC, Kenshalo D, Oweisy H, Hoque AT, Sun D, Herscher LL, Braddon VR, Delporte C, Baum BJ. Safety and efficacy of adenovirus-mediated transfer of the human aquaporin-1 cDNA to irradiated parotid glands of non-human primates. Cancer Gene Ther. 1999;6(6):505–13.
14. Nakao K, Morita R, Saji Y, Ishida K, Tomita Y, Ogawa M, Saitoh M, Tomooka Y, Tsuji T. The development of a bioengineered organ germ method. Nat Methods. 2007;4(3):227–30.
15. Ikeda E, Morita R, Nakao K, Ishida K, Nakamura T, Takano-Yamamoto T, Ogawa M, Mizuno M, Kasugai S, Tsuji T. Fully functional bioengineered tooth replacement as an organ replacement therapy. Proc Natl Acad Sci U S A. 2009;106(32):13475–80.
16. Toyoshima KE, Asakawa K, Ishibashi N, Toki H, Ogawa M, Hasegawa T, Irié T, Tachikawa T, Sato A, Takeda A, Tsuji T. Fully functional hair follicle regeneration through the rearrangement of stem cells and their niches. Nat Commun. 2012;3:784.
17. Ogawa M, Oshima M, Imamura A, Sekine Y, Ishida K, Yamashita K, Nakajima K, Hirayama M, Tachikawa T, Tsuji T. Functional salivary gland regeneration by transplantation of a bioengineered organ germ. Nat Commun. 2013;4:2498.
18. Hirayama M, Ogawa M, Oshima M, Sekine Y, Ishida K, Yamashita K, Ikeda K, Shimmura S, Kawakita T, Tsubota K, Tsuji T. Functional lacrimal gland regeneration by transplantation of a bioengineered organ germ. Nat Commun. 2013;4:2497.
19. Jiménez-Rojo L, Granchi Z, Graf D, Mitsiadis TA. Stem cell fate determination during development and regeneration of ectodermal organs. Front Physiol. 2012;3:107.
20. Pispa J, Thesleff I. Mechanisms of ectodermal organogenesis. Dev Biol. 2003;262(2):195–205.
21. Jaskoll T, Melnick M. Embryonic salivary gland branching morphogenesis. Austin (TX); Madame Curie Bioscience Database [Internet]; 2004.
22. Knosp WM, Knox SM, Hoffman MP. Salivary gland organogenesis. Wiley Interdiscip Rev Dev Biol. 2012;1(1):69–82.
23. Sakai T. Epithelial branching morphogenesis of salivary gland: exploration of new functional regulators. J Med Investig. 2009;56(Suppl):234–8.
24. Hsu JC, Yamada KM. Salivary gland branching morphogenesis – recent progress and future opportunities. Int J Oral Sci. 2010;2(3):117–26.
25. Harunaga J, Hsu JC, Yamada KM. Dynamics of salivary gland morphogenesis. J Dent Res. 2011;90(9): 1070–7.
26. Denny PC, Denny PA. Dynamics of parenchymal cell division, differentiation, and apoptosis in the young adult female mouse submandibular gland. Anat Rec. 1999;254:408–17.
27. Man YG, Ball WD, Marchetti L, Hand AR. Contributions of intercalated duct cells to the normal parenchyma of submandibular glands of adult rats. Anat Rec. 2001;263(2):202–14.
28. Ihrler S, Zietz C, Sendelhofert A, Lang S, Blasenbreu-Vogt S, Löhrs U. A morphogenetic concept of salivary duct regeneration and metaplasia. Virchows Arch. 2002;440(5):519–26.
29. Lombaert IM, MP H. Stem cells in salivary gland development and regeneration. In: Stem cells in craniofacial development and regeneration. Hoboken: Wiley-Blackwell; 2013. p. 271–84.
30. Hayashi Y, Arakaki R, Ishimaru N. Salivary gland and autoimmunity. J Med Investig. 2009;56:185–91.
31. Fox RI, Stern M, Michelson P. Update in Sjögren syndrome. Curr Opin Rheumatol. 2000;12(5):391–8.
32. Nakamura T, Matsui M, Uchida K, Futatsugi A, Kusakawa S, Matsumoto N, Nakamura K, Manabe T, Taketo MM, Mikoshiba K. M3 muscarinic acetylcholine receptor plays a critical role in parasympathetic control of salivation in mice. J Physiol. 2004;558: 561–75.
33. Copelan EA. Hematopoietic stem-cell transplantation. N Engl J Med. 2006;354:1813–26.
34. Segers VFM, Lee RT. Stem-cell therapy for cardiac disease. Nature. 2008;451:937–42.

35. Rotter N, Oder J, Schlenke P. Isolation and characterization of adult stem cells from human salivary glands. Stem Cells Dev. 2008;17(3):509–18.
36. Sugito T, Kagami H, Hata K, Nishiguchi H, Ueda M. Transplantation of cultured salivary gland cells into an atrophic salivary gland. Cell Transplant. 2004;13(6):691–9.
37. Kishi T, Takao T, Fujita K, Taniguchi H. Clonal proliferation of multipotent stem/progenitor cells in the neonatal and adult salivary glands. Biochem Biophys Res Commun. 2006;340(2):544–52.
38. Takahashi S, Schoch E, Walker NI. Origin of acinar cell regeneration after atrophy of the rat parotid induced by duct obstruction. Int J Exp Pathol. 1998;79:293–301.
39. Hisatomi Y, Okumura K, Nakamura K, Matsumoto S, Satoh A, Nagano K, Yamamoto T, Endo F. Flow cytometric isolation of endodermal progenitors from mouse salivary gland differentiate into hepatic and pancreatic lineages. Hepatology. 2004;39(3):667–75.
40. Okumura K, Nakamura K, Hisatomi Y, Nagano K, Tanaka Y, Terada K, Sugiyama T, Umeyama K, Matsumoto K, Yamamoto T, Endo F. Salivary gland progenitor cells induced by duct ligation differentiate into hepatic and pancreatic lineages. Hepatology. 2003;38(1):104–13.
41. Feng J, Van der Zwaag M, Stokman MA, Van Os R, Coppes RP. Isolation and characterization of human salivary gland cells for stem cell transplantation to reduce radiation-induced hyposalivation. Radiother Oncol. 2009;92:466–71.
42. Okumura K, Shinohara M, Endo F. Capability of tissue stem cells to organize into salivary rudiments. Stem Cells Int. 2012;2012:502136.
43. Nanduri LS, Maimets M, Pringle SA, van der Zwaag M, van Os RP, Coppes RP. Regeneration of irradiated salivary glands with stem cell marker expressing cells. Radiother Oncol. 2011;99(3):367–72.
44. Sumita Y et al. Bone marrow-derived cells rescue salivary gland function in mice with head and neck irradiation. Int J Biochem Cell Biol. 2011;43:80–7.
45. Shan Z, Li J, Zheng C, Liu X, Fan Z, Zhang C, Goldsmith CM, Wellner RB, Baum BJ, Wang S. Increased fluid secretion after adenoviral-mediated transfer of the human aquaporin-1 cDNA to irradiated miniature pig parotid glands. Mol Ther. 2005;11(3):444–51.
46. Baum BJ, Alevizos I, Zheng C, Cotrim AP, Liu S, McCullagh L, Goldsmith CM, Burbelo PD, Citrin DE, Mitchell JB, Nottingham LK, Rudy SF, Van Waes C, Whatley MA, Brahim JS, Chiorini JA, Danielides S, Turner RJ, Patronas NJ, Chen CC, Nikolov NP, Illei GG. Early responses to adenoviral-mediated transfer of the aquaporin-1 cDNA for radiation-induced salivary hypofunction. Proc Natl Acad Sci U S A. 2012;109(47):19403–7.
47. Yin H, Nguyen CQ, Samuni Y, Uede T, Peck AB, Chiorini JA. Local delivery of AAV2-CTLA4IgG decreases sialadenitis and improves gland function in the C57BL/6.NOD-Aec1Aec2 mouse model of Sjögren's syndrome. Arthritis Res Ther. 2012;14(1):R40.
48. Wang J, Wang F, Xu J, Ding S, Guo Y. Double-strand adeno-associated virus-mediated exendin-4 expression in salivary glands is efficient in a diabetic rat model. Diabetes Res Clin Pract. 2014;103(3):466–73.
49. Voutetakis A, Bossis I, Kok MR, Zhang W, Wang J, Cotrim AP, Zheng C, Chiorini JA, Nieman LK, Baum BJ. Salivary glands as a potential gene transfer target for gene therapeutics of some monogenetic endocrine disorders. J Endocrinol. 2005;185(3):363–72.
50. Samuni Y, Zheng C, Cawley NX, Cotrim AP, Loh YP, Baum BJ. Sorting of growth hormone-erythropoietin fusion proteins in rat salivary glands. Biochem Biophys Res Commun. 2008;373(1):136–9.
51. Samuni Y, Cawley NX, Zheng C, Cotrim AP, Loh YP, Baum BJ. Sorting behavior of a transgenic erythropoietin-growth hormone fusion protein in murine salivary glands. Hum Gene Ther. 2008;19(3):279–86.
52. Voutetakis A, Zheng C, Metzger M, Cotrim AP, Donahue RE, Dunbar CE, Baum BJ. Sorting of transgenic secretory proteins in rhesus macaque parotid glands after adenovirus-mediated gene transfer. Hum Gene Ther. 2008;19(12):1401–5.
53. Nguyen CQ, Yin H, Lee BH, Chiorini JA, Peck AB. IL17: potential therapeutic target in Sjögren's syndrome using adenovirus-mediated gene transfer. Lab Investig. 2011;91(1):54–62.
54. Wei C, Larsen M, Hoffman MP, Yamada KM. Self-organization and branching morphogenesis of primary salivary epithelial cells. Tissue Eng. 2007;13(4):721–35.
55. Ishida K, Murofushi M, Nakao K, Morita R, Ogawa M, Tsuji T. The regulation of tooth morphogenesis is associated with epithelial cell proliferation and the expression of Sonic hedgehog through epithelial-mesenchymal interactions. Biochem Biophys Res Commun. 2011;405(3):455–61.
56. Ekström J, Khosravani N, Castagnola M, Messana I. Saliva and the control of its secretion. Berlin: Springer; 2012.
57. Matsuo R, Yamamoto T, Yoshitaka K, Morimoto T. Neural substrates for reflex salivation induced by taste, mechanical, and thermal stimulation of the oral region in decerebrate rats. Jpn J Physiol. 1989;39:349–57.
58. Matsuo R, Garrett JR, Proctor GB, Carpenter GH. Reflex secretion of proteins into submandibular saliva in conscious rats, before and after preganglionic sympathectomy. J Physiol. 2000;527:175–84.
59. Proctor GB, Guy HC. Regulation of salivary gland function by autonomic nerves. Auton Neurosci. 2007;133:3–18.
60. Turner RJ, Sugiya H. Understanding salivary fluid and protein secretion. Oral Dis. 2002;8(1):3–11.
61. Matsuo R. Role of saliva in the maintenance of taste sensitivity. Crit Rev Oral Biol Med. 2000;11:216–29.

62. Froehlich DA, Pangborn RM, Whitaker JR. The effect of oral stimulation on human parotid salivary flow rate and alpha-amylase secretion. Physiol Behav. 1987;41(3):209–17.
63. Sasano T, Satoh-Kuriwada S, Shoji N, Sekine-Hayakawa Y, Kawai M, Uneyama H. Application of umami taste stimulation to remedy hypogeusia based on reflex salivation. Biol Pharm Bull. 2010;33(11):1791–5.
64. Ogawa M, Yamashita K, Niikura M, Nakajima K, Toyoshima KE, Oshima M, Tsuji T. Saliva secretion in engrafted mouse bioengineered salivary glands using taste stimulation. J Prosthodont Res. 2014;58(1):17–25.
65. Lamy E, Graca G, Costa GD, Franco C, Silva FC, Baptista ES, et al. Changes in mouse whole saliva soluble proteome induced by tannin-enriched diet. Proteome Sci. 2010;8:65.
66. Cohen S. Isolation of a mouse submaxillary gland protein accelerating incisor eruption and eyelid opening in the new-born animal. J Biol Chem. 1962;237:1555–62.
67. Sreebny LM, Schwartz SS. A reference guide to drugs and dry mouth – 2nd edition. Gerodontology. 1997;14(1):33–47.
68. Wu SM, Hochedlinger K. Harnessing the potential of induced pluripotent stem cells for regenerative medicine. Nat Cell Biol. 2011;13(5):497–505.
69. Cohen DE, Melton D. Turning straw into gold: directing cell fate for regenerative medicine. Nat Rev Genet. 2011;12(4):243–52.
70. Yan X, Qin H, Qu C, Tuan RS, Shi S, Huang GT. iPS cells reprogrammed from human mesenchymal-like stem/progenitor cells of dental tissue origin. Stem Cells Dev. 2010;19(4):469–80.
71. Eiraku M, Takata N, Ishibashi H, Kawada M, Sakakura E, Okuda S, Sekiguchi K, Adachi T, Sasai Y. Self-organizing optic-cup morphogenesis in three-dimensional culture. Nature. 2011;472(7341):51–6.
72. Suga H, Kadoshima T, Minaguchi M, Ohgushi M, Soen M, Nakano T, Takata N, Wataya T, Muguruma K, Miyoshi H, Yonemura S, Oiso Y, Sasai Y. Self-formation of functional adenohypophysis in three-dimensional culture. Nature. 2011;480(7375):57–62.
73. Ozone C, Suga H, Eiraku M, Kadoshima T, Yonemura S, Takata N, Oiso Y, Tsuji T, Sasai Y. Functional anterior pituitary generated in self-organizing culture of human embryonic stem cells. Nat Commun. 2016;7:10351.

Part IV

Therapeutic Considerations for Restoration of Salivary Function

Regulation of Salivary Secretion 11

Guy Carpenter and Polliane Carvalho

Abstract

The production of saliva in conscious humans is under the control of the higher centers of the brain which is upregulated by an autonomic reflex in response to taste, chewing, and smell. The higher centers of the brain maintain the resting rate of salivary secretion which in the healthiest subjects is sufficient to maintain oral health and perform the functions of the mouth such as speaking, swallowing, and preventing the overgrowth of microbial colonies on oral tissues. When asleep, the same higher centers reduce their neural output leading to very low salivary flow, which prevents choking or aspiration of saliva into the lungs leading to pneumonia. An understanding of this complex control of salivary secretion is particularly important for the regeneration of the salivary glands and their functions. Stem cell treatments to replace the salivary tissue are an impressive first step, but the new tissue needs to be under neural control. Inappropriate salivary secretion can be just as much a problem as insufficient salivary flow as demonstrated by drooling in stroke patients and patients on certain antipsychotic medications, who "drown" in their own saliva at night.

11.1 Introduction

Salivary secretion is an autonomic reflex activated mainly by taste, chewing, and smell for some salivary glands [10, 17, 36]. In conscious humans, there is a resting rate (approximately 0.5 ml/min) of salivary secretion into the mouth by the three pairs of major glands (parotid, submandibular, and sublingual) and the hundreds of minor glands. The resting salivary rate is influenced mostly by the higher centers of the brain (such as the hypothalamus and amygdala), which increase their neural input to the salivary centers located in the brainstem during the day but decrease at night and during times of anxiety leading to a dry mouth at those times. Upon stimulation by taste, smell, or chewing, salivary secretion is greatly upregulated two to three times greater than the resting rate. Thus when we put food into the mouth, nerves associated with taste

G. Carpenter, PhD (✉) • P. Carvalho, MSc, PhD
Mucosal and Salivary Biology Division,
King's College London Dental Institute, London, UK
e-mail: guy.carpenter@kcl.ac.uk

S. Cha (ed.), *Salivary Gland Development and Regeneration*, DOI 10.1007/978-3-319-43513-8_11

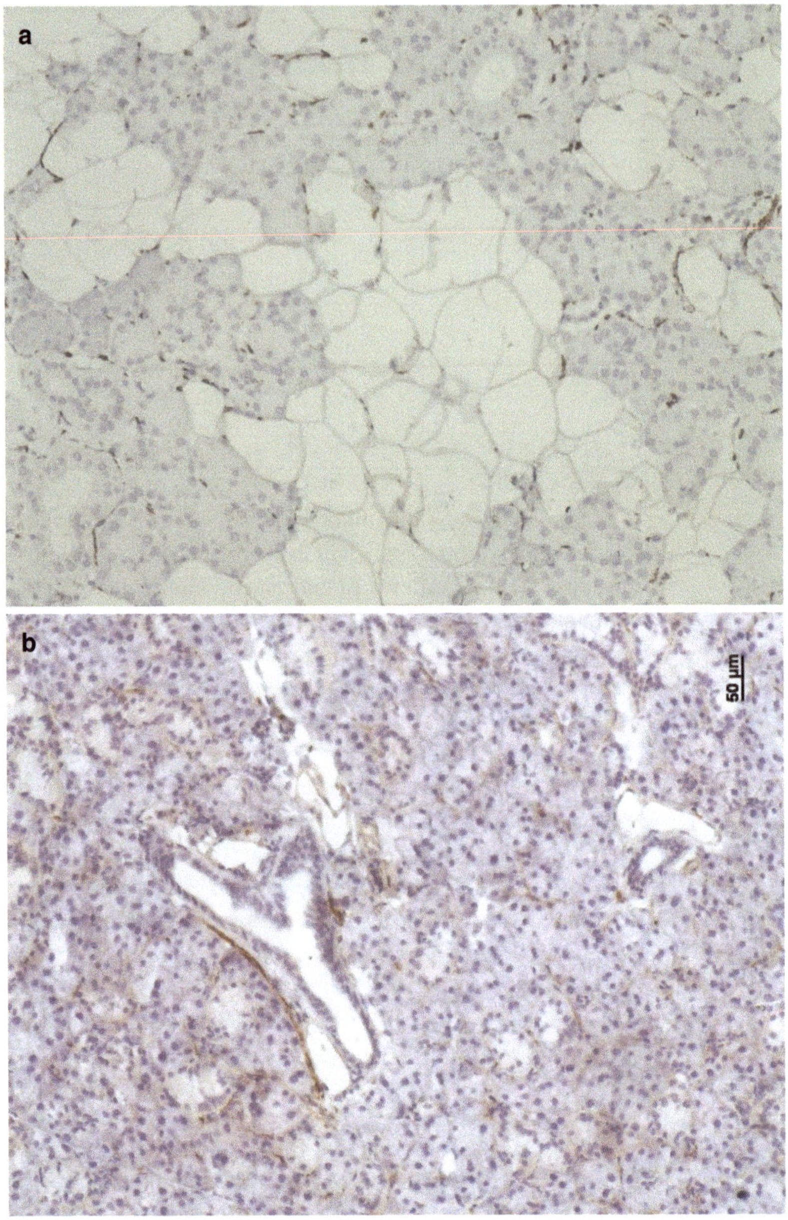

Fig. 11.1 Sympathetic nerve innervation of acinar and ductal cells in the submandibular glands of humans and mice. (**a**) Tyrosine hydroxylase (*brown*) staining (1:100 dilution) of human submandibular glands lightly counter-stained with hematoxylin (*blue*) shows dense sympathetic innervation of the acinar and ductal cells as well as the empty-looking fat cells. The empty appearance of the fat cells is caused by the tissue fixation process. (**b**) Tyrosine hydroxylase (1:100) staining of mouse submandibular gland also demonstrates plentiful sympathetic nerves around acini and ducts. However, there are no fat cells in mouse salivary glands

buds, bare nerve endings in the mucosa, and olfactory receptors in the nose and mechanoreceptors (Ruffini endings) in the mouth are all stimulated and send signals via autonomic afferents back to the salivary centers in solitary tract nucleus [28]. Sympathetic and parasympathetic efferents are then sent to each of the glands to increase salivary secretion [38] (Fig. 11.1a, b).

Salivary flow is controlled separately from salivary protein secretion, which itself is differently regulated depending on the secretory cell type, granule or vesicular secretion route, and even the protein itself [30]. Anaesthetized animals with isolated nerve preparations allowed researchers to show that parasympathetic stimulation *per se* caused a high flow with a low protein concentration-type saliva, whereas sympathetic stimulation evoked a high protein with low-flow saliva [13, 38]. In conscious animals, however, it was found that most salivary secretion is composed of parasympathetic stimulation with smaller amounts of sympathetic stimulation overlaid [7, 28]. In contrast to the rest of the body, the autonomic nerves within the salivary glands work in harmony rather than antagonistically. Similar experiments in humans using adrenergic and cholinergic blocking drugs revealed that a similar situation occurs [2].

11.2 Salivary Secretion by Taste, Chewing, and Smell

Taste buds are mostly located at the back of the tongue within the foliate and circumvallate papilla. Some taste buds occur at the front of the tongue associated with (but not always within) the fungiform papillae, which are the red dots readily apparent on the tongue. The taste maps of the tongue depicting sweet tastes at the front of the tongue and salt at the sides, etc. often reproduced in textbooks are now largely discounted. There is abundant evidence to show most areas of the tongue are able to detect most tastes. There is considerable variation in the number of taste buds between people, which has some correlation with super-taster status – a heightened ability to detect and discriminate tastes [16]. At both circumvallate and foliate papilla, the taste buds located within crypts are bathed in a secretion from serous minor salivary glands. These glands (von Ebner's) have some interesting proteins that have been suggested to be involved in fat detection, such as lingual lipase and lipocalin [22]. However, the output from these glands is so small that it would be highly unlikely that they play a significant role in fat detection or digestion within food. However, it is possible these lipases maintain the environment within the crypts to maintain taste bud acuity [29, 45].

Much progress has been made in characterizing the different channels responsible for the detection of the basic tastes by taste bud cells [9]. Salt taste is transmitted by sodium and possibly potassium channels located on the apical surface of taste bud cells that signal to afferent nerves via ATP molecules, whereas sour taste (which is composed of protons) is detected by a separate channel. Receptors for bitter tastes and glutamate have also been determined [3, 19]. Now that specific receptors have been cloned, more studies are examining the confounding factors of taste receptors, such as age [31] and obesity.

In addition to taste, the other major stimulus for increased salivary secretion is chewing. Mechanoreceptors in the gingival pocket surrounding each tooth are the main receptor for tooth movement related to chewing. Several studies have shown that increases in chewing activity lead to increased salivary secretion [1, 17] although, interestingly, empty chewing (i.e., clenching teeth) does not lead to salivary secretion. Under normal eating conditions, taste and chewing afferent nerve signals are combined to cause, at best, an additive effect on salivary secretion.

Additionally, smells can also stimulate salivary secretion. Olfactory stimuli have been shown to stimulate submandibular/sublingual secretion but not parotid glands [24, 25]. When food is consumed, aerosols are released from the food, probably aided by mixing with saliva, which travel via the retronasal route to the olfactory neuroepithelium in the nose, and contribute flavor signals to the basic tastes detected by the tongue. Indeed, much of the taste of food comes from the olfactory input rather than the taste or chewing that occurs in the mouth. Olfactory stimulation of salivary glands is probably of least importance to salivary secretion stimulated by food in the mouth but does appear to contribute to the mouthwatering phenomenon. This is the subjective feeling of excessive saliva in the mouth often associated with the thought of food. However, few scientists have been able to show a thought or sight evoked secretion of saliva. It would appear that in some situations, smells are apparent which could lead to some secretion and the mouthwatering response. However, in many situations where a mouthwatering response occurs in the absence of food-related smells, it could also be due to facial muscles squeezing on turgid salivary ducts to cause transient flows sufficient to be detected as mouthwatering [18].

11.3 Mechanism of Salivary Secretion

As noted above, the fluid component of saliva is differently regulated to the protein (and to some extent the ionic component) of saliva. Salivary glands are composed of polarized epithelial cells and have two main forms – the acini and the ducts. Often described as resembling a bunch of grapes, the acini are the site of fluid formation, while the ducts modify the saliva and convey it to the mouth. Fluid is mobilized by creating an ionic gradient across the acinar cells (primary saliva) which then is modified by the ducts [43]. The osmotic gradient is created by the selective secretion of chloride ions through the apical membranes of polarized acini. Thus once the parasympathetic nerves from the brain have conveyed the signal to secrete by releasing acetylcholine which binds to muscarinic receptors on the acinar cells, activation of intracellular calcium signaling elicits the opening of chloride channels on the apical side of the acini. Sodium ions follow the chloride ions through an electrochemical attraction so that a higher concentration of sodium chloride exists in the ductal/apical side of the cell compared to the basolateral/interstitial side. This osmotic gradient draws fluid from blood vessels, via the interstitial compartment, toward the apical side and into the ductal system. Water may pass either around the acini through the tight junctions between cells or via the aquaporin channels within the acini [27].

11.4 Neural Connections to the Different Glands

Salivary glands are unique in utilizing parasympathetic and sympathetic innervation in an additive/synergistic manner rather than the more usual antagonist setup found for the regulation of blood flow and other functions in the body. Taste, mechanical, or smell signals generate afferent signals in fibers of the facial (CNVII), glossopharyngeal (CNIX), and trigeminal (CNV) nerves. The nucleus of the solitary tract is innervated by the CNVII and CNIX and sends interneurons to the salivary centers. Interneurons also supply the primary sympathetic salivary centers which are located in the upper thoracic segments of the spinal cord. Efferent nerve fibers from the salivary nuclei conduct signals via the chorda lingual nerve to the submandibular ganglion and onto the submandibular and sublingual glands. The parotid gland is supplied by efferent fibers in the glossopharyngeal (tympanic branch) nerve to the otic ganglion and postganglionic fibers in the auriculotemporal nerve. There also appears to be a contribution to the parotid gland efferent supply from the facial nerve. Minor salivary glands are supplied by parasympathetic nerve fibers in the buccal branch of the mandibular nerve, the lingual nerve, and the palatine nerve.

The salivary reflex is affected by the higher centers of the brain and shows circadian-like variations particularly in the resting salivary flow. This central neural activity appears to contribute towards the lower salivary secretion during sleep and zero flow during anesthesia. Suppression of impulse traffic from the salivary nuclei to salivary glands leading to reduced salivation and dry mouth is most obviously demonstrated during fear and anxiety. However, these effects are less obvious in stimulated flow rates, where the effects of taste and chewing predominate.

Significant advancements in our understanding of the brain have been made possible by functional MRI [40]. By the injection of labeled glucose (or other substrates), the active regions of the brain can be imaged when stimuli such as food or drinks are put in the mouth. Despite some recent advances in understanding of how tastes are perceived, relatively little attention has been paid as to how taste affects the salivary nuclei.

Most people believe that the thought of foods activates salivary secretion, the so-called mouthwatering [20]. However, neither Pavlov nor Lashley found any evidence to support the presence of a conditional salivary reflex in man. fMRI studies have demonstrated the considerable differences between animal and human brains in response to food [41]. Experiments by the author also suggest that just the thought of food does not sustain a stimulated salivary flow and that most mouthwatering experiences are the result of smells evoking submandibular/sublingual salivary flow [18]. Using flow meters, it was possible to detect, particularly when subjects were hungry, small spikes of salivary flow. It was speculated that facial muscles compress the turgid ducts coming from salivary glands to the mouth to cause small transient "flows" of saliva that can be easily perceived by the subject.

11.5 Neurotransmitters and Receptors

Salivary acinar and ductal cells have been well studied for their receptors. Muscarinic receptors are usually cited as most important for fluid secretion, but they also cause a significant degree of protein secretion, mostly mucin and sIgA [8]. Using mouse knockout models, M3 appears most important with smaller contributions from M1 [11, 14] although the in vivo situation is likely to be far more complex with inputs from purinergic [4] and peptidergic neurotransmitters [12].

Acinar cell activation of fluid transport is achieved through increases in intracellular calcium concentration and binding of calcium to ion-transporting proteins. The acinar cell muscarinic receptors are G-protein-coupled receptors; binding of acetylcholine leads to a G-protein/phospholipase C-mediated generation of inositol triphosphate (IP3) from phosphatidylinositol 4,5-bisphosphate. IP3 interacts with IP3 receptors (IP3Rs) on the endoplasmic reticulum causing release of stored calcium. Cytoplasmic calcium levels are tightly controlled by rapid removal of calcium through the actions of plasma membrane and ER calcium pumps. Store-operated calcium entry has been shown to be dependent upon the presence of three proteins, STIM1, Orai1, and TRPC1 channels. Other receptors (α1-adrenoceptor, substance P neurokinin 1 receptor, P2Y receptor, P2X receptors) utilize intracellular calcium signaling mechanisms but may make comparatively minor contributions to salivary fluid secretion under physiological conditions.

11.6 Protein Secretion

Protein secretion, following stimulation by sympathetic and to lesser extent parasympathetic nerves, activates adrenergic receptors on cells and via intracellular cyclic AMP signaling causes the storage granules to migrate toward the apical membrane, fuse, and then release their secretory protein cargo into the ductal lumen. While the storage granule mechanism is the major route by which proteins enter the ductal lumen, non-storage vesicles also transport other proteins such as secretory IgA (sIgA). This is the main antibody in saliva since it is actively transported via a membrane receptor (polymeric immunoglobulin receptor) into saliva, whereas other classes of antibody such as IgG and IgE are unable to bind the membrane receptor and so passively diffuse into saliva. Differences in the secretion of sIgA and other proteins highlight differences between different secretory mechanisms within one cell [7]. However acinar and ductal cells have different secretory proteins and are regulated by different neural impulses [39]. Thus considerable complexity can exist in protein secretion within a single salivary gland.

In humans, during normal conscious reflex secretion, this complexity is less apparent since secretory inputs are processed centrally and so fluid and protein secretion seem to occur together. Thus, from a single gland, such as the parotid, which is the easiest to collect from using a Lashley suction cup, a similar range of proteins are secreted at rest and when stimulated by different taste stimuli although the relative proportions of some proteins (such as sIgA and amylase) may vary [37]. A more detailed study by mass spectrometer methods has revealed that there are some small changes in the composition of proteins [33]. The greatest variation in protein secretion is seen in whole-mouth saliva, which is the combination of all the salivary glands. At rest, submandibular and sublingual glands predominate; when stimulated by taste, parotid is the single most dominant contributor to salivary protein. However, smell or chewing without taste evokes some differences in salivary proteins, most noticeably muc7 and statherin [18].

11.7 Studies of Neural Agonists and Antagonists

α2-Adrenoceptor agonists (e.g., clonidine) and antagonists (e.g., yohimbine) have been demonstrated to act centrally in studies of reflex secretion in human subjects and cholinergically evoked secretion in animal models. α2-Adrenoceptor blockade can increase salivary secretion, while α2-adrenoceptor agonists inhibit secretion. It appears that adrenergic agonists such as amphetamine exert an inhibitory effect on the flow of saliva through the release of noradrenaline from nerves in the medulla causing activation of inhibitory α2-adrenoceptors rather than through a peripheral vasoconstrictive effect. These central effects of amphetamine that cause a dry mouth contrast with its action in the periphery leading to increased secretion of protein by salivary cells and increased salivary protein concentration. The presence of muscarinic receptors on neurons of the salivary nuclei may also partly explain the observed effects on salivary secretion evoked by intracerebro-ventricular injection of pilocarpine or atropine which were found to, respectively, stimulate and inhibit salivation.

The recent use of botulinum toxin for intramuscular paralysis has prompted a number of researchers to use this on salivary glands, principally as a treatment for drooling [26]. Drooling represents the greatest concern of cares of stroke victims, Parkinson's, and other degenerative diseases as the drooling constantly wets clothing and has led some clinicians to deliberately damage the salivary glands by irradiation to effect a remedy [21]. Studies on rabbits have demonstrated that irradiation has atrophic effects on the salivary glands [46] partially by damaging existing cells but also by damaging stem cells leading to reduced repopulation during normal cell turnover [35]. Currently Botox treatment is limited to terminally ill patients due to the risk of whole body neurotoxicity and inhibiting the muscles involved in swallowing thus potentially leading to even greater excessive salivation (sialorrhea)/choking.

11.8 Considerations for Regeneration of Salivary Glands

Initial and recent studies by Coppes and colleagues [32, 35] have demonstrated that stem cell therapy of salivary glands involves an initial short-term recovery of the already existing salivary cells and a longer-term repopulation of the glandular stem cells. Short-term effects demonstrated a recovery of salivary flow in response to autonomimetics demonstrating a functional recovery of the acinar cells. These studies have been a vital step in demonstrating the potential of the treatment and opens further lines of inquiry.

The treatment with stem cells is though fraught with potential problems, the most serious of which is the potential transformation of the injected stem cells into noncancerous growths called teratomas. Even if this risk is extremely low, it has to be balanced against the benefits to patients from increased salivary production.

Injected stem cells appear to help preexisting salivary acinar cells recover function as well as repopulating the stem cell pool for longer-term function. Bone marrow soups [44] or mesenchymal stem cell therapy has multiple growth factors that probably boost the recovery of acinar cells in an atrophic/diseased state. Epithelial stem cells extracted from existing salivary glands (labeled with anti-c-Kit antibodies) are probably required for longer-term repopulation of the stem/progenitor cell pool [35]. In both cases, the endogenous structure of the gland is used to position these cells so that they can contribute to salivary secretion into the mouth.

An alternative approach is to bioengineer the gland in vitro and then transplant the gland into the existing ductal (excretory duct) structures [34]. In most of these studies, an autonomimetic has been used to test the function or ability of the gland to produce saliva. This drug (usually pilocarpine) is injected into the body and reaches the salivary glands by the bloodstream. It then stimulates salivary secretion by directly binding to muscarinic receptors on the acinar cells bypassing any nerve-acinar cell junction. Few papers have considered whether the nerve-acinar cell junction has formed or is even functional.

Any researcher devising therapies for the recovery of salivary glands has to have an appreciation of the complex neural control of salivary glands as detailed above as well as the structural architecture. In particular, the diurnal variation in salivary flow with an upregulation of the resting flow rate during periods of eating and chewing but a downregulation of salivary secretion during sleep will need careful attention. The problems of a high flow at night have been well documented by Ekström and colleagues when describing patients on clozapine – an antipsychotic prescribed to treatment-resistant schizophrenia patients. The night-time sialorrhea causes patients to choke at night with the feeling of "drowning" frequently reported. Clearly then we need to carefully regulate the degree of regeneration of salivary glands whether by stem cells or drugs.

In studies of transplanted salivary glands to avoid irradiation fields in treatment of head and neck cancers [5] or treatment of chronic dry eyes [15, 23], it has become apparent that the transplanted glands can become innervated from nerves attached to local blood vessels [15] leading to some interesting effects whereby salivary secretion increased with exercise and temperature (reflecting increased blood flow). The preferred option that is required is the regenerated salivary gland to be innervated by both parasympathetic and sympathetic nerves that were originally in the gland so that it responds in a manner of reflex to stimuli that initiate increased salivation.

Although some animal studies have shown the ability of salivary gland nerves to regrow into regenerating glands [6], this has not yet been shown in humans. Certainly with disease such as Sjögren's syndrome, it is known that there can be a loss of the fine nerve fibers adjacent to areas of inflammation [42]. Thus, in any treatment of salivary glands, some account has to be taken of the preexisting innervation to determine the likely chances of successful regeneration. It is well established that salivary glands require an intact innervation to maintain their histological appearance and functional capability.

References

1. Anderson DJ, Hector MP, Linden RWA. The effects of unilateral and bilateral chewing, empty clenching and simulated bruxism, on the masticatory-parotid salivary reflex in man. Exp Physiol. 1996;81:305–12.
2. Baum BJ, Wellner RB. Receptors in salivary glands. In: Garrett JR, Ekstrom J, Anderson LC, editors. Neural mechanisms of salivary secretion. Basel: Karger; 1999.
3. Behrens M, Meyerhof W. Bitter taste receptor research comes of age: from characterization to modulation of TAS2Rs. Semin Cell Dev Biol. 2013;24:215–21.
4. Bhattacharya S, Imbery JF, Ampem PT, Giovannucci DR. Crosstalk between purinergic receptors and canonical signaling pathways in the mouse salivary gland. Cell Calcium. 2015;58:589–97.
5. Burghartz M, Ginzkey C, Hackenberg S, Hagen R. Two-stage autotransplantation of the human submandibular gland: first long-term results. Laryngoscope. 2016;126:1551–5.
6. Carpenter GH, Khosravani N, Ekstrom J, Osailan SM, Paterson KP, Proctor GB. Altered plasticity of the parasympathetic innervation in the recovering rat submandibular gland following extensive atrophy. Exp Physiol. 2009;94:213–9.
7. Carpenter GH, Proctor GB, Anderson LC, Zhang XS, Garrett JR. Immunoglobulin A secretion into saliva during dual sympathetic and parasympathetic nerve stimulation of rat submandibular glands. Exp Physiol. 2000;85:281–6.
8. Carpenter GH, Proctor GB, Ebersole LE, Garrett JR. Secretion of IgA by rat parotid and submandibular cells in response to autonomimetic stimulation in vitro. Int Immunopharmacol. 2004;4:1005–14.
9. Chandrashekar J, Hoon MA, Ryba NJP, Zuker CS. The receptors and cells for mammalian taste. Nature. 2006;444:288–94.

10. Chaudhari N, Roper SD. The cell biology of taste journal of cell biology. J Cell Biol. 2010;190: 285–96.
11. Culp DJ, Luo W, Richardson LA, Watson GE, Latchney LR. Both M(1) and M(3) receptors regulate exocrine secretion by mucous acini. Am J Phys Cell Phys. 1996;40:C1963–72.
12. Del Fiacco M, Quartu M, Ekstrom J, Melis T, Boi M, Isola M, Loy F, Serra MP. Effect of the neuropeptides vasoactive intestinal peptide, peptide histidine methionine and substance P on human major salivary gland secretion. Oral Dis. 2015;21:216–23.
13. Garrett JR. The Proper Role of Nerves in Salivary Secretion - A Review. J Dent Res. 1987;66:387–97.
14. Gautam D, Heard TS, Cui YH, Miller G, Bloodworth L, Wess J. Cholinergic stimulation of salivary secretion studied with M-1 and M-3 muscarinic receptor single- and double-knockout mice. Mol Pharmacol. 2004;66:260–7.
15. Gerling G, Garrett JR, Paterson KP, Sieg JP, Collin RO, Carpenter GH, Hakim SG, Lauer I, Proctor GB. Innervation and secretory function of transplanted human submandibular glands. Transplantation. 2008;85:135–40.
16. Hayes JE, Bartoshuk LM, Kidd JR, Duffy VB. Supertasting and PROP bitterness depends on more than the TAS2R38 gene. Chem Senses. 2008;33:255–65.
17. Hector MP, Linden RW. Reflexes of salivary secretion. In: Garrett JR, Ekstrom J, Anderson LC, editors. Neural mechanisms of salivary secretion. Basel: Karger; 1999.
18. Ilangakoon Y, Carpenter GH. Is the mouthwatering sensation a true salivary reflex? J Texture Stud. 2011;42:212–6.
19. Iwatsuki K, Ichikawa R, Uematsu A, Kitamura A, Uneyama H, Torii K. Detecting sweet and umami tastes in the gastrointestinal tract. Acta Physiol. 2012;204:169–77.
20. Jenkins GN, Dawes C. Psychic flow of saliva in man. Arch Oral Biol. 1966;11:1203–4.
21. Kasarskis EJ, Hodskins JST, WH C. Unilateral parotid electron beam radiotherapy as palliative treatment for sialorrhea in amyotrophic lateral sclerosis. J Neurol Sci. 2011;308:155–7.
22. Kawai T, Fushiki T. Importance of lipolysis in oral cavity for orosensory detection of fat. Am J Phys Regul Integr Comp Phys. 2003;285:R447–54.
23. Konitzer J, Gerdan L, Sieg P, Hakim SG, Bruggemann A, Gebhard M, Buchmann I. Salivary gland scintigraphy in patients suffering from severe keratoconjunctivitis sicca Evaluation of the viability of submandibular glands before and after transfer into the temporal fossa. Nuklearmedizin-Nuclear Med. 2015;54:189–95.
24. Lee VM, Linden RWA. An olfactory submandibular salivary reflex in humans. Exp Physiol. 1992a;77: 221–4.
25. Lee VM, Linden RWA. The effect of odors on stimulated parotid salivary flow in humans. Physiol Behav. 1992b;52:1121–5.
26. Lungren MP, Halula S, Coyne S, Sidell D, Racadio JM, Patel MN. Ultrasound-guided botulinum toxin type a salivary gland injection in children for refractory sialorrhea: 10-year experience at a large tertiary children's hospital. Pediatr Neurol. 2016;54:70–5.
27. Ma TH, Song YL, Gillespie A, Carlson EJ, Epstein CJ, Verkman AS. Defective secretion of saliva in transgenic mice lacking aquaporin-5 water channels. J Biol Chem. 1999;274:20071–4.
28. Matsuo R. Central connections for salivary innervations and efferent impulse formation. In: Garrett JR, Ekstrom J, Anderson LC, editors. Neural mechanisms of salivary secretion. Basel: Karger; 1999.
29. Matsuo R. Role of saliva in the maintenance of taste sensitivity. Crit Rev Oral Biol Med. 2000;11:216–29.
30. Melvin JE, Yule D, Shuttleworth T, Begenisich T. Regulation of fluid and electrolyte secretion in salivary gland acinar cells. Annu Rev Physiol. 2005;67:445–69.
31. Mennella JA, Pepino MY, Duke FF, Reed DR. Age modifies the genotype-phenotype relationship for the bitter receptor TAS2R38. BMC Genet. 2010;11:60.
32. Nanduri LSY, Maimets M, Pringle SA, Van Der Zwaag M, Van Os RP, Coppes RP. Regeneration of irradiated salivary glands with stem cell marker expressing cells. Radiother Oncol. 2011;99:367–72.
33. Neyraud E, Sayd T, Morzel M, Dransfield E. Proteomic analysis of human whole and parotid salivas following stimulation by different tastes. J Proteome Res. 2006;5:2474–80.
34. Ogawa M, Oshima M, Imamura A, Sekine Y, Ishida K, Yamashita K, Nakajima K, Hirayama M, Tachikawa T, Tsuji T. Functional salivary gland regeneration by transplantation of a bioengineered organ germ. Nat Commun. 2013;4:2498.
35. Pringle S, Maimets M, Van Der Zwaag M, Stokman MA, Van Gosliga D, Zwart E, Witjes MJH, De Haan G, Van Os R, Coppes RP. Human salivary gland stem cells functionally restore radiation damaged salivary glands. Stem Cells. 2016;34:640–52.
36. Proctor GB. The physiology of salivary secretion. Periodontology 2000. 2016;70:11–25.
37. Proctor GB, Carpenter GH. Chewing stimulates secretion of human salivary secretory immunoglobulin A. J Dent Res. 2001;80:909–13.
38. Proctor GB, Carpenter GH. Regulation of salivary gland function by autonomic nerves. Auton Neurosci-Basic Clin. 2007;133:3–18.
39. Proctor GB, Carpenter GH. Salivary secretion: mechanism and neural regulation. Monogr Oral Sci. 2014;24:14–29.
40. Rolls ET. Taste, olfactory and food texture reward processing in the brain and obesity. Int J Obes. 2011;35:550–61.
41. Small DM. Taste representation in the human insula. Brain Struct Funct. 2010;214:551–61.
42. Sorensen CE, Larsen JO, Reibel J, Lauritzen M, Mortensen EL, Osler M, Pedersen AML. Associations between xerostomia, histopathological alterations, and autonomic innervation of labial salivary glands in men in late midlife. Exp Gerontol. 2014;57:211–7.

43. Thaysen JH, Thorn NA, Schwartz IL. Excretion of sodium, potassium, chloride and carbon dioxide in human parotid saliva. Am J Physiol. 1954;178:155–9.
44. Tran SD, Liu YN, Xia DS, Maria OM, Khalili S, Wang RWJ, Quan VH, Hu S, Seuntjens J. Paracrine effects of bone marrow soup restore organ function, regeneration, and repair in salivary glands damaged by irradiation. PLoS ONE. 2013;8:e61632.
45. Voigt N, Stein J, Galindo MM, Dunkel A, Raguse JD, Meyerhof W, Hofmann T, Behrens M. The role of lipolysis in human orosensory fat perception. J Lipid Res. 2014;55:870–82.
46. Xu H, Shan XF, Cong X, Yang NY, Wu LL, Yu GY, Zhang Y, Cai ZG. Pre- and Post-synaptic Effects of Botulinum Toxin A on Submandibular Glands. J Dent Res. 2015;94:1454–62.

Salivary Gland Gene Therapy in Experimental and Clinical Trials

12

Michael Passineau

Abstract

Salivary gland gene therapy presents an opportunity to reprogram the organ on the molecular level and achieve unprecedented therapeutic advancements. This chapter will review the basic biology of gene transfer, with emphasis on those vector systems that have performed well in the salivary gland in animal models. Various therapeutic applications of salivary gland gene therapy will be discussed, including radiation-induced xerostomia and Sjögren's syndrome. The concept of salivary glands as endogenous bioreactors for systemic gene therapeutics in monogenetic and acquired diseases will also be reviewed.

A brief history of the field, with regard to animal models, clinical translational studies, and ultimately a successful phase I/II clinical trial, will be presented. The merits and limitations of the several animal models of salivary gland gene therapy will be reviewed. The chapter concludes with a discussion of human salivary gland gene therapy clinical trials, completed and ongoing, and will point out congruence and discord between preclinical animal studies and clinical trials. Salivary gland gene therapy is now established as safe and therapeutically effective in humans, and the near future of this field will be focused on making this technology practical for outpatient use and broadly disseminating it into the practice of oral and dental medicine.

12.1 Overview

Gene therapy may be broadly defined as the act of delivering a genetic sequence to a target cell or tissue to effect changes in gene expression. Typically, this involves utilizing a *vector* containing an expression cassette comprised of a promoter, open-reading frame for the gene to be expressed (referred to as the "transgene"), and a

M. Passineau, PhD
Gene Therapy Program, Allegheny Health Network, Pittsburgh, PA, USA
e-mail: mpassine@wpahs.org

S. Cha (ed.), *Salivary Gland Development and Regeneration*, DOI 10.1007/978-3-319-43513-8_12

polyA sequence. The presence of these three minimalist elements results in the expression of the transgene within the target cell/tissue at levels proportional to the activity of the promoter in the target. While transgene expression is the classic example of gene therapy, other approaches include delivery of silencing RNA, microRNA, and, increasingly, gene editing.

Gene therapy is fundamentally different from traditional pharmacology and even biopharmacology in that rather than manipulating the existing cellular machinery, gene therapy allows the manipulation of the composition of the cellular machinery itself. Accordingly, gene therapy theoretically expands the armamentarium of therapeutic options beyond what is available through conventional pharmacology, and the two approaches could potentially be synergistic. When considering salivary gland disorders, gene therapy is particularly attractive due to the relative paucity of conventional treatment options for salivary gland disorders.

Historically, gene therapy as a broad field has been very slow to meet its initial promise, and by far the most important factor limiting the mainstreaming of gene therapy has been the difficulty of achieving vector delivery to the target cell/tissue. As detailed below, cellular entry represents a fundamental challenge for gene therapy vectorology, but so does macroscopic delivery of the vector to the target tissue. In particular, delivery of gene therapy vectors to internal, solid organs is inherently challenging, since this requires either surgical intervention or intravenous administration of a vector capable of efficiently trafficking to the target tissue without unwanted accumulation in off-target tissues. This latter consideration has proven to be a major drawback of intravenous administration of gene therapy vectors, particularly viral vectors.

The salivary gland has unique attributes that make it an intriguing and practical site for gene therapy, obviating some of the above concerns. These considerations have been previously elaborated [1] and include the direct accessibility of the organ through bloodless intraoral cannulation of the salivary ducts (Wharton's and Stensen's). This exceptionally simple and safe accessibility has facilitated careful and comparative studies of vectorology in the salivary gland and enabled translation of salivary gland gene therapy from a proof of concept in 1991 [2] to a successful phase I human gene therapy clinical trial in 2006. This chapter will review the roughly two decades between those milestone events, as well as speculating on the future of this promising new approach to salivary gland therapeutics, repair, and regeneration.

12.2 Gene Delivery Technology

The fundamental challenge facing the gene therapist is, simply put, *how to get a gene drug from the outside of a target cell into the cytoplasm.* While only 7 nm of cell membrane stands in the way of this objective, that barrier has proven exceptionally difficult to traverse, to the extent that the gene therapist will often describe the cell membrane as "the longest 7 nm in nature." In marked contrast to prokaryotic cells, eukaryotic cells repel foreign nucleic acids, based both upon passive biophysical principles (DNA is strongly hydrophilic, precluding its diffusion through the cell membrane) and active immunological barriers, both extracellular and intracellular. In one often-repeated statement to Time magazine in 1999 [3], Inder Verma remarked, "There are only three problems in gene therapy: delivery, delivery and delivery."

Dr. Verma's statement accurately reflected the driving force behind the remarkable adherence to Gartner's curve that the gene therapy field has observed. On the positive side, this singular challenge of delivery has not diminished the theoretical potential of gene therapy over the past two decades, and indeed this challenge has driven rational and often successful research designed to meet it. What is now clear is that the diversity of gene therapy applications precludes generalized applications of gene therapy techniques. What is needed rather is a gene therapy "toolbox" comprised of dozen of vectors, devices, and techniques, each of which may address only a few or even a single disease state. This principle has been well demonstrated in salivary gland gene therapy over the past two decades, and this chapter will pay particular attention to the refinement of the salivary gland gene therapy "toolkit" into its present form.

12.2.1 Viral Vectors

The only biological entity in nature capable of efficient transport of nucleic acid across eukaryotic cell membranes is the virus. Indeed, the life cycle of the virus depends upon successful transmembrane transfer of genetic material into the host cell cytoplasm. For this reason, the early days of gene therapy research were dominated by viral vectors, and as of data from July 2015 (http://www.wiley.com//legacy/wileychi/genmed/clinical/), >70 % of all human gene therapy clinical trials have utilized a viral vector.

Viral vectors have two key attributes that determine their suitability for a given gene therapy application, such as gene transfer to the salivary gland (see Fig. 12.1, left panel): (1) proteins in the capsid or envelope of the virus mediate cell binding through receptor/ligand interactions, and thus cell binding affinity can vary dramatically from one cell type to another, and (2) endosome escape is a key feature of the viral life cycle but results in deposition of antigenic viral capsid proteins on MHC receptors, triggering host response to the infected cell. Both of these attributes must be managed in such a way as to match the viral vector choice to the particular application. For example, the canonical adenovirus type 5 (Ad5) is strongly immunogenic in the host, but this can actually be an advantage in gene therapy applications in oncolysis or immunization. In the salivary gland, the most promising applications of gene therapy are for chronic conditions, and the anti-Ad5 host response is very undesirable.

In the salivary gland, a very helpful study published early on surveyed the efficacy of several viral vectors for gene transfer to the salivary gland [4]. This work demonstrated that the only viral vectors capable of robust transduction of the salivary gland are adenovirus and adeno-associated virus (AAV), and these vectors have formed the sole basis of viral-mediated salivary gland gene therapy since 2002. Unfortunately, it is known that AAV does not transduce acinar cells [5], and thus the field does not yet have a viral vector that is well-suited for any gene therapy application requiring gene transfer to the acini of the salivary gland. It is this author's opinion that an alternate AAV serotype will ultimately be discovered that can transduce acinar cells (source: unpublished data from John Chiorini, PhD), but given the variability in viral tropism for the salivary glands of different species, any AAV candidates for acinar cell gene transfer will need to be empirically validated in humans, a complex and challenging task.

12.2.2 Nonviral Vectors

Nonviral vectors can potentially obviate both of the disadvantages of viral vectors in that target cell affinity can be engineered directly into a synthetic construct (often referred to generally as "nanoparticles"; see Fig. 12.1, right panel), and lack of viral protein antigens can evade host immune recognition. The downside of nonviral vectors is that they lack the viral mechanisms that mediate escape from the endosome. Endosomal escape has proven extraordinarily difficult to engineer artificially, and without robust endosome escape, gene transfer efficiency is low due to the vector remaining entombed within the endosome. There has been one report of an endosomolytic nanoparticle capable of effective siRNA to the salivary gland of the rodent [6], but questions remain as to the relevance of this technology for accomplishing transgene expression. The general consensus of the gene therapy literature thus far regarding nanoparticles is that they are often capable of delivering siRNA to diverse targets but are far less effective in accomplishing expression of an exogenous transgene.

Ultrasound-assisted gene transfer (UAGT; see Fig. 12.1, center panel) was proposed in 2010 [7] as a novel method for accomplishing gene transfer to the salivary gland of animal models. This technique circumvents endosomal escape altogether by relying upon transient disruption of the cell membrane and direct transmembrane transit of a

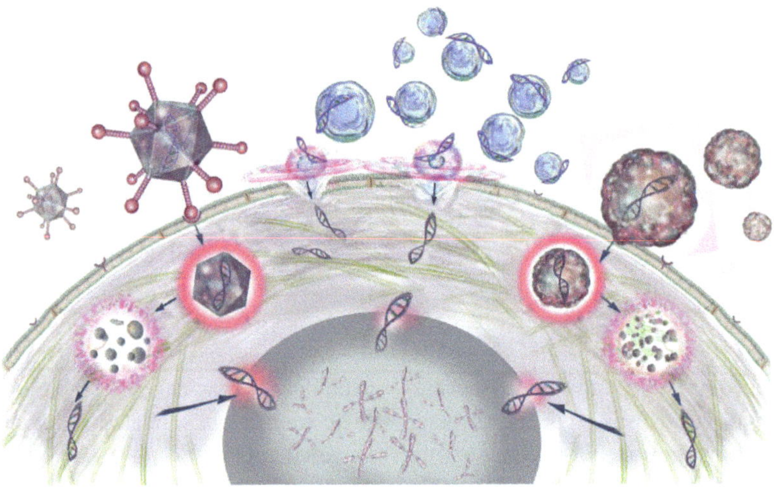

Fig. 12.1 Basic biology of exogenous gene transfer to a target cell. The cell membrane prohibits the passage of genetic material. Ligand/receptor interactions by viral vectors (*left*) or nanoparticles (*right*) can mediate endocytosis of the vector, but this leads to a destination in an endosome, not the cytoplasm. Viral mechanisms (*left*) lead to endosomal disruption, but viral antigens remain and are presented on MHC receptors. Nanoparticles are engineered to evade MHC activation, but endosomal escape has proven difficult to artificially engineer. Sonoporation (*center*) involves the direct transmembrane transit of the nonviral vector via transient pores in the cell membrane produced by fluid-phase cavitation. In all instances, once the genetic material has been deposited in the cytoplasm, transgene expression can occur (© 2013 Michael J Passineau, PhD)

naked DNA vector. The technique relies upon biophysical phenomenon referred to as "sonoporation" in which the DNA vector is associated with ~2.5 μm microbubbles comprised of a lipid bilayer surrounding a perfluoropropane gas core. These microbubbles resonate in a 1 MHz acoustic field, and if the acoustic field is of sufficient power, the microbubbles are violently destroyed, resulting in fluid cavitation, which disrupts the cell membrane. Sonoporation alone appears to have minimal effects upon cellular homeostasis, and no damage to the gland is apparent after UAGT, either by histological or proteomic analysis [8].

12.3 Gene Repair

Gene repair refers to the use of gene transfer to *correct* a deleterious genetic polymorphism in the target cell. The gene repair paradigm was the founding principle of gene therapy many decades ago, driven primarily by the well-characterized gene/disease relationship that exists between CFTR mutations and cystic fibrosis [9]. It is not fair to credit CF as the genesis of the gene therapy research field all on its own, because many other monogenetic disease states were pursued from the earliest days of gene therapy research. Nevertheless, CF is the prototype gene therapy application and, ironically, one of the most difficult to address, with no effective gene therapy treatment yet manifest, despite almost 30 years of research.

Salivary gland gene therapy has never employed the gene repair paradigm, chiefly because of the extremely low impact of monogenetic diseases of the salivary gland. However, an intriguing application of salivary gland gene therapy has been proposed that attempts to address systemic monogenetic diseases by repurposing a portion of the salivary gland into an "endogenous bioreactor" [10, 11] for systemic delivery of biomolecules deficient in these disease states. A variety of therapeutic biomolecules have been expressed in the salivary glands of various animal models and been shown to circulate systemically, with examples including erythropoietin [10, 11], human growth hormone [10–12], α-galactosidase A [13], and GLP-1 [14].

The clinical translation of the salivary gland bioreactor paradigm faces two principal challenges. First, the sorting of the transgene between the apical (exocrine-directed) and basal (endocrine-directed) compartments of the acinar cell cannot be predicted and varies between species. A substantial body of careful research, almost all of it carried out at the NIDCR, has sought to understand and manipulate these two sorting pathways, but the results have not yet produced clear-cut principles likely to apply to humans [15–19]. Thus, the animal models, particularly rodent models, are limited in their ability to predict whether a transgene intended for endocrine circulation after expression in the salivary glands of humans would indeed traffic as intended. Heretofore, no clinical trials have been approved to study this gene therapy strategy in humans, and until such a trial can empirically address these sorting issues, the idea remains extremely intriguing but unrealized.

The second issue that confounds the use of salivary glands as endogenous bioreactors is the imprecision of the relationship [vector dose/systemic transgene circulating]. The principles of pharmacology as they relate to biopharmaceuticals dictate that dosing must be maintained within a relatively narrow window in order to maximize efficacy while avoiding intolerable side effects. With an exogenously delivered biological agent, this dosing can be tightly controlled, but producing the biological agent endogenously simply does not allow this level of precision. For this reason, the concept of producing a growth hormone such as HGH, endogenously, might not be workable due to the risk of overdose and the tendency of transgene expression to degrade over time. The notable exception might be orphan diseases such as lysosomal storage diseases, where even a small amount of circulating enzyme can be beneficial, and increased enzyme is only additive to the therapeutic benefit. This might also be the case in a schema like the expression of GLP-1 in type 2 diabetes mellitus where the salivary gland provides a baseline level of the therapeutic that could reduce (but not eliminate) the need for exogenous administration of conventional pharmaceuticals.

12.4 Genetic Medicine

Genetic medicine as a gene therapy strategy differs from gene repair in that the therapeutic effect is achieved not by replacement of a defective gene but by the use of a transgene to manipulate the physiology of the target cell. In this paradigm, the context is rarely inherited monogenetic disease (where a gene repair strategy would presumably be more direct) but rather complex acquired disease. As mentioned above, monogenetic inherited diseases of the salivary gland are very rare, so genetic medicine is by far the more important paradigm to consider in the near-to-intermediate future of salivary gland gene therapy.

The majority of genetic medicine research in the salivary gland has focused on radiation-induced xerostomia, likely due to the consistency and reasonably direct clinical relevance of irradiated animal models. Sjögren's syndrome (SS) has also attracted the interest of the salivary gland gene therapy community, owing to its very high prevalence, but the clinical relevance of the animal models is far more problematic. Animal models of salivary gland dysfunction, and their limitations, will be discussed in the following section. Recently, an intriguing application of genetic medicine involves the ectopic synthesis of a hormone called PYY, normally produced by endocrine cells of the gut epithelium, in the salivary gland. This approach has been shown to modulate taste [20] and induce satiety [21], with potential applications to the treatment of obesity and some forms of anorexia.

Animal studies utilizing genetic medicine approaches to radiation-induced xerostomia have been impressively varied but fall into one of two categories: (1) protection of the gland from the predictable radiation insult or (2) functional restoration of a salivary gland where damage is already manifest. The latter application has heretofore been focused exclusively on a single transgene, aquaporin-1 (AQP1) that presumably localizes circumferentially in the cell membrane of surviving ductal cells (and possibly acinar cells, which may survive in small numbers) and facilitates transmembrane flux [22] of interstitial fluid into the intraductal labyrinth. The first report of this genetic medicine treatment paradigm utilizing the archetype adenoviral type 5 vector was in 1997 [23], and since that time, this therapeutic approach has been upscaled from rodents to miniature swine [24], replicated with an AAV vector in rodents [25] and later miniswine [26], carried out with nonviral ultrasound-assisted gene transfer in miniswine [27], and finally shown both safety and efficacy in a phase I dose-escalation human clinical trial [28]. The field-wide implications of this milestone clinical trial are discussed at the end of this chapter.

Radioprotective gene therapy strategies have been more diverse but are much less advanced down the clinical translational pathway. Examples include Tousled kinase [29, 30], human keratinocyte growth factor [31, 32], vascular endothelial growth factor and/or fibroblast growth factor [33, 34], and heat shock protein 25 [35]. The mechanisms by which these various treatments exert their therapeutic effect are well understood in some cases, less so in others. All of these biological therapies appear to be safe and could potentially be candidates for human clinical trials. However, the cost/benefit analysis of using an adenoviral vector, which is itself inflammatory, must be considered. Alternatives, such as AAV or sonoporation might be considered for clinical trials but only after more work is done to optimize the transgene expression dynamics of each therapeutic candidate to maximize protection and minimize unintended biological consequences.

SS is the most prevalent salivary gland disease, affecting roughly 0.5–3 % of the general population, with a 1:9 male to female ratio. Thus, SS presents the greatest single opportunity for clinical impact using salivary gland genetic medicine strategies. Given the sophisticated state of salivary gland gene therapy relative to gene therapy applications in other organs and tissues, there is reason to be optimistic that this hope may be realized in the coming decade(s). However, there are two principal hurdles to exploiting gene therapy to disrupt and/or reverse SS: (1) since a major element of primary SS is inflammatory, the prospects for using a virus, even AAV, to treat this disease locally within the salivary gland are doubtful, and (2) despite decades of research, the molecular etiology of local salivary gland dysfunction in SS is poorly understood, meaning that the molecular targets for a gene therapy strategy in humans are not at all clear. The advent of UAGT as a nonviral platform for salivary gland gene therapy may obviate the former concern, but the latter remains unresolved.

Gene therapy is fundamentally a method for altering the intracellular programming of the target cell and as such presents nearly unlimited versatility. However – and to extend the metaphor of software – since the "program" of the pathobiology in SS is not understood, there is no basis upon which to directly act upon it with gene therapy. At the highest level, it might be effective to simply utilize transgenes with broad anti-inflammatory activity, based upon what is known about the disease at the histological level. Even starting with this basic premise, a meaningful animal model must be engaged before a clinical translational strategy can approach clinical trials, and it is principally animal models that have hindered the progress of gene therapy for SS.

In the main, it is this author's opinion that an animal model with direct relevance to translating SS gene therapy to humans does not yet exist. This is not for lack of effort, as evidenced by a recent and very helpful review by Park et al. [36] that inventories more than a dozen mouse models of SS. Missing from the Park et al. review are several additional animal models where induction of SS-like disease is itself accomplished by viral gene transfer [37–39]. The phenotypes of these mouse models are variable, with respect to salivary and lacrimal gland manifestations, as well as systemic manifestations and autoantibodies. Similarly, a number of gene therapy studies have been carried out targeting the salivary glands of these mouse models, in particularly the C57BL/6.NOD-Aec1Aec2 model [40–44]. These studies do suggest the potential of gene therapy for SS, but their relevance to the human condition is not clear, and at this point it is difficult to imagine a successful Investigational New Drug (IND) application for a human gene therapy clinical trial based upon any of this evidence.

So what role can mouse models play in unraveling the complex molecular choreography of human SS and more importantly in providing the rationale for a human clinical trial involving gene transfer to the salivary gland in SS? Perhaps one answer lies in working backward to mouse models rather than forward. The availability of salivary gland biopsies from SS patients through various tissue banking efforts provides a rich resource for genomic and transcriptomic studies of the molecular pathobiology of SS in the affected salivary gland, already yielding insights on the human condition [45, 46]. Since the only relevant targets for gene therapy are the human ones, it may be best to de-emphasize the importance of SS-like phenotype in mice and rather focus gene therapy efforts going forward on mouse models that allow demonstration of clear-cut modulation of genetic targets known to be relevant to the human condition.

A final potential application of salivary gland-based genetic medicine strategies bears mentioning, although it has not yet been exploited in a peer-reviewed research manuscript. It is theoretically possible to use gene therapy to alter the protein composition of saliva for applications focused on the oral cavity itself. One can envision the introduction of proteins or peptides with specific activity against intractable or opportunistic dental pathogens, such as *Candida albicans* or *Aggregatibacter actinomycetemcomitans*. In this author's opinion, this novel methodology for chronic oral disease is extremely promising and warrants greater attention.

12.5 Experimental Models of Salivary Gland Disease and Gene Therapy

The great promise of gene therapy is that it presents therapeutic opportunities that are simply not possible with traditional pharmacotherapies. However, with this new paradigm come additional risks, some known, and some yet unknown. Gene therapy is not a new field and has been an active area of research for more than four decades, with the first successful human gene therapy intervention now almost three decades in the past (1990). Despite this history of promise and setbacks, gene therapy has blossomed during the second decade of this century, with several industry-sponsored approval applications now pending before the FDA and the EMA. Some of the trials and tribulations of the gene therapy field, such as the gradual waning of viral-mediated RPE65 gene repair in congenital blindness [47], were perhaps predictable. Others, including the persistence of AQP1 expression in the human salivary gland following viral-mediated gene therapy [48], were not. As the field traverses this critical juncture in its history, it is absolutely clear that high-quality, large animal preclinical models of gene therapy are one key to assuring smooth clinical translation of candidate gene therapies. In this regard, the salivary gland gene therapy field enjoys a distinct advantage, as discussed below.

One of the major advantages of salivary gland gene therapy is the accessibility of the organ itself, via intraoral cannulation of the salivary duct (parotid or submandibular). Technically, the cells of the salivary gland are epithelium, and the tight junctions between these cells, combined with the encapsulation of the organ, make the salivary gland a cutaneous structure. Delivery of a vector to the salivary gland via cannulation avoids communication with the systemic circulation and avoids many of the complexities of systemic toxicity that have complicated other applications of gene therapy. Fundamentally, cannulating the salivary duct of large animals (e.g., miniswine) or even that of a rodent is very similar to the actual situation that would be faced in humans, further increasing the relevance of animal models to clinical translation.

The first models of salivary gland gene therapy were rodents, as expected, and their low cost, as well as the availability of transgenic modification (in mice), makes rodents the mainstay of research development in this field. The primary drawback of these animals is the very small size and relative fragility of the salivary duct, making the technique extremely challenging from a technical standpoint and also prone to variability. A second issue that limits rodent models is the major difference in salivary gland structure and function between rodents and humans. In rodents, the submandibular gland (SMG) is the largest gland and plays a much greater role in saliva production than the parotid gland. In humans, the roles of the parotid and submandibular are reversed. Thus, caution must be exercised when interpreting the results of gene therapy interventions in the SMG of rodents. Nevertheless, with the notable exception of exocrine/endocrine sorting, most gene therapy insights gained in the SMG of rodents do faithfully upscale to larger animal models.

Between rodents and humans, a large animal model of gene therapy is highly desirable, if not essential. Whereas primates are indispensable as preclinical models in other areas of research, they present very considerable challenges, primarily in ethical and cost considerations. Fortunately, primates have proven unnecessary for translation of salivary gland gene therapy to clinical trials and may even be less informative than other animal models [49]. To a lesser degree, canines also present challenges as preclinical models, primarily due to their status as companion animals. Fortunately, the pig has proven to be nearly ideal as a preclinical model of salivary gland gene therapy and has proven sufficient for clinical translation of the first-in-man salivary gland gene therapy [24]. The parotid glands of pigs are similar in overall size and location to humans (see Fig. 12.2), and while pigs are highly intelligent animals, their status as an agricultural product obviates ethical concerns that plague canine or primate studies. The principal challenge in the use of pigs as models of salivary gland gene therapy is the forceful growth kinetics of the domestic farm swine, a trait that has been bred into these animals over millennia but is cumbersome for chronic studies. For reasons of convenience, the miniature pig has been established as the penultimate preclinical model of salivary gland gene therapy after an impressive body of collaborative work by Songlin Wang in Beijing and Bruce Baum in Bethesda characterized the essential elements of this animal model [50, 51].

12.6 Clinical Trials of Gene Therapy

As of the date of this writing, clinicaltrials.gov lists only two gene therapy studies involving the salivary gland, one completed, and the other approved but not yet recruiting. Both trials utilize the same therapeutic philosophy, expressing a water channel called aquaporin-1 (AQP1) in the salivary glands of patients whose salivary glands have been damaged by radiotherapy for head-and-neck cancer. Since acinar cells are known to be destroyed in the context of radiation-induced xerostomia, it is presumed that AQP1 achieves expression in the surviving ductal cells, driving transcellular fluid flux from the interstitial space into the intraductal labyrinth.

The first trial, NCT00372320, is usually referred to as the "AdAQP1" trial and involved a phase I dose-escalation paradigm to evaluate the safety of adenoviral gene therapy in the human salivary gland [28]. As a first-in-man study, the primary outcome measure was safety, but secondary measures included both objective (salivary flow) and subjective (xerostomia) metrics of therapeutic efficacy. It is difficult to overstate the importance of this successful trial as an inflection point for the field, as it established proof-of-principle for the safety and efficacy of salivary gland gene therapy, allowing subsequent studies to focus on practicality of this approach for dissemination to the oral medicine clinical community.

Two important observations from the AdAQP1 trial are worth noting. First, the dose-escalation strategy demonstrated that there is an ideal dosing range for efficacy, above and below which is ineffective and possibly harmful [28]. The second observation, specifically addressed in a later report [48], was the *duration* of therapeutic effect in the human patients, which was unexpected in that it substantially exceeded the transient effects seen in preclinical large animal studies [24]. This issue will need to be considered in future clinical trials, and while it is premature to draw broad conclusions, this single observation gives reason to hope that salivary gland gene therapy in humans may produce therapeutic effects that last for at least several months.

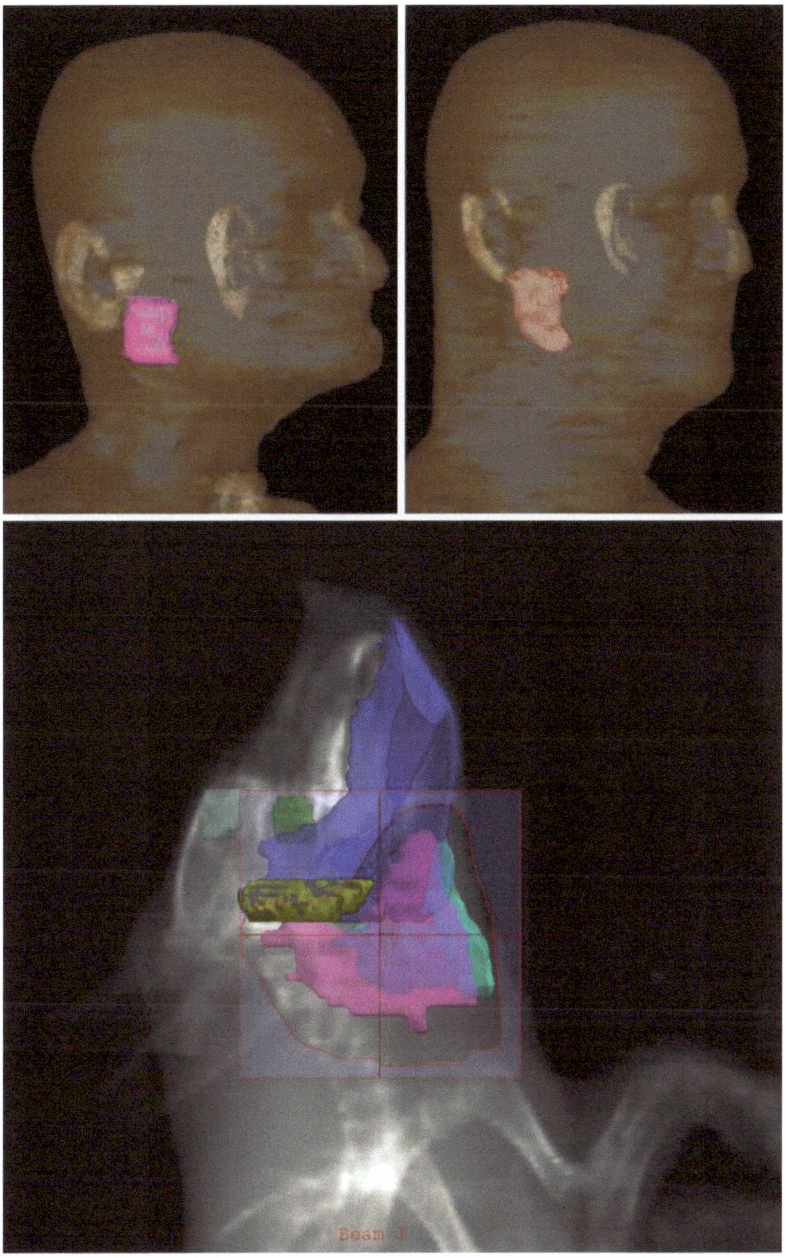

Fig. 12.2 Comparison of parotid gland position and relative size and humans versus miniswine. *Upper* shows digital radiography reconstruction (DRR) of parotid glands from two de-identified patients from the Allegheny Cancer Center (Pittsburgh, PA). The *lower* shows (DRR) of a miniswine subject. The right parotid is shown in magenta. Note that human and swine images are not referenced to the same scale. (*Upper*, reproduced from [18], *lower* © 2014 Olivier Gayou, PhD)

The second trial, NCT02446249, builds upon the success of the AdAQP1 clinical trial, while attempting to improve upon what is presumed to be its principal weakness, the highly immunogenic adenoviral vector. In this follow-up trial, AAV2 will be used to express the AQP1 transgene, and preclinical studies suggest that the duration of therapeutic efficacy could be much longer than that seen with adenovirus [26]. AAV has been used safely in other human clinical trials, and this trial holds great promise for long-lasting palliative therapy in radiation-induced xerostomia.

Looking into the future, it is now clear that salivary gland gene therapy is a promising new modality for treating salivary gland dysfunction in radiation-induced xerostomia and may soon find widespread application in this condition [52]. However, this patient population represents but a small niche of patients suffering from xerostomia and hyposalivation, with age-related xerostomia and SS representing tens of millions in the developed world and presumably hundreds of millions of individuals worldwide. Since these conditions are chronic but not lethal, the key to effective therapy will be durable transgene expression, either with single treatment or (more likely) with a therapeutic strategy that allows for serial readministration. If the latter is required, questions remain as to the approach that might be taken with viral vectors, as even AAV generates host immune response with repeated dosing. Nonviral alternatives, such as UAGT or nanoparticles, might also be considered but have not yet been evaluated in the salivary glands of humans. With clinical data demonstrating the safety of salivary gland gene therapy now firmly in hand, it is important that clinical development of these various modalities be accelerated in order to mainstream salivary gland gene therapy into the practice of oral and dental medicine.

References

1. Perez P, Rowzee AM, Zheng C, Adriaansen J, Baum BJ. Salivary epithelial cells: An unassuming target site for gene therapeutics. Int J Biochem Cell Biol. 2010;42:773–7.
2. Mastrangeli A, O'Connell B, Aladib W, Fox PC, Baum BJ, Crystal RG. Direct in vivo adenovirus-mediated gene transfer to salivary glands. Am J Phys. 1994;266:G1146–55.
3. Leon J, Park A. Time. 1999;153:68–70, 73.
4. Shai E, Falk H, Honigman A, Panet A, Palmon A. Gene transfer mediated by different viral vectors following direct cannulation of mouse submandibular salivary glands. Eur J Oral Sci. 2002;110:254–60.
5. Katano H, Kok MR, Cotrim AP, Yamano S, Schmidt M, Afione S, Baum BJ, Chiorini JA. Enhanced transduction of mouse salivary glands with aav5-based vectors. Gene Ther. 2006;13:594–601.
6. Arany S, Benoit DS, Dewhurst S, Ovitt CE. Nanoparticle-mediated gene silencing confers radioprotection to salivary glands in vivo. Mol Ther. 2013;21:1182–94.
7. Passineau MJ, Zourelias L, Machen L, Edwards PC, Benza RL. Ultrasound-assisted non-viral gene transfer to the salivary glands. Gene Ther. 2010;17:1318–24.
8. Geguchadze R, Wang Z, Zourelias L, Perez-Riveros P, Edwards PC, Machen L, Passineau MJ. Proteomic profiling of salivary gland after nonviral gene transfer mediated by conventional plasmids and minicircles. Mol Ther Methods Clin Dev. 2014;1:14007.
9. Riordan JR, Rommens JM, Kerem B, Alon N, Rozmahel R, Grzelczak Z, Zielenski J, Lok S, Plavsic N, Chou JL, et al. Identification of the cystic fibrosis gene: cloning and characterization of complementary DNA. Science. 1989;245:1066–73.
10. Voutetakis A, Kok MR, Zheng C, Bossis I, Wang J, Cotrim AP, Marracino N, Goldsmith CM, Chiorini JA, Loh YP, Nieman LK, Baum BJ. Reengineered salivary glands are stable endogenous bioreactors for systemic gene therapeutics. Proc Natl Acad Sci U S A. 2004;101:3053–8.
11. Voutetakis A, Bossis I, Kok MR, Zhang W, Wang J, Cotrim AP, Zheng C, Chiorini JA, Nieman LK, Baum BJ. Salivary glands as a potential gene transfer target for gene therapeutics of some monogenetic endocrine disorders. J Endocrinol. 2005;185:363–72.
12. Racz GZ, Zheng C, Goldsmith CM, Baum BJ, Cawley NX. Toward gene therapy for growth hormone deficiency via salivary gland expression of growth hormone. Oral Dis. 2015;21:149–55.
13. Passineau MJ, Fahrenholz T, Machen L, Zourelias L, Nega K, Paul R, MacDougall MJ, Mamaeva O, Steet R, Barnes J, Kingston HM, Benza RL. Alpha-galactosidase a expressed in the salivary glands partially corrects organ biochemical deficits in the fabry mouse through endocrine trafficking. Hum Gene Ther. 2011;22:293–301.
14. Voutetakis A, Cotrim AP, Rowzee A, Zheng C, Rathod T, Yanik T, Loh YP, Baum BJ, Cawley NX. Systemic delivery of bioactive glucagon-like peptide 1 after adenoviral-mediated gene transfer in the murine salivary gland. Endocrinology. 2010;151:4566–72.
15. Voutetakis A, Zheng C, Wang J, Goldsmith CM, Afione S, Chiorini JA, Wenk ML, Vallant M, Irwin RD, Baum BJ. Gender differences in serotype 2 adeno-associated virus biodistribution after administration to rodent salivary glands. Hum Gene Ther. 2007;18:1109–18.
16. Yan X, Voutetakis A, Zheng C, Hai B, Zhang C, Baum BJ, Wang S. Sorting of transgenic secretory proteins in miniature pig parotid glands following adenoviral-mediated gene transfer. J Gene Med. 2007;9:779–87.

17. Samuni Y, Cawley NX, Zheng C, Cotrim AP, Loh YP, Baum BJ. Sorting behavior of a transgenic erythropoietin-growth hormone fusion protein in murine salivary glands. Hum Gene Ther. 2008;19:279–86.
18. Samuni Y, Zheng C, Cawley NX, Cotrim AP, Loh YP, Baum BJ. Sorting of growth hormone-erythropoietin fusion proteins in rat salivary glands. Biochem Biophys Res Commun. 2008;373:136–9.
19. Voutetakis A, Zheng C, Metzger M, Cotrim AP, Donahue RE, Dunbar CE, Baum BJ. Sorting of transgenic secretory proteins in rhesus macaque parotid glands after adenovirus-mediated gene transfer. Hum Gene Ther. 2008;19:1401–5.
20. La Sala MS, Hurtado MD, Brown AR, Bohorquez DV, Liddle RA, Herzog H, Zolotukhin S, Dotson CD. Modulation of taste responsiveness by the satiation hormone peptide yy. FASEB J. 2013;27: 5022–33.
21. Acosta A, Hurtado MD, Gorbatyuk O, La Sala M, Duncan D, Aslanidi G, Campbell-Thompson M, Zhang L, Herzog H, Voutetakis A, Baum BJ, Zolotukhin S. Salivary pyy: a putative bypass to satiety. PLoS ONE. 2011;6:e26137.
22. Delporte C, Hoque AT, Kulakusky JA, Braddon VR, Goldsmith CM, Wellner RB, Baum BJ. Relationship between adenovirus-mediated aquaporin 1 expression and fluid movement across epithelial cells. Biochem Biophys Res Commun. 1998;246:584–8.
23. Delporte C, O'Connell BC, He X, Lancaster HE, O'Connell AC, Agre P, Baum BJ. Increased fluid secretion after adenoviral-mediated transfer of the aquaporin-1 cdna to irradiated rat salivary glands. Proc Natl Acad Sci U S A. 1997;94:3268–73.
24. Shan Z, Li J, Zheng C, Liu X, Fan Z, Zhang C, Goldsmith CM, Wellner RB, Baum BJ, Wang S. Increased fluid secretion after adenoviral-mediated transfer of the human aquaporin-1 cdna to irradiated miniature pig parotid glands. Mol Ther. 2005;11: 444–51.
25. Braddon VR, Chiorini JA, Wang S, Kotin RM, Baum BJ. Adeno-associated virus-mediated transfer of a functional water channel into salivary epithelial cells in vitro and in vivo. Hum Gene Ther. 1998;9:2777–85.
26. Gao R, Yan X, Zheng C, Goldsmith CM, Afione S, Hai B, Xu J, Zhou J, Zhang C, Chiorini JA, Baum BJ, Wang S. Aav2-mediated transfer of the human aquaporin-1 cdna restores fluid secretion from irradiated miniature pig parotid glands. Gene Ther. 2011;18: 38–42.
27. Wang Z, Zourelias L, Wu C, Edwards PC, Trombetta M, Passineau MJ. Ultrasound-assisted nonviral gene transfer of aqp1 to the irradiated minipig parotid gland restores fluid secretion. Gene Ther. 2015;22: 739–49.
28. Baum BJ, Alevizos I, Zheng C, Cotrim AP, Liu S, McCullagh L, Goldsmith CM, Burbelo PD, Citrin DE, Mitchell JB, Nottingham LK, Rudy SF, Van Waes C, Whatley MA, Brahim JS, Chiorini JA, Danielides S, Turner RJ, Patronas NJ, Chen CC, Nikolov NP, Illei GG. Early responses to adenoviral-mediated transfer of the aquaporin-1 cdna for radiation-induced salivary hypofunction. Proc Natl Acad Sci U S A. 2012;109: 19403–7.
29. Palaniyandi S, Odaka Y, Green W, Abreo F, Caldito G, De Benedetti A, Sunavala-Dossabhoy G. Adenoviral delivery of tousled kinase for the protection of salivary glands against ionizing radiation damage. Gene Ther. 2011;18:275–82.
30. Timiri Shanmugam PS, Dayton RD, Palaniyandi S, Abreo F, Caldito G, Klein RL, Sunavala-Dossabhoy G. Recombinant aav9-tlk1b administration ameliorates fractionated radiation-induced xerostomia. Hum Gene Ther. 2013;24:604–12.
31. Zheng C, Cotrim AP, Sunshine AN, Sugito T, Liu L, Sowers A, Mitchell JB, Baum BJ. Prevention of radiation-induced oral mucositis after adenoviral vector-mediated transfer of the keratinocyte growth factor cdna to mouse submandibular glands. Clin Cancer Res. 2009;15:4641–8.
32. Zheng C, Cotrim AP, Rowzee A, Swaim W, Sowers A, Mitchell JB, Baum BJ. Prevention of radiation-induced salivary hypofunction following hkgf gene delivery to murine submandibular glands. Clin Cancer Res. 2011;17:2842–51.
33. Guo L, Gao R, Xu J, Jin L, Cotrim AP, Yan X, Zheng C, Goldsmith CM, Shan Z, Hai B, Zhou J, Zhang C, Baum BJ, Wang S. Adltr2ef1alpha-fgf2-mediated prevention of fractionated irradiation-induced salivary hypofunction in swine. Gene Ther. 2014;21: 866–73.
34. Cotrim AP, Sowers A, Mitchell JB, Baum BJ. Prevention of irradiation-induced salivary hypofunction by microvessel protection in mouse salivary glands. Mol Ther. 2007;15:2101–6.
35. Lee HJ, Lee YJ, Kwon HC, Bae S, Kim SH, Min JJ, Cho CK, Lee YS. Radioprotective effect of heat shock protein 25 on submandibular glands of rats. Am J Pathol. 2006;169:1601–11.
36. Park YS, Gauna AE, Cha S. Mouse models of primary sjogren's syndrome. Curr Pharm Des. 2015;21: 2350–64.
37. Nguyen CQ, Yin H, Lee BH, Carcamo WC, Chiorini JA, Peck AB. Pathogenic effect of interleukin-17a in induction of sjogren's syndrome-like disease using adenovirus-mediated gene transfer. Arthritis Res Ther. 2010;12:R220.
38. Vosters JL, Landek-Salgado MA, Yin H, Swaim WD, Kimura H, Tak PP, Caturegli P, Chiorini JA. Interleukin-12 induces salivary gland dysfunction in transgenic mice, providing a new model of sjogren's syndrome. Arthritis Rheum. 2009;60: 3633–41.
39. Vosters JL, Yin H, Roescher N, Kok MR, Tak PP, Chiorini JA. Local expression of tumor necrosis factor-receptor 1:Immunoglobulin g can induce sali-

vary gland dysfunction in a murine model of sjogren's syndrome. Arthritis Res Ther. 2009;11:R189.
40. Nguyen CQ, Yin H, Lee BH, Chiorini JA, Peck AB. Il17: potential therapeutic target in sjogren's syndrome using adenovirus-mediated gene transfer. Lab Investig. 2011;91:54–62.
41. Roescher N, Vosters JL, Yin H, Illei GG, Tak PP, Chiorini JA. Effect of soluble icam-1 on a sjogren's syndrome-like phenotype in nod mice is disease stage dependent. PLoS ONE. 2011;6:e19962.
42. Lee BH, Carcamo WC, Chiorini JA, Peck AB, Nguyen CQ. Gene therapy using il-27 ameliorates sjogren's syndrome-like autoimmune exocrinopathy. Arthritis Res Ther. 2012;14:R172.
43. Vosters JL, Roescher N, Illei GG, Chiorini JA, Tak PP. Taci-fc gene therapy improves autoimmune sialadenitis but not salivary gland function in non-obese diabetic mice. Oral Dis. 2012;18:365–74.
44. Wu C, Wang Z, Zourelias L, Thakker H, Passineau MJ. Il-17 sequestration via salivary gland gene therapy in a mouse model of sjogren's syndrome suppresses disease-associated expression of the putative autoantigen klk1b22. Arthritis Res Ther. 2015;17:198.
45. Ice JA, Li H, Adrianto I, Lin PC, Kelly JA, Montgomery CG, Lessard CJ, Moser KL. Genetics of sjogren's syndrome in the genome-wide association era. J Autoimmun. 2012;39:57–63.
46. Burbelo PD, Ambatipudi K, Alevizos I. Genome-wide association studies in sjogren's syndrome: What do the genes tell us about disease pathogenesis? Autoimmun Rev. 2014;13:756–61.
47. Kaiser J. Gene therapy for blindness may fade with time. Science. 2015. http://www.sciencemag.org/news/2015/05/gene-therapy-blindness-may-fade-time
48. Zheng C, Baum BJ, Liu X, Goldsmith CM, Perez P, Jang SI, Cotrim AP, McCullagh L, Ambudkar IS, Alevizos I. Persistence of haqp1 expression in human salivary gland cells following adhaqp1 transduction is associated with a lack of methylation of hcmv promoter. Gene Ther. 2015;22:758–66.
49. Voutetakis A, Zheng C, Cotrim AP, Mineshiba F, Afione S, Roescher N, Swaim WD, Metzger M, Eckhaus MA, Donahue RE, Dunbar CE, Chiorini JA, Baum BJ. Aav5-mediated gene transfer to the parotid glands of non-human primates. Gene Ther. 2010;17:50–60.
50. Li J, Zheng C, Zhang X, Liu X, Zhang C, Goldsmith CM, Baum BJ, Wang S. Developing a convenient large animal model for gene transfer to salivary glands in vivo. J Gene Med. 2004;6:55–63.
51. Li J, Shan Z, Ou G, Liu X, Zhang C, Baum BJ, Wang S. Structural and functional characteristics of irradiation damage to parotid glands in the miniature pig. Int J Radiat Oncol Biol Phys. 2005;62:1510–6.
52. Baum BJ. Radiation-induced salivary hypofunction may become a thing of the past. Oral Dis. 2016; 22:81–4.

Surgical Management of Salivary Gland Disease

13

Varun V. Varadarajan and Peter T. Dziegielewski

Abstract

The study of salivary gland tissue and the surgical management of salivary gland pathology are fundamental to the practicing otolaryngologist-head and neck surgeon. Traditional surgical intervention for both neoplastic and nonneoplastic disease of the salivary glands includes sialadenectomy, superficial or complete parotidectomy, minor procedures involving the salivary ducts, and procedural interventions for xerostomia and sialorrhea. Recent surgical advances of the salivary glands and ducts such as minimally invasive, endoscopic, and robotic techniques have augmented the surgeon's armamentarium for managing salivary gland disease. Novel techniques such as salivary gland transfer are also being pioneered. The mechanisms of salivary gland function remain an active research topic, and future applications may include regeneration of functional salivary gland tissue. This chapter briefly reviews the basic surgical anatomy and physiology of the major and minor salivary glands and describes traditional indications for surgical intervention. The recent advances in salivary gland surgery are described, and the chapter concludes by highlighting recent discoveries in the field of salivary gland regeneration. The implications of these advances for the head and neck surgeon and the potential future of surgical management of salivary gland pathology are discussed.

13.1 Introduction and Historical Perspective

The anatomic study of major salivary glands is documented as early as the second century AD when Galen described anatomic relationships of the major salivary glands [1, 2]. Detailed anatomic depictions were not available in the western world until the fifteenth to sixteenth centuries when anatomists including Andreas

V.V. Varadarajan, MD • P.T. Dziegielewski, MD FRCS(C) (✉)
University of Florida Department of Otolaryngology, Gainesville, FL, USA
e-mail: varun.varadarajan@ent.ufl.edu; Peter.Dziegielewski@ent.ufl.edu

S. Cha (ed.), *Salivary Gland Development and Regeneration*, DOI 10.1007/978-3-319-43513-8_13

Vesalius, Realdus Columbus, William Harvey, Bartholomaeus Eustachius, and others popularized systematic human body dissection. Vesalius is the first anatomist to use the term "salivary glands" and attribute their presence to the secretion of saliva in his writing *De humani corporus fabrica* in 1543. A more sophisticated anatomic understanding of the salivary glands was not attained until the seventeenth century. Nicholas Stenson first described the parotid gland duct in his writing *De glandulis oris et novis earandum vasis* in 1661. Thomas Wharton is credited with the discovery of the submandibular gland duct, which he described in his writing *Adenographia sive glandularum totius corporis descriptio* in 1656. In this writing he also describes what we believe to be the sublingual gland and ducts. Caspar Bartholin further characterized the anatomy of the sublingual gland and duct system in 1685 [1]. Anatomists further characterized the structural relationships, histology, and physiologic function over the ensuing centuries. Traditional surgical interventions were developed to address a range of pathologies including neoplastic, nonneoplastic, obstructive, inflammatory, infectious, and iatrogenic conditions. The twentieth century allowed further development and refinement of salivary gland surgery. Janes was the first surgeon to describe a process for the intraoperative identification of the facial nerve during parotid surgery in 1940 [3]. The development and widespread availability of computerized tomography and magnetic resonance imaging allowed the medical and surgical community to a gain a sophisticated understanding of the salivary gland anatomy. The structure and function of the salivary glands continue also to play an important role in regenerative medicine; functional salivary gland regeneration is an active research topic. Researchers aim to replicate and regenerate the complex histologic organization and function of the human salivary glands in an attempt to potentially allow salivary gland preservation and regrowth.

This chapter will begin by providing an overview of the anatomic and physiologic principles of the salivary glands. We will then review the indications for salivary gland surgery including neoplastic and nonneoplastic disease and briefly discuss the most commonly described surgical approaches to the major salivary glands. The recent advances in salivary gland surgery such as sialendoscopy, salivary gland transfer, and minimally invasive surgery will then be discussed. The chapter will conclude by highlighting recent discoveries in the field of functional salivary gland regeneration and discuss the implications of these advances for the head and neck surgeon.

13.1.1 Anatomic and Physiologic Principles of Major and Minor Salivary Glands

Salivary glands are accessory digestive glands and begin their development during the 6th week of gestation. Epithelial buds invaginate from oral ectoderm into connective tissue mesenchyme. There are three paired major salivary glands (parotid, submandibular, and sublingual glands) and presumably 100s of minor salivary glands. The site of invagination defines the location of the ductal orifice. Tunnels created by ectodermal outpouchings proliferate and branch, creating tubules and acini that ultimately form the structure of the salivary glands. The parotid gland develops first among the major salivary glands followed by the submandibular and sublingual glands. The parotid gland develops around the branches of the facial nerve and is the last to become encapsulated by connective tissue fascia. The associated lymphatic vessels develop after the submandibular and sublingual glands become encapsulated but before parotid gland encapsulation [2–7]. The result of this unique aspect of embryogenesis is the presence of lymphatic channels and lymph nodes within the parotid gland.

13.1.1.1 Parotid Gland and Facial Nerve Anatomy

The parotid gland is the largest of the major salivary glands and is located between the external auditory canal and the mandibular ramus. It is classically described as wedge shaped and extends superficially over the masseter muscle. The parotid tail is a posterior and inferior extension into the neck over the sternocleidomastoid muscle. The parotid gland fascia encapsulates glandular tissue, blood vessels, and lymphatic tissue. The parotid gland is bordered medially by the parapharyngeal space and medial pterygoid muscle, laterally by subcutaneous fat and dermis, superiorly by the zygomatic arch, inferiorly by the styloid process and associated muscles and ligaments, anteriorly by the mandibular ramus and masseter muscle, and posteriorly by the external auditory canal. The styloid process, stylohyoid muscle, and digastric muscle separate the gland from the vessels and nerves of the parapharyngeal space. The medial aspect of the gland, which contacts these structures, is termed the "deep lobe" of the parotid gland although the gland is technically unilobular. The facial nerve is considered by many to be the dividing structure between the superficial and deep lobes of the parotid gland [3–6, 8].

The parotid gland duct, known as the Stensen's duct (named after Nicholas Stenson), is 4–6 cm in length and arises from the anterior aspect of the parotid gland [1, 5]. The Stensen's duct travels in the anterior direction lateral to the masseter muscle. The buccal branch of the facial nerve often travels parallel to the duct. The duct ultimately makes a 90° medial turn (the "masseteric bend") to pierce the buccinator muscle and opens into the buccal mucosa at the level of the second maxillary molar tooth. In 21 % of the human population, accessory parotid tissue is found in proximity to the duct and ductal orifice [4, 9, 10].

The connective tissue fascia that encapsulates the parotid gland is contiguous with the superficial layer of the deep cervical fascia. The fascia sends septations into the parotid tissue. The parotid gland is separated from the submandibular gland by the stylomandibular ligament, which is a continuation of the fascia of the posterior belly of the digastric muscle. The gland has fibrous attachments to the anterior wall of the external auditory canal, mastoid process, and the fascia of the sternocleidomastoid [5].

The transverse facial artery branch of the external carotid artery serves as the arterial supply to the parotid gland, and the transverse facial vein provides venous drainage into the retromandibular vein. The embryological development of the lymphatic tissues prior to parotid gland encapsulation leads to the presence of intraparotid and periparotid lymph nodes and lymphatic channels that drain the forehead, scalp, periorbital regions, auricles, and external auditory canals. Intraparotid lymph nodes also serve as lymphatic drainage to the posterior aspects of the nasopharynx and soft palate [2–6, 8]. This has clinical implications in head and neck malignancy in the abovementioned sites with lymph node metastasis that may require parotidectomy despite no primary salivary gland disease.

Associated nerves are the facial nerve and its branches, the auriculotemporal nerve, and the great auricular nerve. The parasympathetic innervation to the parotid gland stimulates saliva secretion. Preganglionic parasympathetic fibers originate from the inferior salivary nucleus and travel along the glossopharyngeal nerve to the otic ganglion via the lesser superficial petrosal nerve. The auriculotemporal nerve carries sensation from the otic ganglion to the parotid gland. The auriculotemporal nerve is a branch of the mandibular division of the trigeminal nerve; it exits the skull base at foramen ovale and travels anteriorly and laterally from the skull base and infratemporal fossa to the external auditory canal. Sympathetic stimulation originates from the superior cervical ganglion; postganglionic fibers travel to the parotid gland via the external carotid artery [2, 4–6].

The great auricular nerve originates from cervical rootlets C2–C3 and is a branch of the cervical plexus. This nerve branches from the cervical plexus at *Erb's point* and courses superiorly from the posterior aspect of the sternocleidomastoid to the superficial aspect of the parotid gland. The great auricular nerve supplies sensation to the skin overlying the parotid gland, the mastoid and mandibular angle, and the inferior and posterior aspects of the auricle. This nerve may be sacrificed during a parotidectomy [3–6, 8].

The facial nerve is intimately associated with the parotid gland tissue, and a discussion of the surgical anatomy of the parotid gland is incomplete without describing the course of the facial nerve. The main trunk of the nerve exits the stylomastoid foramen and provides branches to the posterior belly of the digastric muscle, posterior auricular muscle, and the stylohyoid muscle before entering the parotid gland. The nerve enters the gland approximately 1 cm after exiting the temporal bone [4–6]. At this point, the nerve divides into superior temporofacial and inferior cervicofacial divisions at the *pes anserinus;* 13.3 % of patients have three divisions [11]. The five terminal branches of the nerve from superior to inferior are the temporal, zygomatic, buccal, marginal mandibular, and cervical branches. Communicating branches between these terminal branches are very common, and the terminal branching is variable. Identification of the facial nerve is critical in parotidectomy. Several classic anatomic relationships have been used to localize the main trunk of the facial nerve. The "tragal pointer" is a deep extension of conchal cartilage that is an anatomic landmark; the nerve is located 1 cm inferior and medial to the tragal pointer [4, 9, 12]. The nerve is located posterior and lateral to the base of the styloid process. The main trunk is also located 6–8 mm deep to the tympanomastoid suture line of the temporal bone exiting the stylomastoid foramen. The nerve is also located superior and deep to the proximal attachment of the posterior belly of the digastric muscle. Facial nerve dissection is discussed elsewhere in this text. The mastoid cortex and air cells can be removed to identify the facial canal if the above methods do not allow identification of the nerve. Anterograde dissection and further skeletonization of the nerve starting from the main trunk allow safe removal of parotid gland tissue. If a distal branch is found before the main trunk, retrograde dissection can also be performed to trace the nerve to the main trunk [3, 4, 6]. Further description of parotidectomy is described later in this chapter.

13.1.1.2 Submandibular Gland Anatomy

The submandibular gland is the second largest paired major salivary gland and is located within the submandibular triangle in the neck. The gland is associated with neck level IB lymph nodes and extends medial and deep to the inferior border of the posterior mandibular body. The gland curves over the posterior border of the mylohyoid muscle, which anatomically divides the gland into two lobes. The superficial lobe is located in the posterolateral sublingual space, while the larger deep lobe is located inferior to the mylohyoid muscle. Like the parotid gland, the fibrous encapsulation of the submandibular gland derives from the superficial layer of the deep cervical fascia [2–6, 13]. The submandibular duct is termed the *Wharton's duct* (named after anatomist Thomas Wharton) [1]. The duct extends from the medial aspect of the gland and extends anteriorly to open into the oral cavity lateral to the lingual frenulum. The duct courses between the mylohyoid and hyoglossus muscles. The opening is at the apex or on the walls of the papilla on the anterior floor of mouth. The duct is approximately 5 cm in length and is between 0.5 and 1.5 mm in diameter [4, 14]. The lingual nerve curves around the inferior border of the duct from a lateral to anteromedial direction to provide sensory innervation to the anterior 2/3rds of the tongue. The arterial supply of the submandibular gland is via the glandular branch of the facial artery branch of the external carotid artery. The facial artery travels deep to the digastric and stylohyoid muscle to pass into a groove on the posterior and deep surface of the gland. The artery courses both anteriorly and superiorly to the superior aspect of the gland until it curves over the facial notch of the mandibular body to then ascend over the lateral aspect of the mandibular body anterior to the masseter muscle. Venous drainage is provided by the facial vein which travels superficial to the submandibular gland and drains into the common facial vein [2, 4–6]. Figure 13.1 depicts the anatomic relationships of the structures to the submandibular gland.[4].

Like the parotid gland, the sympathetic innervation to the submandibular gland is provided by postganglionic sympathetic fibers originating from the superior cervical ganglion, which

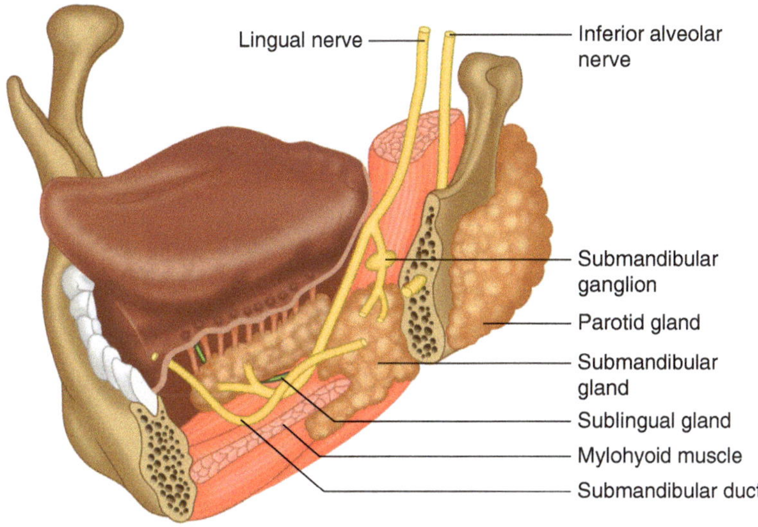

Fig. 13.1 Anatomic relationships of the submandibular gland to its adjacent structures (Reprinted from Ref. 4)

travel along the external carotid artery branches. The parasympathetic innervation originates in the superior salivatory nucleus, and travels down the facial nerve via the nervus intermedius and ultimately joins the lingual nerve via the chorda tympani nerve to synapse in the submandibular ganglion. Postganglionic fibers synapse onto glandular cells [2, 4].

The marginal mandibular branch of the facial nerve is closely associated with the submandibular gland and is often found coursing anteriorly within 1–2 cm of the angle of the mandible [3–6]. The nerve loops below the mandible and has a variable course and superior-inferior position to the inferior border of the mandible. The facial vein is deep to this nerve; the vein can be ligated and reflected superiorly from the gland during submandibular gland surgery to protect the nerve. This maneuver has been termed the *Hayes-Martin maneuver* after the well-known head and neck surgeon [3, 13]. Several surgical approaches have been described for submandibular gland surgery including transcervical, submental, retroauricular, or intraoral approaches [3]. The lateral cervical approach is most often described and allows direct access to the gland; this technique is again described later in this chapter.

13.1.1.3 Sublingual Gland Anatomy

The sublingual gland is the smallest of the paired major salivary glands. This gland is located in the sublingual space in between the mylohyoid muscle and the oral cavity floor mucosa. The genioglossus muscle is medial to the gland, while the mandible is lateral to the gland. The Wharton's duct also travels within this space along with the terminal branches of the lingual and hypoglossal nerves. The sublingual gland is located laterally to these structures. This gland is approximately 3 cm in length and oval in shape and has no fibrous capsule. The sublingual gland may have a major drainage duct *(Bartholin duct)* and minor drainage duct but drains into the oral cavity along the sublingual fold via 8–20 small ducts termed the *ducts of Rivinus* [3–6].

The arterial supply of the sublingual gland is mainly from the sublingual branches from the lingual branch of the external carotid artery. There are also branches from the submental branch of the facial artery. The lingual and facial veins provide venous drainage to the sublingual gland. The sympathetic and parasympathetic innervation to the gland is similar to the submandibular gland as described above. Postganglionic parasympathetic nerves originate in the submandibular ganglion.

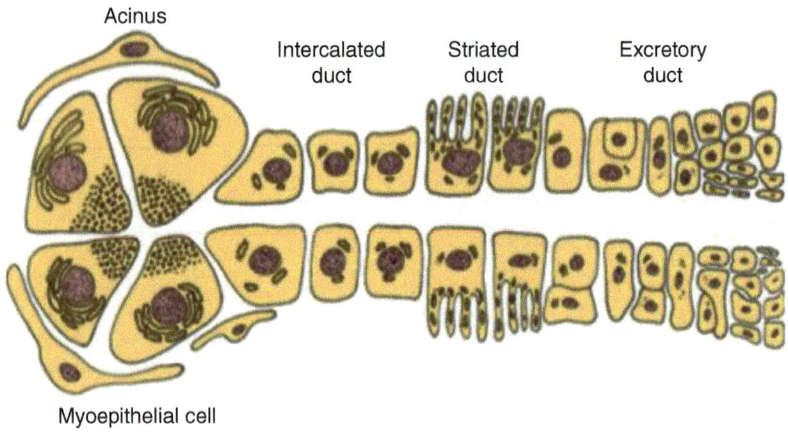

Fig. 13.2 Basic salivary gland unit (Reprinted from Ref. 8)

The lingual gland can be surgically approached in a transoral fashion with direct incision into floor of mouth into the sublingual space [3, 5, 6].

13.1.1.4 Minor Salivary Glands

There are 100–1000 minor salivary glands that are distributed throughout the oral cavity, oropharynx, larynx, tracheobronchial tree, and nasal cavity. The arterial supply, venous and lymphatic drainage, and innervation depend on the anatomic location of the minor salivary glands. Postganglionic fibers from the submandibular gland innervate the minor salivary glands of the inferior oral cavity and oropharynx. Palatine nerves supply postganglionic fibers from the pterygopalatine ganglion to the superior oral cavity and palate [2, 4].

13.1.2 Salivary Gland Physiology

The basic histologic architecture of all salivary glands consists of a branching duct system that terminates at the salivary acini. The acinus is the site of production of saliva and is surrounded and supported by myoepithelial cells which contract to express saliva into the ducts, myofibroblasts, extracellular matrix and stromal cells, immune cells, vascular endothelial cells, and nerve cells. The acinus contains many acinar cells that produce saliva into the acinar lumen [2, 4, 7]. Acinar cells are bipolar, pyramidal shaped cells which secrete fluid and proteins from their apical surface. The acinus expresses saliva into the secretory duct, which consists of intercalated and striated duct. Myoepithelial cells also surround the intercalated ducts. The intercalated ducts consist of cuboidal shaped cells and continue as striated ducts, which contain columnar cells with microvilli on their luminal surface. The acinus and these proximal ductal components are together considered the *secretory end piece* and are organized into lobules [2]. These ducts drain into excretory and collecting ducts, which consist of a bicellular layer (apical flat epithelial cells and basal columnar cells) and lie outside of the lobules. This structural organization varies between glands. Figure 13.2 depicts the basic salivary gland unit [8].

The ducts of Rivinus are the collecting ducts of the sublingual gland, while the Stenson and Wharton's ducts are the terminal collecting ducts of the parotid and submandibular glands, respectively. The ducts serve as transport conduits while also modifying saliva composition. The medullary brainstem salivary center is a major central neural control center for salivation; however, there are multiple other stimuli for saliva secretion including taste and olfaction and the mechanical act of mastication. Salivation can be increased or decreased as a side effect of medications and can be affected by systemic medical conditions [2, 4].

The average human produces between 1 and 1.5 L of saliva daily. The minimal human salivary flow rate is at least 0.1 mL per min when unstimulated and at least 0.2 mL/min when stimulated although the average range of salivary flow rates are 0.3 mL/min when unstimulated to 7 mL/min as the maximum stimulated flow rate [2, 4, 15, 16]. The salivary glands have different viscosity of saliva that reflects the histologic subtype of acinar cells within its lobules. The parotid gland consists of mostly serous subtype acini and secretes watery saliva. ~25 % of the daily saliva production is from the parotid gland. The sublingual gland and minor salivary glands consist of mostly mucous acini and secrete viscous saliva. These glands together comprise of 2–4 % of the daily saliva production. The submandibular gland acini are a mixture of serous and mucous types, and therefore, the gland secretes an intermediate viscosity saliva which contributes ~70 % of the daily saliva production [2, 17]. Mucinous cells are found surrounding the lumen of the acini while serous acinar cells are organized at the end of the acinus to form a *serous demilune* [2]. Viscosity is also affected by the stimulating factor for saliva production. Parasympathetic innervation stimulates a less viscous and watery type of saliva while sympathetic stimulation produces a thick, low volume, viscous saliva. Parasympathetic innervation uses the neurotransmitter acetylcholine binding to muscarinic receptors to stimulate salivation. Sympathetic stimulation uses the neurotransmitter norepinephrine binding to adrenergic receptors [4, 8, 15].

Saliva consists of electrolytes, proteins, and other molecules. The acinus generates the fluid component of saliva in the form of an isotonic solution. Sodium and chloride ions are resorbed in the proximal ductal network, while potassium and bicarbonate ions are secreted. Electrolyte reabsorption and secretion involves active transport processes. The majority of the protein component is secreted at the level of the acinus; however, the secretory duct also contributes protein molecules. The end product is a hypotonic solution with pH 6–7. Salivary flow rates also affect electrolyte composition as slower flow rates allow more time for sodium and chloride resorption. However, increased flow rates stimulate increased bicarbonate secretion. Potassium is unaffected by flow rates [2, 4, 15, 17].

13.2 Traditional Surgical Interventions

13.2.1 Nonneoplastic and Inflammatory Salivary Gland Disease

Nonneoplastic salivary gland diseases include wide differential diagnosis including infectious, inflammatory, obstructive, traumatic, and radiation-induced etiologies. These disease processes more commonly involve the major salivary glands and may involve either the salivary gland parenchyma or the ducts. Some conditions may be a condition of a systemic disease process. Presentation may be acute, chronic, or recurrent and may be present in both adult and pediatric populations.

13.2.1.1 Acute Suppurative Sialadenitis

Acute suppurative sialadenitis is a condition in which retrograde bacterial contamination of the salivary ducts from the microflora of the oral cavity causes an acute infection of the ducts and salivary gland. The submandibular gland is the most common (the original description is correct as far as I know. In addition, the following sentences make more sense if the parotid gland is listed as the most common gland) salivary gland affected by bacterial sialadenitis due to increased viscosity and the decreased concentration of the antibacterial lysosomes, IgA antibodies, and sialic acid in parotid gland vs. the submandibular, sublingual, and minor salivary glands. Submandibular and sublingual gland saliva also contains glycoproteins that have been shown to competitively inhibit bacterial attachment on the epithelium of salivary ducts [18–21]. Suppurative bacterial sialadenitis has been associated with patients undergoing major abdominal or hip surgery in their postoperative hospital course. These infections may also be associated with a preexisting malignancy or head and neck infection [19,

22]. The inciting events to a bacterial infection of the salivary glands are reduced flow and salivary stasis due to obstruction due to a sialolith, foreign body, injury, or other factors that reduce flow. Predisposing conditions include dehydration, periodontal disease, immunodeficiency, diabetes mellitus, neurodegenerative disease, systemic autoimmune conditions, cystic fibrosis, radiation injury, chemotherapy, and medications that cause reduced salivary flow as a side effect [18–20].

The clinical presentation begins with pain and rapid, diffuse enlargement of the salivary gland. Palpation reveals tenderness, warmth, and induration. A stone may be identified with bimanual palpation and may be the predisposing factor for recurrent infections. Purulence may be expressed from the papilla of the involved by pressing on the gland and sent for culture. Polymicrobial infections are common; however, *Staphylococcus aureus* is reported to be the most causative organism. *Streptococcus pneumonia, Streptococcus pyogenes, Streptococcus viridans, Haemophilus influenzae,* and *Escherichia coli* are other aerobic organisms that have been cultured. Anaerobic organisms responsible for infections include *Bacteroides* species, *Peptostreptococcus, Prevotella* species, *Fusobacterium* species, and *Burkholderia pseudomallei* [18, 19, 21]. Computerized tomography (CT) and ultrasound are used to evaluate for an abscess or a sialolith. Sialography is contraindicated due to the risk of exacerbation of the infection.

The treatment of acute bacterial sialadenitis consists of antibiotics, frequent gland massage, sialogogues, hydration, electrolyte repletion, and the application of heat packs. The causative factor must be addressed if identified. Medications that reduce salivary flow must be discontinued. Broad-spectrum antibiotic therapy directed against gram-positive organisms and anaerobes can be narrowed once culture results are available. Sialendoscopy is contraindicated during an acute infection but can address sialoliths or other obstructive etiologies after the infection is treated; this technique is discussed in another section of this chapter. If treatment does not improve symptoms within 2–3 days, an abscess or antibiotic resistance must be considered [18–20, 23]. In the event of an abscess, diagnostic imaging (e.g. CT scan) is helpful to confirm location and extent of the purulence collection before an incision and drainage is performed. Image-guided drainage with CT or ultrasound is a minimally invasive method but may not be an option in some institutions. If surgical drainage is required for a parotid abscess, a modified Blair incision can be used for exposure, and blunt dissection toward the abscess is oriented in the direction of the facial nerve branches to avoid injury. The parotid fascia must be incised parallel to the facial nerve branches. For a submandibular abscess, the transcervical approach is the most direct method; the marginal mandibular nerve must either be protected with the Hayes-Martin maneuver or identified and protected. A surgical drain may be placed [18].

13.2.1.2 Viral Sialadenitis

Viral sialadenitis is similar to bacterial sialadenitis and is thought to develop more commonly by the hematogenous route more often than retrograde ductal migration. Mumps is the most common form of viral sialadenitis. The infectious agent is paramyxovirus which typically affects the parotid gland. The infection historically occurred most frequently in children and the incidence has decreased after routine vaccination. Recently, an increasing number of young adults are being diagnosed with the infection [18, 19]. Viral sialadenitis is nonsuppurative unless a bacterial superinfection occurs. After infection via the respiratory tract, the virus enters an incubation period of several weeks. The clinical presentation includes a nonspecific viral prodrome of fever, myalgia, and malaise. Salivary gland enlargement presents within the first week and is often bilateral. Other manifestations of mumps include orchitis, myocarditis, and aseptic meningitis. The virus may rarely cause sensorineural hearing loss. Acute viral sialadenitis (usually parotitis) can also be caused by cytomegalovirus, coxsackie viruses A and B, lymphocytic choriomeningitis virus, enteric cytopathic human orphan virus, and influenza virus. The submandibular gland is rarely involved. Diagnosis can be made with viral serology or isolation of the virus through cerebrospinal fluid. Symptomatic treatment is usually sufficient for acute viral sialadenitis, and surgery is rarely indicated [18, 19]..

Human immunodeficiency virus (HIV) can cause diffuse enlargement of the salivary glands (most often the parotid gland), which is referred to as HIV-associated salivary gland disease (HIV-SGD). Associated symptoms can include xerostomia and lymphadenopathy. Treatment includes antiviral drugs, sialogogues, and oral hygiene. HIV can also predispose patients to benign lymphoepithelial cysts, Kaposi sarcoma, and lymphoma of the salivary glands. The lymphoepithelial cysts can usually be managemed with needle aspiration for symptomatic but temporary relief, or sclerotherapy. Surgical intervention with extracapsular dissection or superficial parotidectomy is reserved for refractory disease [18, 19].

13.2.1.3 Sialadenitis in the Pediatric Population

Neonatal suppurative parotitis is an uncommon but reported condition that occurs most often in male and preterm neonates; dehydration appears to be the inciting factor. Clinical presentation consists of fever, irritability, anorexia, failure to thrive, gland swelling, and erythema of the overlying skin. Bilateral glands may be involved. Infection may originate from oral flora or hematogenous dissemination of bacteria. A number of pathogens can be responsible, *S. aureus* being the most common. *E. coli, Pseudomonas aeruginosa,* and group B *Streptococcus* species have been reported. Antibiotics are the mainstay of therapy with drainage or surgical intervention for refractory cases [18, 19, 24].

Juvenile recurrent parotitis is a nonsuppurative inflammatory condition in which the parotid gland periodically enlarges with associated tenderness, fever, and malaise. It is the most common salivary gland disease of childhood after mumps. The condition may be unilateral or less commonly bilateral. The peak incidence is between 3 and 6 years of age. Episodes occur every 3–4 months. The etiology of juvenile recurrent parotitis is unclear, and multiple etiologies have been proposed, including congenital duct malformation (ectasia), immunologic deficiencies, and infectious causes (*Staphylococcus* and *Streptococcus* species*)*. Strictures and ductal abnormalities causing obstruction can develop [18, 19, 25, 26]. Treatment is similar to acute sialadenitis and consists of gland massage, sialogogues, hydration, and application of heat. Antistaphylococcal antibiotic therapy can be started after a culture is obtained from the parotid duct. Conservative treatment is almost always sufficient although parotidectomy can be considered in refractory cases. Ductal ligation and tympanic neurectomy have also been described; however, these are rarely performed [19, 25, 26]. Angioplasty balloon catheters have also been used for stricture dilations by interventional radiology under fluoroscopic control [25, 27–29]. Sialendoscopic techniques may be used to address strictures; this technique is discussed later in this chapter.

13.2.1.4 Chronic Sialadenitis

Chronic sialadenitis is a condition in which there are recurrent episodes of inflammation and pain in the major salivary glands; the parotid gland is the most commonly involved gland. Symptoms are worse with eating. Similar to acute sialadenitis, salivary stasis, reduced salivary flow rates, ductal obstruction (with a sialolith or other foreign body), systemic disease, or dehydration are possible predisposing factors. Sialolithiasis is the most common cause [18, 19]. Repeated episodes of acute sialadenitis cause permanent structural changes including acinar destruction, ductal ectasia, and fibrosis. the gland becomes enlarged exacerbations, and saliva is difficult to express from the duct. Xerostomia and change in salivary content (altered electrolyte composition, increased immunoglobulins with IgG predominance, albumin, transferrin, increased lysozyme concentrations) develop in long-standing disease [18, 19]. Gland atrophy can occur, and firm, fibrotic areas of the gland may be palpated. These firm areas must be ruled out for malignancy. Ductal strictures can form and cause obstruction. CT and ultrasound can help to further characterize gland structure and identify non-palpable sialoliths, while sialography (traditional and MRI sialography) can characterize ductal architecture. Treatment includes massage, sialogogues, heat, and hydration. Several procedural interventions have been described for symptomatic management including ductal papilla dilation, sialodochoplasty, ductal steroid injection, ductal ligation, and tympanic neurectomy [18, 19]. Interventional radiology techniques to dilate ductal strictures under fluoroscopy have been reported [27–29]. Surgical extirpation of the gland can be considered when all other treatment modalities fail to sufficiently relieve symptoms. Sialendoscopy (described below) is a developing treatment modality that can delay or prevent the need for open surgical intervention.

Benign lymphoepithelial lesions (LE lesions) can develop in the setting of long-standing chronic disease. LE lesions are characterized by a lymphoplasmacytic infiltrate, acinar atrophy, and ductal metaplasia leading to the development of epimyoepithelial islands [19, 30]. This condition is well described in patients with Sjögren's syndrome and has been termed *Mikulicz's disease.* LE lesions are usually asymptomatic enlargements unless they become infected which may require drainage or surgical removal. Kuttner's tumor (chronic sclerosing sialadenitis) is a similar process occurring in the submandibular gland characterized by a firm, painless swelling associated with areas of gland atrophy. Kuttner's tumor differs histologically (lymphoid infiltrate and discrete tubular structures with regularly aligned nuclei) from LE lesions. Patients with benign LE lesions and Kuttner's tumor must be monitored for development of ductal carcinoma [18, 19, 30, 31].

13.2.1.5 Sialolithiasis

Sialolithiasis is the development of calculi in the salivary gland ductal system. Sialolithiasis accounts for 50 % of major salivary gland diseases [27, 32]. They occur most frequently in the submandibular gland (80 %) followed by the parotid gland (20 %) and sublingual gland (1 %) [18, 19, 33]. Minor salivary gland stones are rare and are most often in the upper lip or buccal mucosal glands. Sialolithiasis occurs more frequently in men. The calculi are composed of calcium phosphate and calcium carbonate and are mixed with organic molecules including glycoproteins, mucopolysaccharides, and cellular debris [19, 33]. The nidus for calculi development is believed to be an inorganic substance that allows salt precipitation in the setting of salivary stasis or reduced flow [18, 19, 34]. Due to the more alkaline and viscous properties of the submandibular gland saliva, the submandibular duct is reported to be the most susceptible to sialolith formation. The duct is also long and saliva flows against the force of gravity. Calculi occur in chronic sialadenitis patients as well as patients with gout. Calculi may be the predisposing factor in acute suppurative sialadenitis. Symptoms include postprandial pain and swelling as well as a history of acute suppurative sialadenitis [18, 19].

Sialography, CT, and ultrasound and MRI sialography can be used for diagnosis although calculi less than 2 mm may be missed by imaging [18]. Plain films are more useful for submandibular stones, which are usually radio-opaque unlike parotid stones. Virtual MRI endoscopy is a new modification of MRI that allows a three-dimensional endoscopic view of the ductal system. Treatment may be conservative and consists of gland massage, sialogogues, hydration, and observation for spontaneous passage. This is often successful for small (<2 mm) sialoliths [18, 19, 34, 35]. Procedural interventions are considered for refractory cases; the best approach depends on the location, size, and shape of the sialolith. Transoral sialolith removal can be attempted; however, gland extirpation may be required. Stenson's duct calculi can be approached transorally if the calculus is medial to the masseter muscle. Shockwave lithotripsy and sialendoscopy (discussed below) are being increasingly used. Combined approaches with endoscopy and either transoral or external approaches have been shown to be successful [18, 19]. Interventional radiology techniques under fluoroscopy have been described as well; the first sialolith removed via basket under fluoroscopy was reported by Kelly in 1991 [29]. Coronary angioplasty balloon, embolectomy catheters, and wire loop snares have also been used to remove stones. Capaccio's literature review revealed that fluoroscopic guided sialolith removal was reasonable for mobile stones in proximal and middle submandibular ductal system as well as parotid stones [27–29] .

13.2.1.6 Granulomatous Diseases

Granulomatous diseases of the head and neck may involve the salivary glands and the lymphatic networks associated with the glands. Granulomatous infections can invade salivary gland parenchyma in advanced cases. The most commonly discussed granulomatous infectious diseases of the head and neck are tuberculous and nontuberculous mycobacterial disease, cat scratch disease, toxoplasmosis, and actinomycosis. Noninfectious granulomatous disease includes sarcoidosis and Sjögren's syndrome.

Mycobacterium tuberculosis is the pathogen associated with tuberculosis, which can manifest as cervicofacial lymphadenopathy. Although salivary gland involvement is rare, it is reported in immigrants from underdeveloped countries as well as immunocompromised patients. Infection can be primary by way of ductal migration from the oral or oropharyngeal saliva or lymphoid tissue or can be secondary with either lymphatic or hematogenous spread. The intraglandular lymph nodes of the parotid gland may become sites of latent infection. The parotid gland is the most common gland affected. Submandibular gland infection is more common in systemic and disseminated tuberculosis. The infection can present as an inflammatory lesion that mimics sialadenitis or can present as a mass that masquerades as a neoplasm. Diagnosis involves purified protein derivative (PPD) skin test, chest x-ray, and fine needle aspiration (FNA) of lesions. FNA cytology may reveal characteristic granulomatous inflammation with epithelioid histiocytes. Samples may be sent for acid fast staining. In cases in which the diagnosis is uncertain or is resistant to antibacterial therapy, the involved glands are excised [18–20].

PPD skin test may be negative in nontuberculosis mycobacterium (NTM) infections that more commonly present with cervicofacial lymphadenopathy. These infections are usually localized without systemic signs or symptoms. The most common NTM infections are caused by *M. kansasii, M. scrofulaceum, M. avium-intracellulare*, and *M. bovis*. These infections are encountered in children less than 5 years of age, and the pathogens are carried in soil, water, and food products (including milk) and domestic or wild animals. Clinical presentation is classically described as a neck mass with rapid enlargement, violaceous overlying skin changes, and resistance to initial antibiotic therapy. Cervical lymphadenopathy is common. The infection may progress to an abscess that may spontaneously drain and form a sinus tract. Diagnosis with FNA biopsy is controversial and carries the risk of fistula tract formation. Cultures take up to 6 weeks to result and may be negative. Antibiotic therapy may also require weeks to months of treatment and may not be effective. Complete gland excision is therefore considered and can serve as definitive treatment. If the parotid gland is involved, superficial and/or deep parotidectomy with facial nerve preservation must be performed [18, 19].

Cat scratch disease is a local infection that originates at the scratch site with ensuing granulomatous lymphadenitis in the draining lymph nodes. *Bartonella henselae* is the pathogen and is a gram-negative bacillus that is usually spread to the skin laceration from the scratch or bite of a household cat. The upper extremity is the most common site of infection followed by the head and neck. Head and neck infection can involve the lymph nodes associated with the parotid gland or the submandibular gland. The infection starts as a pustule at the site of scratch or bite and progresses to local and regional lymphadenopathy over 1–2 weeks. Erythema and lymphadenitis frequently develops and may progress to abscess formation with spontaneous drainage. Antibody detection for *B. henselae* or PCR detection is used for diagnosis. Bacilli may be visible in tissue specimens with Warthin-Starry silver staining. Culture requires 6 weeks due to the slow growth of the organism. The infection is usually self-limiting, and antibiotic therapy is reserved for patients with advanced or systemic spread of disease. Surgical excision, like tuberculous disease, is reserved for infections that fail to resolve. Resolution may take several months. Parinaud's oculoglandular syndrome is an atypical presentation of cat scratch disease characterized by unilateral granulomatous conjunctivitis with ipsilateral cervicofacial or salivary gland lymph node involvement. Parotid involvement with facial nerve palsy has been reported [18, 19].

Toxoplasmosis is caused by the organism *Toxoplasma gondii* and rarely involves the salivary glands. Domestic cats are the host for this organism. The pathogen is transferred through ingestion of infected meat or through cat feces. Hematogenous dissemination can spread the disease to the intraparotid lymph nodes or the periparotid lymph nodes. Antibiotic therapy is usually sufficient even in advanced cases; surgery is reserved for large suppurative lesions [18, 19].

Actinomycosis is caused by the organism *Actinomycosis* species (*A. israelii, A. bovis,* and *A. naeslundii),* a gram-positive, nonacid fast bacilli. The microscopic appearance is similar to mycobacteria and fungi given the branching, filamentous appearance. *A. israelii* is commonly found as part of the oropharyngeal lymphoid tissue flora and in carious dentition [19]. Cervicofacial infection is the most common presentation and is caused by invasion of the organism after trauma or poor oral hygiene. Retrograde ductal migration may explain salivary gland infection although direct invasion into parotid or submandibular gland tissue is also possible. Infection of the salivary gland is characterized by painless enlargement of the gland with chronic purulent drainage. The disease can progress to form cutaneous drainage tracts. Fibrotic changes and soft tissue destruction cause induration of the gland upon palpation. Microscopic examination of tissue samples or swabs reveals the characteristic sulfur granules in the presence of branching filamentous, gram-positive rods. Long term (minimum 6 weeks) antibiotic therapy is sufficient for limited disease, but surgical extirpation is required in the presence of fistulous tracts and for cases refractory to antibiotics [18–20].

13.2.1.7 Noninfectious Inflammatory Disease

Sjögren's syndrome (SS) is an autoimmune disorder characterized by autoimmune destruction of exocrine glands. B- and T-cell-mediated damage causes symptoms including xerostomia, dry eyes (foreign body sensation in the eye), dysphagia, and enlargement of the salivary glands. Sjögren's syndrome can be primary or secondary (associated with another autoimmune disorder). The disease is more commonly seen in women during the fourth and fifth decade of life. Exam reveals xerostomia, dental caries, and possible oral candidiasis. Systemic manifestations include arthritis, pneumonitis, skin rash, myositis, and other complaints. Ocular exam may reveal decreased tear secretion (may be evaluated with Schirmer test), lacrimal gland enlargement, enlarged conjunctival vessels, corneal damage, and pericorneal injection. These findings are characteristic of keratoconjunctivitis sicca. The Imaging by CT or MRI can reveal calcification in involved salivary glands. Sialography may reveal sialectasis. Histopathology reveals lymphocytic infiltration of gland tissue starting with the ducts and progressing to destroy and replace acinar tissue, which in turn reduces the salivary gland function. Laboratory tests include the detection of autoantibodies against RNA/protein complexes Ro (SS-A) and La (SS-B) in addition to rheumatoid factor (RF) and ANA (antinuclear antibody). SS patients are also at increased risk for developing lymphoma. Diagnosis is aided by biopsy of a minor salivary gland of the labial mucosa. Biopsy may be performed in the clinic setting: several lobes of minor salivary gland tissue must be sampled (collected, or biopsied, if authors like) and examined. Diagnostic criteria have been established and involve the presence of signs and symptoms of keratoconjunctivitis sicca, symptoms of xerostomia and signs of decreased salivary gland function, salivary gland biopsy results, and presence of Ro and La antibodies. The presence of another autoimmune disorder such as systemic lupus erythematous or rheumatoid arthritis suggests secondary SS. Treatment includes symptomatic treatment to protect the eyes and the teeth with eye lubricants, eye patches, saliva substitutes, dental care and oral hygiene, and pilocarpine [18, 19, 30, 36].

Sarcoidosis is a granulomatous disorder with a wide range of systemic manifestations involving multiple organ systems. Common presenting symptoms include cough, dyspnea, weight loss, erythema nodosum, arthralgias, and myalgias. Salivary gland involvement is rare but presents as gland swelling. Uveoparotid fever is a manifestation of sarcoidosis characterized by uveitis, parotid gland enlargement, and facial paralysis. The parotid gland enlargement can last months but is self-limited. Minor salivary gland biopsy, as in SS, may aid diagnosis. Corticosteroids are used for treatment in uveoparotid fever and are effective for resolution of the facial paralysis [18, 19, 37].

13.2.1.8 Radiation-Induced Sialadenitis

Radiation-induced xerostomia is a well-known complication of radiotherapy for head and neck cancer. Radiation dosages greater than 20–30 Gy predispose glands to lipid peroxidase injury, enzyme spillage, and cell lysis [38]. Injury begins with an acute inflammatory reaction that leads to acinus destruction with continued irradiation. Strictures and kinks can form in the ducts and can cause duct obstruction. Increased incidence of pleomorphic adenomas and malignant salivary gland neoplasms have been reported in patients with radiation exposure.

Iodine-131 treatment for thyroid malignancy may cause dose-dependent sialadenitis. The sodium-potassium-chloride transporter in salivary gland tissue concentrates radioactive iodine to levels that can cause parenchymal damage. The parotid glands are most commonly affected followed by the submandibular glands. Sialendoscopy (described below) has revealed ductal stenosis, mucous plugs, and other findings of chronic inflammation. Sialendoscopy has been used to provide symptomatic relief by allowing ductal irrigation and/or steroid instillation [18, 39, 40].

13.2.1.9 Trauma

Traumatic injury to the salivary glands, ducts, or associated nerves requires surgical exploration and repair. Penetrating or laceration injuries to the parotid gland place the duct and facial nerve at risk. Blunt trauma may cause hematomas that require drainage to prevent fibrosis or superinfection. Penetrating injuries posterior to the anterior border of the masseter muscle must be evaluated for ductal injury due to the proximity of the duct to the skin. A probe may be placed transorally and the duct may be assessed through the wound. The proximal end of a lacerated duct may be identified by gland massage, which should demonstrate secretion of saliva in the wound. If the duct is transected, a salivary stent or catheter is placed in the duct, and primary end-to-end anastomosis is performed. The catheter or stent is left in place to allow healing for 2 weeks [18]. The duct may also be rerouted and sutured into the oral cavity. Serial dilations may be required after repair to prevent strictures and stenosis. Salivary gland parenchyma lacerations can be closed primarily with interrupted sutures. Sialoceles or a salivary cutaneous fistula may develop shortly after the repair. Serial drainage and a pressure dressing can conservatively manage these conditions. Botox injections have been used to decrease the salivary production and allow fistula resolution. If the fistula fails to resolve, ductal injury must be suspected [41]. The ductal system can be evaluated with sialography or MRI sialography. If the above management fails to resolve the fistula, gland excision can be considered [18, 42]. Ductal injuries are less common in the submandibular and sublingual glands; however, the approach to repair is similar as described above.

Patients with penetrating facial trauma must also be assessed for facial nerve injury. Physical exam may reveal weakness or complete paralysis. Nerve stimulation can further characterize the injury. Facial nerve injuries that lie posterior to a line drawn from the lateral canthus to the mental foramen must be repaired immediately. If the injury lies anterior to this line, the injury can likely be observed for recovery [18].

13.2.1.10 Cysts and Ranula

Cystic lesions of the salivary glands occur most often in the parotid gland and may be congenital or acquired. Congenital cysts include dermoid cysts, branchial cleft cysts (typically first branchial cleft cyst), and congenital duct cysts. Dermoid cysts consist of keratinizing squamous epithelium with associated dermal appendages; these cysts must be completely excised [18, 43]. First branchial cleft cysts are rare and typically present within the parotid gland. They are classified as type I (ectodermal derived duplication of the external auditory canal) or type II (ectoderm and mesoderm derived cyst or fistula). These lesions may become repeatedly infected in which case they must be excised when there is no active infection to allow clear dissection of the cyst and its tract [44]. The tracts are always intimately associated with the facial nerve and superficial parotidectomy with facial nerve monitoring, and preservation is often required. Congenital duct cysts can be diagnosed and further characterized by sialography. Intervention is not warranted unless the cyst becomes infected [18].

Acquired cysts of the salivary glands include posttraumatic cysts, postinfectious cysts, neoplasms, benign LE cysts (described above), mucoceles, and sialectasis with duct obstruction (due to sialoliths vs. other etiology). Unless the cyst is associated with a neoplasm or becomes infected, no intervention is warranted as long as the cyst is asymptomatic.

Mucoceles form due to extravasation of mucous; mucous retention cysts are true cysts and are lined by epithelium. Both of these phenomena usually occur in minor salivary glands on the labial mucosa, buccal mucosa, and ventral tongue. Treatment for symptomatic mucoceles or retention cysts is accomplished by complete excision or marsupialization [18]. A ranula is a large mucocele that arises from the sublingual gland from a ruptured duct or acinus. It presents as a cystic mass on the floor of the mouth. If the ranula continues to increase in size, it can dissect through a congenital dehiscence of the mylohyoid muscle or in between the mylohyoid and hyoglossus muscles into the submandibular space and present as a neck mass [45–47]. This is referred to as a "plunging ranula." Surgical intervention involves either marsupialization of a small ranula, surgical excision of the ranula, or attempts at inducing fibrosis that would prevent reformation. Methods to induce fibrosis include laser vaporization and sclerosing agents [18, 45, 46, 48]. An outpatient method of inducing fibrosis involves placing several sutures into the ranula to allow drainage with subsequent suture removal once adequate fibrosis has been achieved (Fig. 13.3). This relatively new technique has been termed "micro-marsupialization." The concept was introduced in 1995; however, its safety and efficacy have been under recent investigation [49, 50].

An imperforate submandibular or sublingual duct orifice may also present as an intraoral cystic swelling. These congenital sialoceles may mimic ranulas but have a true epithelial lining. Although marsupialization is the classic treatment for sialoceles, simple sialodochostomy has been reported as a safe and effective method of treatment for congenital sialocele associated with an imperforate submandibular or sublingual duct [51].

13.2.2 Neoplasm

The major salivary glands originate from epithelial invaginations from oral ectoderm during the 6th week of gestation, as described above. The ingrowths develop into the ductal system. The acinus drains into the intercalated duct, which in turn drains into the striated duct followed by the excretory duct [2].

There are two theories of tumorigenesis for neoplasms of the salivary glands: the multicellular theory and the bicellular reserve theory [8, 52]. The multicellular theory states that each type of neoplasm is derived from a differentiated cell of origin within the salivary gland (Warthin's tumors and oncocytic tumors arise from striated duct, acinic cell tumors arise from acinic cells, etc.). The bicellular theory states that all primary salivary gland neoplasms originate from the basal or stem cell of either the excretory duct or the intercalated duct cells (adenomatous tumors such as pleomorphic adenomas and oncocytic tumors originate from the intercalated duct, while tumors with an epidermoid component such as squamous cell carcinoma and mucoepidermoid carcinoma originate from the excretory duct) [8, 52]. Several etiologic factors have been linked with salivary gland neoplasms including environmental factors such as radiation exposure (Warthin's tumor), viruses (EBV and lymphoepithelial carcinoma), tobacco use (Warthin's tumor), exposure to silica dust and nitrosamines, diet, and genetic factors [8, 13].

The majority of salivary gland tumors (approximately 70 %) arise in the parotid and the majority of parotid gland tumors are benign (approximately 80 %). Ten percent of tumors arise in the submandibular gland, and the ratio of benign to malignant tumors is similar to the parotid gland. Twenty percent of salivary gland

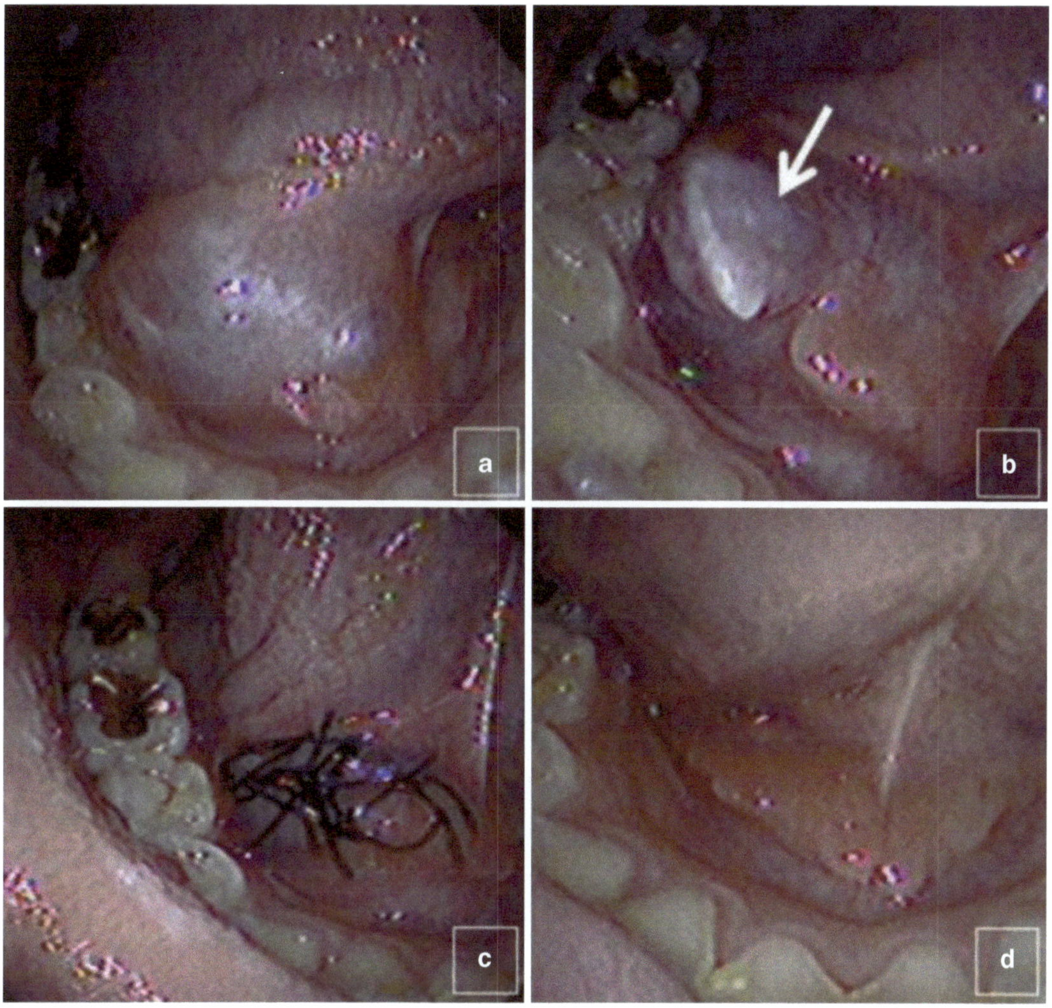

Fig. 13.3 Photographs depicting suture marsupialization of a right-sided intraoral ranula. In image (**a**), the lesion is depicted in the right floor of the mouth; this lesion was masupialized with suture but recurred. Image (**b**) depicts the recurrent intraoral ranula located posterior to the initial lesion. In image (**c**), repeat suture marsupialization has been performed. Image (**d**) demonstrates the resolution of the lesion; the sutures have been removed 2 weeks after placement (Reprinted from Ref. 50)

tumors arise in minor salivary glands, and 50–75 % of these tumors are malignant [13, 53]. Most salivary gland tumors in adults are benign. Salivary gland tumors in the pediatric population are far less common than in adults; however, the majority of pediatric salivary gland tumors are malignant. Other lesions that may present in the salivary glands of the pediatric population include hemangiomas, vascular malformations, and lymphatic malformations [8, 13, 24].

Tumors of the parotid gland present as painless swelling; the rate of enlargement is often slow for benign tumors. Obstruction of the duct may cause rapid swelling or predispose the gland to sialadenitis. Cutaneous malignancy of the scalp or facial skin may also metastasize to the intraparotid or periparotid lymph nodes. Benign tumors are typically mobile and well defined. Tumors may originate from the superficial or deep lobe of the parotid and may present on the face or in the neck or may occupy the parapharyngeal space and present as intraoral swelling. Malignant tumors are more likely to be fixed to surrounding tissues and cause facial nerve paresis. Malignant tumors are more likely to be associated with regional and cervical lymphadenopathy [8, 13, 54, 55].

Tumors of the submandibular gland are less common but present as a mobile mass in the submandibular triangle of the neck. Malignant lesions may be fixed to surrounding structures and may cause tongue weakness or numbness from perineural spread. Lower lip weakness may suggest involvement of the marginal mandibular branch of the facial nerve. Tumors of the sublingual gland are rare and may present as a floor of mouth mass. Clinical presentations of minor salivary gland tumors depend on the location of the gland; the most common sites of presentation are the palate and the parapharyngeal space [8, 13, 54, 55].

Diagnosis of a parotid or submandibular gland tumor can be obtained by fine needle aspiration (FNA) biopsy. This may be guided by ultrasound if the mass is indiscrete or difficult to visualize. Complications of FNA biopsy include local infection, hemorrhage, infarction, fibrosis, and tumor seeding. Mukunyadzi et al. noted that FNA biopsy with a 25 G needle is not only safe but also allowed the surgeon to obtain an adequate diagnosis without tumor seeding [8, 56].

The extent and type of imaging of salivary gland tumors depends on the size, location, and suspicion for malignancy. Small and well-defined and palpable tumors may require ultrasound evaluation or may not require any imaging; however, larger tumors and suspicion for malignancy require workup with CT, ultrasound with color Doppler, positron emission topography, or even MRI in advanced tumors with perineural invasion [8, 13, 54, 55].

13.2.2.1 Benign Salivary Gland Neoplasms

Pleomorphic Adenoma

Pleomorphic adenoma, which is also known as "benign mixed tumor," is the most common (65–75 %) salivary gland tumor. It is most often found in the parotid gland followed by the submandibular gland and the minor salivary glands. Pleomorphic adenoma is a slow-growing and painless tumor that contains both mesenchymal and epithelial components; the mesenchymal stroma varies between tumors [8, 13, 55]. These tumors may originate in the superficial lobe of the parotid and can extend to the deep lobe into the parapharyngeal space. Pleomorphic adenoma of the minor salivary glands can occur on the palate, the labial mucosa (more commonly the upper lip), or parapharyngeal space. The consistency of the tumor is typically smooth and rubbery in texture. Encapsulation is present; however, it may be incomplete with "pseudopod" extensions of the tumor [8, 13]. Due to the risk of recurrence and the presence of pseudopod extensions, complete surgical resection with a margin of normal tissue is performed. This may entail partial or superficial parotidectomy with facial nerve preservation. Rarely, pleomorphic adenoma has been reported to metastasize to the bone, lungs, skin, and other regions of the head and neck. Recurrence or metastasis is attributed to either leaving residual tumor in the surgical bed or rupture of the tumor during excision. Malignant transformation is rare and is termed "carcinoma ex pleomorphic adenoma" [8, 13].

Warthin's Tumor

Warthin tumor, also known as papillary cystadenoma lymphomatosum, is the second most common (5–10 %) benign neoplasm of the salivary glands. The most common site for Warthin's tumor is the parotid gland. Smoking is a known risk factor and the tumor is more common in men. The tumor may be multicentric; it may present bilaterally in up to 12 % of patients [57]. The tumor is slow growing, painless, and smooth in appearance with a well-defined capsule. A cross-section often reveals multiple cystic spaces with brown mucoid fluid. Histology is characteristic of a projection of a double-layered papillary epithelium with lymphoid stroma into cystic spaces. Treatment of Warthin's tumor is complete surgical excision. Recurrence may be attributed to undiagnosed multicentricity [8, 13, 55].

Oncocytoma

Oncocytomas represent 1 % of salivary gland tumors that present almost exclusively in the parotid gland. It presents as a painless mass and is firm, encapsulated, and rubbery in consistency. Histologically, the tumor contains granular eosinophilic cells with abundant, hyperplastic mitochondria and indented nuclei [13]. Complete resection in an extracapsular fashion is sufficient treatment. Oncocytomas of the minor salivary glands may be locally invasive and have potential to destroy adjacent tissues despite their benign nature. Surgical excision is the preferred treatment [8, 13, 55].

Basal Cell Adenoma

Basal cell adenoma represents 2–3 % of salivary gland tumors and occurs most often in the parotid gland but has been reported in the submandibular and minor salivary glands. It typically affects patients in their seventh to eighth decade of life and affects women more commonly than men. Basal cell adenomas are encapsulated and can present in four distinct histological patterns (solid, tubular, trabecular, and membranous). Minor salivary gland basal cell adenomas may lack a capsule. Surgical resection is the treatment of choice. The membranous subtype is nodular in appearance and can display multicentricity, which increases the risk of recurrence after surgical resection. Basal cell adenomas appear similar histologically to adenoid cystic carcinoma; however, they do not invade surrounding tissues or adjacent nerves [8, 13, 55].

Other Benign Neoplasms

There are a number of other benign tumors of the salivary glands such as canalicular adenomas (presents in minor salivary glands), oncocytic papillary cystadenoma (most often in larynx), myoepithelioma, sialadenoma papilliferum, inverted ductal papilloma, and others that are outside the scope of this chapter. The treatment of the majority of these tumors is surgical resection [8, 13, 55].

13.2.2.2 Malignant Salivary Gland Tumors

Mucoepidermoid Carcinoma

Mucoepidermoid carcinoma is the most common malignant salivary gland neoplasm (approximately 30–35 % of all malignant salivary gland neoplasms) [13]. The most common site for mucoepidermoid carcinoma is the parotid gland. Although the most common malignant tumor of the parotid gland is mucoepidermoid carcinoma, it is the second most common malignant neoplasm of the submandibular gland after adenoid cystic carcinoma. Mucoepidermoid carcinoma usually occurs after the third decade of life with a female predominance [13]. These tumors are classified as either high grade or low grade based on histological findings. Low-grade tumors contain mucoid as well as epidermal cell components and rarely metastasize, while high-grade tumors are predominated by epidermoid cells and have a high propensity to metastasize. Low-grade tumors are usually small, can be encapsulated, and contain mucinous fluid. High-grade tumors are usually solid and may have no encapsulation. The prognosis is worse for high-grade mucoepidermoid carcinoma. Surgical resection is recommended for low-grade tumors; the neck is not treated in the clinically N0 neck due to the low incidence of nodal metastasis [13, 54]. High-grade tumors are treated with complete surgical resection, and elective neck dissection is usually performed due to the higher rate (21 %) of occult nodal metastasis [58]; this is often followed by adjuvant radiation therapy.

Adenoid Cystic Carcinoma

Adenoid cystic carcinoma is the second most common parotid gland malignancy but is the most common malignancy of the submandibular gland and the minor salivary glands. The tumor may be partially encapsulated (or without a capsule) and appears histologically as basaloid epithelium arranged in cribriform, solid (worst prognosis), and tubular patterns with an eosinophilic stroma. Although adenoid cystic carcinoma is a slow-growing tumor, the tumor infiltrates surrounding tissue and demonstrates perineural invasion with resultant facial nerve palsy. Local recurrences after resection and distant metastasis to the lung are not uncommon. Surgical resection with postoperative radiation is typically recommended. Occult metastasis is rare and elective neck dissection for the N0 neck is not performed. If the neck is clinically positive, the overall survival is lower [13, 54].

Acinic Cell Carcinoma

Acinic cell carcinoma most commonly affects women after the fourth decade of life and most often presents in the parotid masses. The tumor can be multicentric and can present bilaterally in the parotid masses. The tumor is well encapsulated and contains both serous acinar cells and acinar cells with a clear appearing cytoplasm. Several histologic patterns are possible (cystic, papillary, vacuolated, follicular), and the cells stain positive on periodic acid-Schiff (PAS) stain. Treatment is surgical resection; adjuvant radiation is performed with facial nerve involvement, neck metastasis, skin involvement, or other poor prognostic indicators. For histologically high-grade lesions, elective neck dissection is performed. Local recurrences can present years after treatment [13, 54, 55].

Other Malignant Neoplasms

Other malignant salivary gland neoplasms include adenocarcinoma (minor salivary glands and parotid gland), polymorphous low-grade adenocarcinoma, carcinoma ex-pleomorphic adenoma (derived from pleomorphic adenoma), primary squamous cell carcinoma (most often in submandibular gland), undifferentiated carcinomas, salivary duct carcinomas, sarcomas, lymphomas, and others. The treatment of these lesions is complete surgical resection with neck dissection for high grade lesions [13, 54, 55].

13.2.2.3 Surgery of the Parotid Gland

The facial nerve branches divide the parotid into arbitrary superficial and deep "lobes" as it courses through the parotid gland parenchyma. Benign neoplasms and low-grade, well-encapsulated malignancies in the superficial lobe are treated with a superficial parotidectomy with preservation of the facial nerve branches. A total parotidectomy involves resecting the deep parotid tissue as well; this is reserved for malignancies of the deep lobe, high-grade tumors, or tumors with nodal metastasis. Cutaneous malignancies of the scalp or the face with nodal metastasis to the parotid gland or high risk for nodal metastasis also require parotidectomy [3, 6, 13, 54].

The facial nerve can be preserved if the tumor has not invaded the neural tissue; intraoperative frozen sections can assess for tumor margins in the nerve tissue. Nerve grafting should be performed for sacrificed nerves. Neck dissections are performed in the clinically positive neck, and elective neck dissections are performed in the clinically N0 neck for high-grade tumors [13, 54]. The extent and specific indications for neck dissections are outside the scope of this chapter.

The most common approach to the parotid gland is via a modified facelift (modified Blair) incision or a preauricular incision that curves along a skin crease into the neck approximately 2 cm below the angle and border of the mandible. A skin flap is raised anteriorly in a level superficial to the parotid fascia until the masseter muscle is encountered. The greater auricular nerve is identified and preserved if possible in the event that nerve grafting is required. The posterior branch of the nerve can usually be preserved to mainatin sensation to the ear lobe; the anterior branch is often sacrificed. The tail of the parotid is dissected from the sternocleidomastoid muscle. The posterior belly of the digastric muscle can be identified here and can serve as a landmark for facial nerve identification. Blunt dissection is then performed to separate the tragal cartilage from the parotid gland tissue. This reveals the tragal pointer, which guides the surgeon in localizing the facial nerve (1 cm medial). If the tumor interferes with identifying the nerve, a distal branch may be traced in a retrograde fashion to find the main nerve trunk [3, 5, 6, 13, 54]. Fibrosis from prior surgery, radiation, or other anatomic distortion may prevent adequate identification of the nerve; the mastoid cavity may then be drilled to find the intratemporal facial nerve, which is then followed to its extratemporal course. The nerve is identified and traced anteriorly to its main branches, separating the superficial and deep lobes of the gland [3, 13, 54]. Janes was the first surgeon to describe the identification process for the facial nerve trunk in 1940 [3, 59]. The nerve may then be mobilized if the deep lobe of the gland is to be removed. The posterior auricular artery, external carotid artery, and retromandibular vein may be encountered during superficial

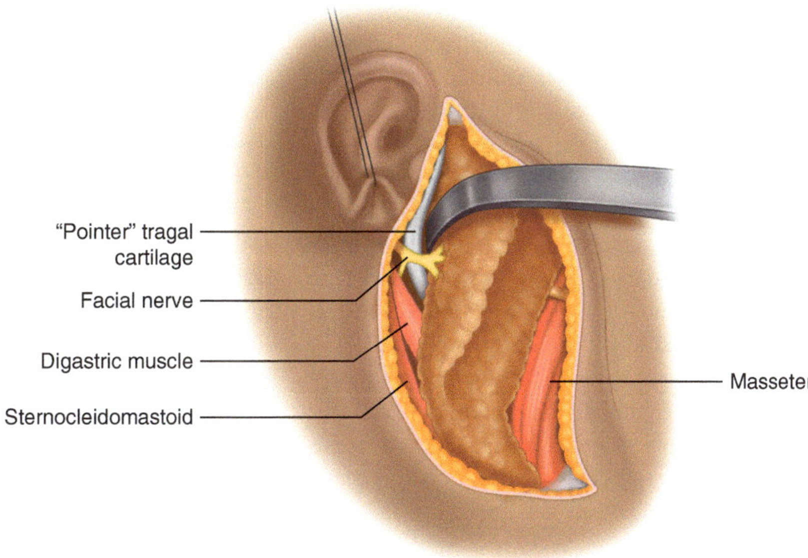

Fig. 13.4 Surgical anatomy of the parotid gland and relationship to facial nerve (Reprinted from Ref. 54)

parotidectomy, and the internal carotid artery and internal jugular vein are likely encountered during deep lobe removal [3, 13]. Figure 13.4 depicts the surgical approach to the facial nerve during a parotid gland dissection [54].

Intraoperative frozen sections may be sent to assess for extent of disease. If the tumor invades the facial nerve, the nerve may need to be traced proximally into the temporal bone to obtain negative margins. Nerve reconstruction after nerve sacrifice is performed by primary repair or with nerve grafting using the great auricular nerve or the sural nerve from the lower extremity [13, 54].

Large parotid tumors may extend into the parapharyngeal space. Tumors described as "dumbbell tumors" may involve both the superficial and deep lobes as they straddle the mandibular ramus and stylomandibular ligament. Parapharyngeal space tumors can be removed either through a transoral approach or via transcervical approach, which may require division of the styloid process, stylomandibular ligament, stylohyoid ligament, and associated muscles, or even mandibulotomy for increased access. The transoral approach is usually reserved for well-encapsulated benign tumors due to the limited access and exposure [3, 13, 54, 60].

Complications of parotid gland surgery can be classified as early and late complications. Early complications include bleeding, hematoma or seroma, infection, skin flap necrosis, trismus due to inflammation or fibrosis of the masseter muscle, development of a sialocele, and facial nerve paralysis. Facial nerve paralysis is usually temporary, and permanent facial nerve paralysis occurs in less than 4 % of parotidectomies for benign disease in which the nerve was identified and preserved [13, 61, 62]. The nerve may be stretched, compressed, or injured due to thermal energy from electrocautery, or ischemia from extensive dissection. Postoperative edema of the nerve may contribute to paresis and some surgeons administer postoperative steroids to reduce edema. Continuous facial nerve monitoring with EMG is used to allow intraoperative nerve stimulation and to warn the surgeon when the nerve is in close proximity [3, 13].

Late complications include Frey's syndrome, tumor recurrence, and poor cosmesis due to either scarring or loss of tissue bulk. Frey's syndrome is a well-known complication and is also referred to as "gustatory sweating" or "auriculotemporal nerve syndrome" [3, 8, 13, 62, 63]. Frey's syndrome was first described in 1853 by Baillarger, and the pathophysiology was described by Frey in 1923 [64, 65]. This complication is thought to occur due to aberrant reinnervation of nerve fibers from postganglionic parasympathetic fibers of the parotid gland (which use acetylcholine as a neurotransmitter) to the sweat glands and transected postganglionic sympathetic fibers to the sweat glands (which also use acetylcholine as a neurotransmitter). This causes sweating and flushing of the cheek skin as a parasympathetic response during salivation. Using a thicker skin flap and less extensive parotid dissection may reduce the incidence of Frey's syndrome. Using fascial flaps, muscle flaps, or synthetic material as a barrier has also been described [3, 13, 63]. 10 % of patients have symptomatic Frey's syndrome [13]. Symptomatic treatments include; botox injections, topical antiperspirants, topical anticholinergics, or tympanic neurectomy. Postoperative radiation has decreased the incidence of Frey's syndrome [3]. Sialoceles are subcutaneous saliva collections that can be managed with observation, needle aspiration, or a pressure dressing [3]. Botox injections may decrease salivation to promote resolution [3]. Frey's syndrome was first described in 1853 by Baillarger, and the pathophysiology was described by Frey in 1923 [64, 65].

13.2.2.4 Surgery of the Submandibular Gland

Surgical resection of the submandibular gland is typically confined to the submandibular triangle unless an extensive malignant neoplasm extends to surrounding structures. The submandibular gland is typically approached in a transcervical fashion, although submental, transoral, retroauricular, and endoscopic-assisted/endoscopic robot-assisted approaches have been described. The transcervical approach improves direct access to the gland. An incision is made 1.5–2 cm below the inferior border of the mandible along a neck skin crease. A subplatysmal flap is raised superiorly, and the marginal mandibular nerve is identified and preserved or simply preserved with the Hayes-Martin maneuver as described previously. The fascia investing the submandibular gland is incised to expose the gland. The facial artery is encountered and ligated. The mylohyoid muscle is then retracted anteriorly to expose the anterior aspect of the gland. The nerve to the mylohyoid may be encountered and can be preserved. The lingual nerve is identified and mobilized from the gland. The hypoglossal nerve is often identified at this point and preserved. Wharton's duct is identified and ligated [3, 8, 13]. Figure 13.5 depicts the classic lateral cervical approach to the submandibular gland [13].

Complications of submandibular gland surgery include bleeding or hematoma, seroma, infection, scarring, injury to the marginal mandibular branch of the facial nerve, injury to the lingual nerve, or injury to the hypoglossal nerve. Temporary lower lip paresis can occur with similar injury mechanisms to the facial nerve during parotid surgery. Tongue weakness and tongue hypesthesias can occur from hypoglossal nerve and lingual nerve injury, respectively [3, 8, 13].

13.2.2.5 Surgical Approach to the Sublingual Gland

The sub lingual gland is approached in a transoral fashion. A linear incision can be made parallel to the ipsilateral mandible. The gland can be bluntly dissected from adjacent structures such as the Wharton's duct and the lingual nerve. The superior and medial aspects of the gland can be dissected such that the gland can be peeled from the sublingual space with blunt dissection. Injuries to the lingual nerve and Wharton's duct may occur when attempting gland removal. A floor of mouth hematoma may occur and compromise the patient's airway if hemostasis is not adequately acquired at the time of surgery [3].

13.3 Advances in Salivary Gland Surgery

13.3.1 Sialendoscopy

Sialendoscopy is a relatively novel technique that provides visualization into the salivary ducts via a small-caliber endoscope. The technology was originally invented for the purpose of diagnosis but is currently used to treat a variety of nonneoplastic salivary gland pathologies while allowing gland preservation [66, 67]. The endoscope is

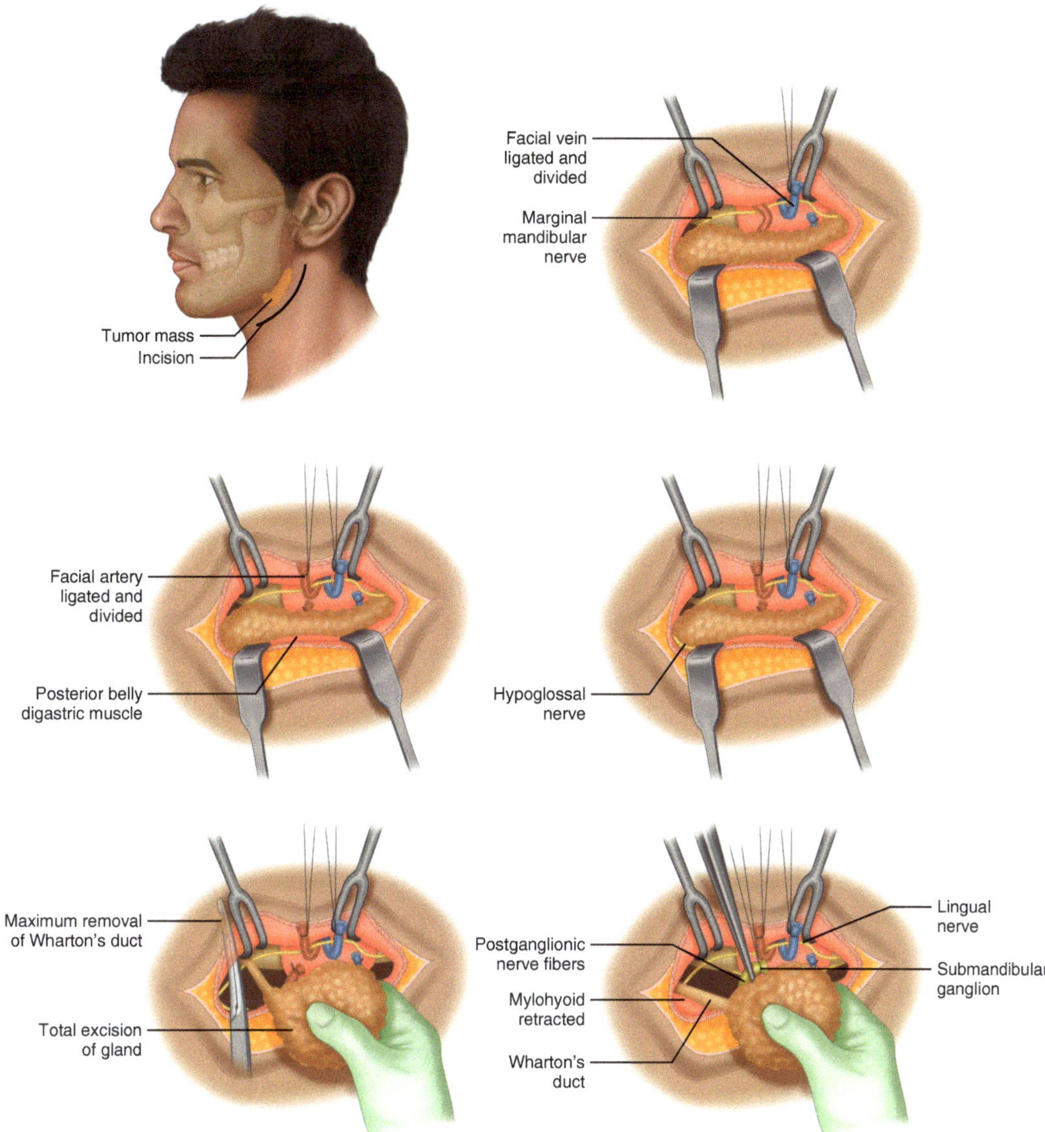

Fig. 13.5 Surgical approach to the submandibular gland depicting the Hayes-Martin maneuver (Reprinted from Ref. 13)

passed into the Stenson's duct or Wharton's duct, and saline is irrigated through the endoscope to fill the lumen and distend the salivary ductal tree. Katz was the first to describe salivary endoscopy and endoscopic anatomy in 1991 [10]. The masseteric bend of Stensen's duct was characterized by endoscopic anatomy. The optical resolution has since improved; the sialendoscopists of today can perform procedures to treat a variety of conditions.

The endoscopes can range in size from 0.8 to 1.6 mm in diameter although some studies recommend limiting the caliber to 1.2 mm to avoid iatrogenic injury [67]. The endoscopes most commonly used today are semirigid although Katz first described the use of a flexible endoscope [35]. Atienza and López-Cedrún recently performed a systematic review of the management of obstructive salivary disorders with sialendoscopy and concluded that sialendoscopy is both safe and

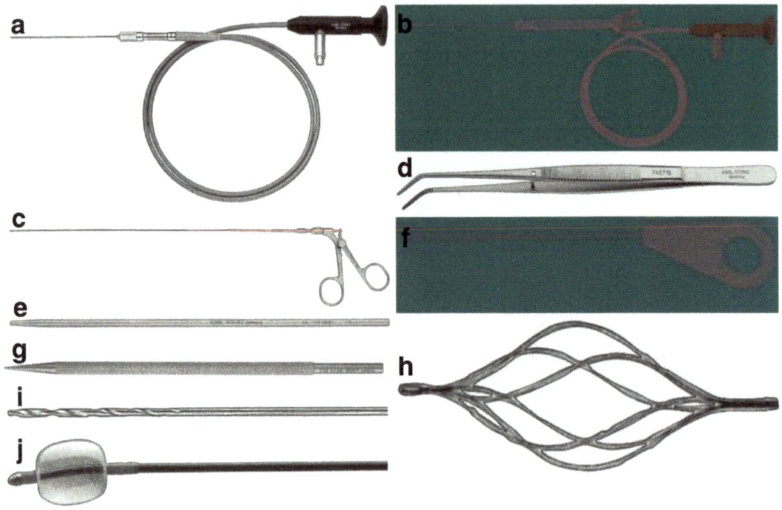

Fig. 13.6 Instruments used in Sialendoscopy. (**a**) sialendoscope, modular; (**b**) sialendoscope, all in one; (**c**) biopsy and grasping forceps; (**d**) forceps; (**e**) bougies; (**f**) probes; (**g**) dilatator; (**h**) stone extractor; (**i**) microdrill; (**j**) balloon catheter (Courtesy of Karl Storz Ref. [68])

effective for the treatment of obstructive salivary gland disorders. Four thousand one hundred thirty-four sialendoscopic procedures were performed in Atienza's review [67]. Figure 13.6 depicts typical instruments used for sialendoscopy [68].

Sialendoscopy and sialendoscopic interventions are typically outpatient procedures and can be performed either under local or general anesthesia. The sialendoscope is introduced into the duct although serial dilation of the papilla and duct with lacrimal probes may be required for insertion. The lumen is irrigated with normal saline which distends the ductal tree and allows both visualization and room for the endoscope and instruments to pass. One port is required for saline irrigation. Larger endoscopes contain an instrumentation port for forceps, wire baskets, micro-drills, balloons, stents, or laser interventions. Medications such as steroids may also be instilled into the lumen. Patients are instructed to massage their glands postoperatively. Stents are typically removed several weeks after the procedure [35, 66, 67].

Atienza's review revealed that sialolithiasis is involved in 66 % of the patients that undergo interventional sialendoscopy [67]. Marchal recommends that sialoliths less than 3 mm in the parotid gland and less than 4 mm for the submandibular gland can safely be removed endoscopically [34]. Laser lithotripsy or combined endoscopic and transoral maneuvers such as papillotomy, sialolithotomy, or ductal dissection can be performed for larger stones. Stones that are 8 mm or greater typically require a combined surgical approach [27, 34, 67]. Figure 13.7 depicts endoscopic images during endoscopic sialolith removal [35].

Inflammatory disorders such as radiation- and radioiodine-induced sialadenitis and autoimmune sialadenitis may also be treated with sialendoscopy irrigation with instillation of steroids [67]. Strictures can be found in diseases such as Sjögren's syndrome, radiation-induced sialadenitis, and juvenile recurrent parotitis; balloon dilation and steroid instillation may be performed for these conditions. 1 mm balloons that dilate to 3 mm are available for stricture dilation. Acute sialadenitis is a contraindication for sialendoscopy as the risk for ductal perforation increases in this setting [34, 35, 66, 67]. Sialendoscopy has also been used in the pediatric population for juvenile recurrent parotitis, Sjögren's syndrome, and other acquired or congenital strictures [69]. Atienza's review of sialendoscopy for treatment of obstructive disorders reports a success rate of 76 % for all sources of obstruction. In this systematic review, success was defined by resolution of obstruction with no symptoms upon patient follow-up. The success rate increases to 96 % when sialendoscopic intervention was combined with another surgical approach (papillotomy, transoral incisions, incisions through parotid fascia, external incisions) [67].

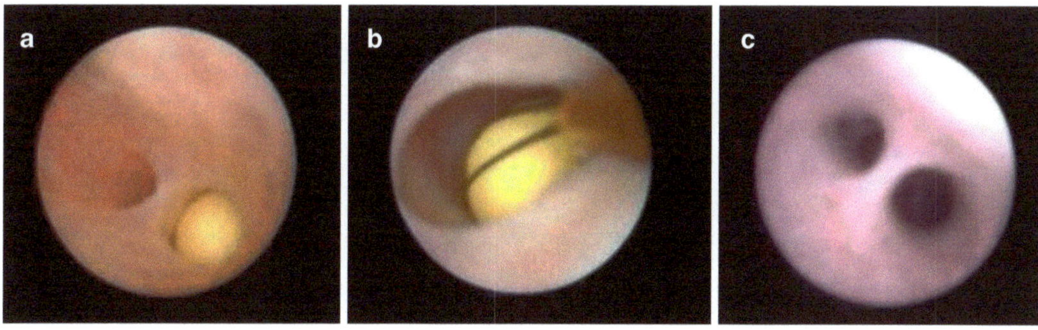

Fig. 13.7 Endoscopic sialolith removal using a wire basket. (**a**) Sialendoscopic view of sialolith in Wharton's duct; (**b**) sialolith engaged in wire basket; (**c**) view of duct after sialolith removal (Reprinted from Ref. 35)

The most common complication of sialendoscopy is post-procedural glandular swelling (typically resolves within 48 h), perforation of the duct, and injuries to adjacent blood vessels and nerves. Other complications reported have been postoperative stenosis due to structural failure of the duct or papilla; the studies in Atienza's review often cited the use of a stent to compensate for this complication. Failure to remove stones or failure of equipment (such as the wire basket for stone retrieval) was also reported. 4.6 % of glands were excised after a sialendoscopic procedure in Atienza's review [67].

13.3.2 Minimally Invasive and Robotic Surgery

Endoscopic surgery has had limited use in neck surgery due to the lack of anatomic space for instrumentation and the need for high insufflation pressures for the neck [70]. Multiple minimally invasive approaches to the submandibular gland have been reported, and several authors report the benefits of minimal scarring with an endoscopic-assisted submandibular sialadenectomy through a number of different incisions in the neck, hairline, retroauricular, facelift incision, and modified facelift approaches [71–78]. A video-assisted approach to submandibular gland sialadenectomy may yield excellent results with minimal scarring; this approach has yet to become a widely accepted and routinely practiced technique at most major institutions. Figure 13.8 depicts an endoscopic view during an endoscopic-assisted transoral submandibular sialadenectomy [75].

In contrast, the role of robotic surgery has become increasingly significant since its first application in the field of otolaryngology in 2002 at the Medical College of Georgia [70, 79]. Advantages of surgical robotics include increased precision, three-dimensional magnification, improved articulation, and possibly improved surgical ergonomics [80]. Transoral robotic surgery is becoming an established part of the head and neck oncologic surgeon's armamentarium. Robotic capabilities in head and neck surgery are continuing to be developed, and the robotic surgical procedures have been well documented in the oral cavity, oropharynx, larynx, skull base and otologic procedures, thyroidectomy, and salivary gland excision [70, 80, 81].

Robotic surgery has been described for salivary gland excision, sialolith excision, and ranula excision [81]. Terris et al. used a cadaver model to perform endorobotic submandibular gland excision in 6 cadavers and 11 total glands; they reported faster procedure times compared to neck endoscopic surgery alone (Fig. 13.9) [82].

Lee et al. performed a prospective study comparing robot-assisted and endoscopic-assisted submandibular sialadenectomy [83]. They concluded that the early postoperative outcomes were comparable and that patients in both cohorts were satisfied with their cosmesis. Although more convenient for the surgeon, the robot did not give the surgeon any clinical advantage over the endoscope.

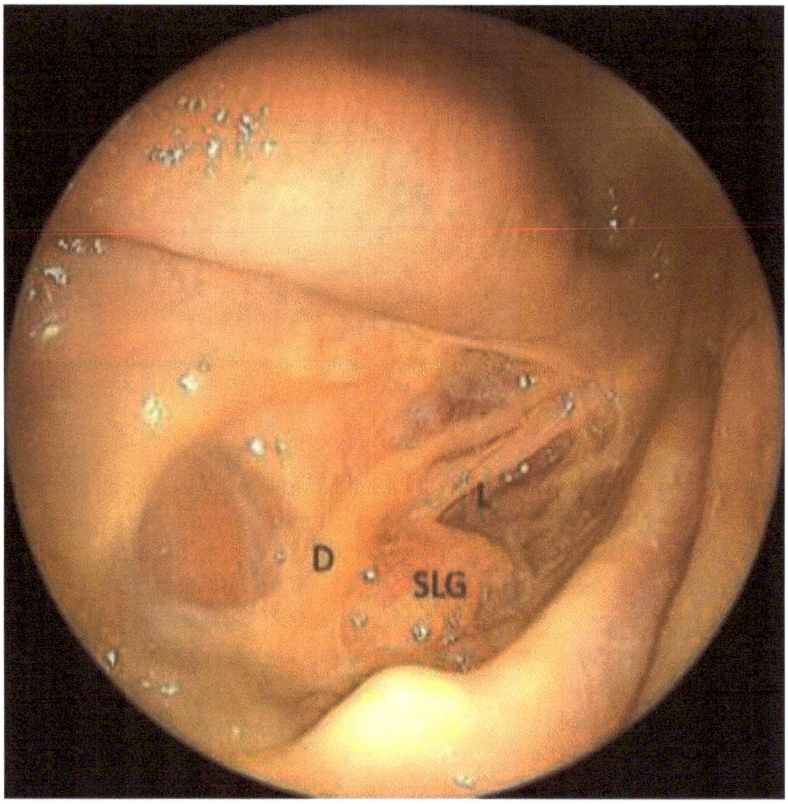

Fig. 13.8 Intra-operative photograph of endoscopic-assisted approach to transoral left submandibular sialadenectomy. *D* submandibular duct, *L* lingual nerve, *SLG* sublingual gland (Reprinted from Ref. 75)

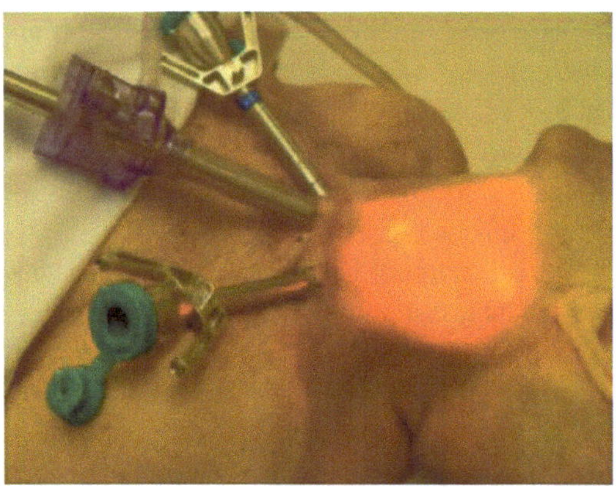

Fig. 13.9 Photograph depicting trochar placement for endorobotic submandibular gland excision (Reprinted from Ref. 82)

Walvekar reported the first case in which a submandibular gland megalith (19 × 11 mm) was removed using a combination approach with sialendoscopy to localize and trap the sialolith while transoral robotic surgery was used to remove the sialolith [84]. Razavi published a case series in 2015 describing robot-assisted sialolithotomy with sialendoscopy (RASS) for the management of large (>5 mm) hilar submandibular gland sialoliths [85]. Twenty-two patients underwent this procedure, and success (defined as gland preservation with absence of symptom recurrence) was reported in 100 % of the subjects. This cohort was compared to a historical cohort in Razavi's study that consisted of patients that underwent sialolithotomy via a combined sialendoscopy/traditional transoral approach for which the success rate was 75 %. Although further investigation and prospective studies are warranted, these results suggest that the safety and efficacy of robot-assisted sialolithotomy is excellent. Surgical robotics may eventually become an important adjunct to sialendoscopy. Walvekar et al. also reported the first removal of a floor of mouth ranula using the surgical robot [86].

13.3.3 Procedural Interventions for Xerostomia

Xerostomia affects the majority of patients who undergo primary or adjuvant radiation therapy for head and neck cancer. Salivary glands are radiosensitive and gland destruction leads to hyposalivation [87, 88]. Decreased saliva causes the patient to experience xerostomia, dysphagia, and dysarthria while also predisposing the patient to dental caries and local oral and dental infections. Traditional therapy for postradiation xerostomia is a combination of strict oral hygiene, saliva substitutes, fluoride agents, pilocarpine, and sialogogues. Amifostine has been used as a cytoprotectant [89].

13.3.3.1 Acupuncture

Acupuncture is an adjuvant alternative medicine modality extensively described in the treatment of xerostomia due to radiation as well as Sjögren's syndrome. Although the mechanisms are not clearly elucidated, acupuncture has been shown to increase salivary flow in patients with Sjögren's syndrome as well as radiation xerostomia [88, 90, 91]. Acupuncture does not appear to be a widely available or accepted treatment given the lack of standardization in its technique as well as prospective randomized trials evaluating its efficacy. The existing studies consist of small sample sizes with a variety of technical variations including needle location, needle stimulation, needle depth, number of treatments, and frequency of treatments [88]. Li et al. acknowledged the lack of standardization and proposed an acupuncture protocol for patients with radiation-induced xerostomia [91]. Zhuang et al. performed a systematic review of the literature depicting acupuncture as a treatment modality for radiation-induced xerostomia and acknowledged that there is insufficient evidence to determine its safety or efficacy [88]. Furness et al.'s Cochrane review concluded that there is low quality evidence that acupuncture affects xerostomia symptoms greater than placebo [92]. Therefore, routine use of acupuncture for radiation-induced xerostomia is not recommended at this time.

13.3.3.2 Salivary Gland Transfer

Salivary gland transfer is a relatively new technique that has been developed to address postradiation xerostomia as well as dry eyes and keratoconjunctivitis sicca [93–95]. Autologous transplantation of both major and minor salivary glands has been described for these indications [96, 97]. In patients undergoing radiotherapy for head and neck cancer, salivary gland transfer may be performed at the time of surgical intervention in anticipation of postoperative radiation. Jha et al. reported the first submandibular gland transfer to the submental space for shielding prior to radiotherapy [98]. Wu et al. performed a systematic review of the literature containing 369 patients who underwent submandibular gland transfer before radiotherapy in the included studies [99]. Both stimulated and unstimulated salivary flow rates were noted to be much higher in patients who underwent the intervention vs. patients who either received pilocarpine or no other intervention. They concluded that submandibular gland transfer is highly effective in the prevention of postradiation xerostomia without serious adverse effects.

Major and minor salivary gland transfer techniques have also been performed for the purposes of eye lubrication in the setting of dry eyes and keratoconjunctivitis sicca [96, 97, 100]. Figure 13.10 depicts a schematic diagram of the four possible salivary gland transfers for xerophthalmia [101]. The composition of saliva and tears is fairly similar, and the digestive component of saliva and presence of amylase have not been found to be destructive to the ocular surface [96]. Mucosal grafts containing salivary gland tissue for dry eyes were first described by Murube in 1998 [102]; labial minor salivary glands were described to significantly reduce dry eye symptoms despite the minor differences in composition and increased viscosity. The graft is sutured to the undersurface of the eyelid. The grafts appear to be 90 % viable although further prospective studies are warranted [97].

Limitations noted in major gland transfer include potential gland necrosis, hypersecretion, and donor site morbidity such as facial nerve injury. Surgical options described in animal and

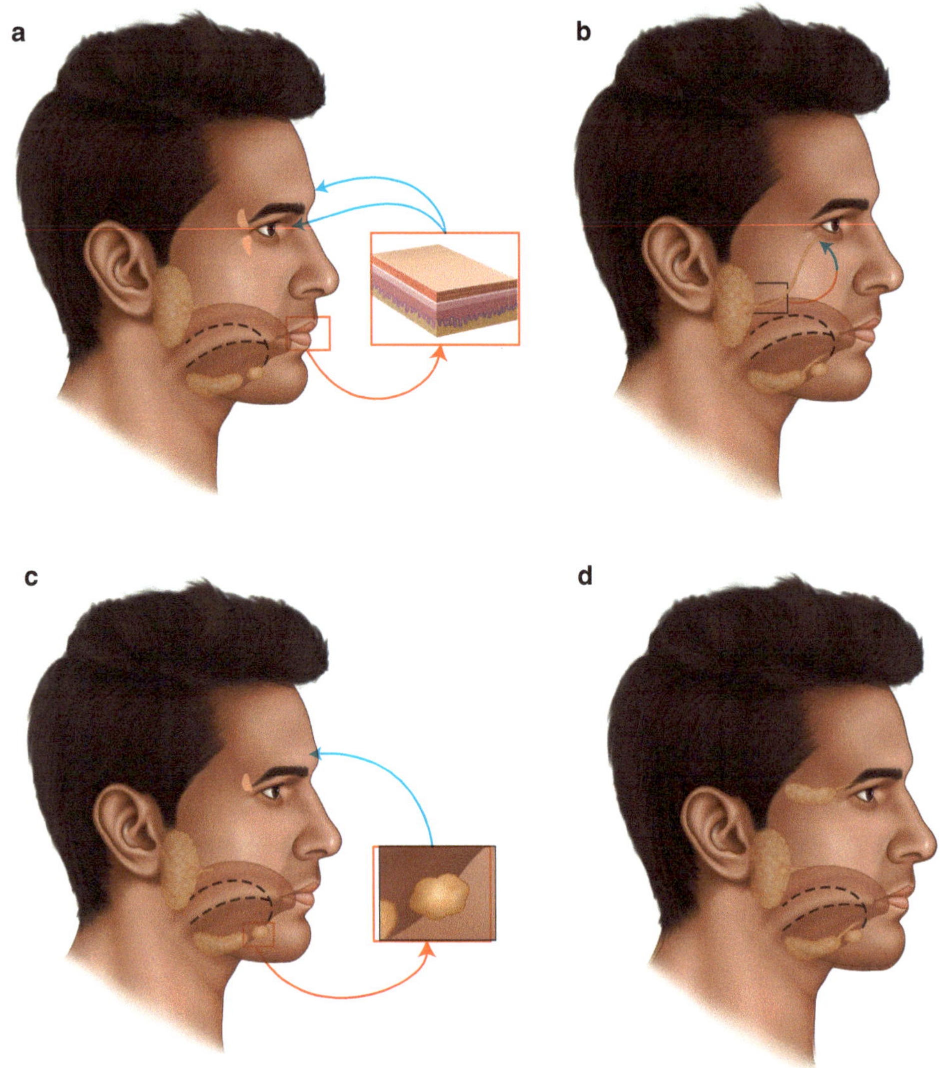

Fig. 13.10 Diagrammatic illustration depicting three different surgical techniques of major salivary gland transfer for xerophthalmia. (**a**) Transplantation of the minor salivary glands; (**b**) Transposition of the parotid gland duct; (**c**) sublingual gland transplantation; (**d**) submandibular gland transplantation. (Reprinted from Ref. 102)

human studies include transposition of Stenson's duct to the inferior fornix, free transplantation of the sublingual gland to the conjunctival fornix without microvascular anastomosis, and free transplantation of the submandibular gland with or without microvascular anastomosis and implantation of Wharton's duct into the upper temporal fornix [96, 97]. Parotid gland transfer or Stensen's duct transposition has been reported to produce copious amounts of tearing (saliva) and epiphora as a gustatory response. Complications include blepharitis, corneal calcifications, and increased bacterial load in the conjunctival sac in canine studies [96, 103]. Epiphora may lead to increased eye wiping and subsequent keratitis. Sublingual gland transplantation is not performed due to the high rate of necrosis given that the gland is transplanted as a free graft without vascular supply [96].

Submandibular gland transplantation with microvascular anastomosis appears to have the most advantages to the ophthalmologic surgeon. This procedure was first described by Murube-del

Castillo in 1986, and several other authors have replicated this technique [94, 96, 102, 104–108]. The seromucinous nature of submandibular gland secretions simulate the seromucinous lacrimal secretions. The gustatory reflex of epiphora is not present due to intraoperative denervation during the procedure [96]. Major salivary gland transplantation to the ocular tissues requires a team of ophthalmologists as well as head and neck surgeons. After resection of the submandibular gland along with the duct, a surrounding cuff of mucosa at the papilla, and facial artery and vein glandular branches, the vessels are anastomosed to branches of the superficial temporal artery or preauricular vessels. The gland is placed in a pocket created in the temporalis muscle, and the duct is tunneled subcutaneously to the conjunctival fornix. Prospective studies have revealed improved Schirmer's test, fluorescein break-up time, use of artificial tears, and discomfort. Epiphora and epithelial edema were common complications for successful transplantations [96]. Geerling and Sieg note that of the three major salivary glands, autologous submandibular gland transplantation is the only procedure that can currently be recommended in humans [96].

13.3.4 Procedural Interventions for Sialorrhea

Sialorrhea is the term for excessive salivation in both children and adults with neurologic impairment. Patients with neurologic impairment suffer from a defect in their oral and oropharyngeal phases of swallowing which causes pooling of saliva. Other causes for sialorrhea include oral inflammation, gastroesophageal reflux, medication side effects and toxins, or anatomic abnormalities of the oral cavity and oropharynx (tonsillar hypertrophy, macroglossia) [109]. Hypersecretion of saliva in combination with poor oropharyngeal and facial muscle control and dysphagia leads to pooling of saliva in the oral cavity, oropharynx, and larynx. Patients with sialorrhea often suffer from dehydration, chapped lips, and are socially marginalized due to the odor and appearance of excess saliva. Nonsurgical options include physical therapy, medications (glycopyrrolate, scopolamine), botox injection, treatment of gastroesophageal reflux, and radiation to the glands [109, 110]. Surgical options include gland excision, tympanic neurectomy, and duct ligation. A combination of the above procedures involving multiple glands may be performed. Salivary duct repositioning or rerouting is also a well-known technique to address sialorrhea. The parotid or submandibular ducts are rerouted to the posterior oropharynx or elsewhere in the oral cavity mucosa to avoid more definitive procedures such as gland excision. Duct rerouting has allowed preservation of salivation with reduction of drooling [110]. Hockstein states that the most definitive surgical therapy is bilateral parotid duct ligation with bilateral submandibular gland excision [109]. Reed et al. performed a meta-analysis of the surgical management of drooling and noted that there is no single procedure that is agreed upon as the most effective [110]. Large, directly comparative studies depicting the safety and efficacy of the above procedures are required to identify a procedure that may be universally performed by otolaryngologists and maxillofacial surgeons.

13.3.4.1 Salivary Duct Repositioning

Salivary duct repositioning is used to address sialorrhea and xerophthalmia and to prevent postoperative salivary duct obstruction from oral cancer resection or salivary calculi. Parotid salivary duct repositioning is described often in the literature as a surgical procedure to address sialorrhea. The procedure is most often described in pediatric populations, and submandibular duct repositioning is the most commonly described. Puraviappan et al. performed a prospective study in which the efficacy of submandibular duct relocation was assessed in eight children with cerebral palsy using a visual analogue score by the patient's parents. They reported that seven of eight patients had significant reduction in drooling and reported parent satisfaction in all patients [111]. De et al. reported outcomes for submandibular duct relocation for 56 pediatric patients; drooling was significantly reduced in 49 cases, and parental satisfaction was noted to be high.

The main complication reported was ranula formation in five cases. They conclude that duct repositioning is a significant means to improve quality of life in pediatric patients with sialorrhea [112]. Panarese et al. reported outcomes for 37 pediatric patients and noted that 76.5 % of patients had long-term control of sialorrhea, and the authors also concluded that the procedure is safe and successful and improves quality of life in the majority of patients [113]. Uppal et al. performed a retrospective review of 23 neurologically impaired children and noted an overall improvement in drooling in 20 patients (13 patients with complete cessation of drooling); reported complications were ranula, submandibular gland swelling (three transient, two which required gland excision). Three patients were reported to have a poor outcome, and they noted that these patients had the most severe oral-motor dysfunction [114]. Katona et al. performed submandibular duct relocations on 14 young adults and children using high-frequency radiosurgery techniques; 79 % of patients achieved a satisfactory decrease in sialorrhea. Katona et al. also reported decreased operative time with high-frequency radiosurgery and endorsed its safety and efficacy [115].

Salivary Duct Repositioning for Xerophthalmia

Parotid duct relocation, as described in the previous section, has also been used to treat xerophthalmia due to several etiologies (autoimmune, inflammatory). The duct is rerouted either transorally or extraorally to the conjunctival fornix; an external approach was first described by Filatov and Chevaljev in 1951 [96, 116–118]. Zhang et al. reported outcomes on 40 cases in which parotid duct transposition was performed for xerophthalmia, 82.5 % of patients had tearing postoperatively, and vision was improved in 72.5 % of patients [119]. The etiologies for dry eye in this studies included Stevens-Johnson syndrome, ocular pemphigoid, and alkali eye burn. They concluded that parotid duct transplantation is a simple and easy procedure that should be considered in patients with dry eyes caused by Stevens-Johnsons syndrome but not in ocular pemphigoid (these patients failed to improve). Complications reported may include duct obstruction, dislocation, and ductal contraction, which have the potential to cause entropion or ectropion. Gustatory epiphora is a manifestation due to the nature of the parasympathetic innervation as described previously. Prospective studies are warranted to further evaluate the efficacy and rate of complications.

Salivary Duct Repositioning for Head and Neck Cancer

Salivary duct repositioning has also been used in the setting of oral cancer. Salivary gland swelling and pain may occur in the postoperative period, which may be confused for a tumor recurrence. Duct relocation may prevent this potential false-positive diagnosis and decrease postoperative salivary gland colic. Stenson's duct rerouting has been reported as a means for gland preservation without compromising cancer resection [120, 121]. The salivary duct can be repositioned even in the setting of oral cancer reconstruction with a free flap by routing the duct through the free flap [122]. Sakakibara reported that repositioning of Wharton's duct could lower the likelihood of postoperative obstructive complications. Mehta et al. performed parotid duct relocation in buccal mucosa cancer resection in 562 patients and reported a markedly reduced incidence of postoperative sialocele and parotitis [121].

13.4 The Future: Salivary Gland Regeneration

The minimally invasive and gland-sparing procedures described above are relatively recent innovations in the history of head and neck surgery. The future of salivary gland surgery is promising and may build on the principles of gland sparing techniques. Regenerative medicine is a rapidly emerging field of research and will certainly impact the surgical management of salivary gland disease. The molecular and genetic mechanisms of salivary gland biogenesis and development are becoming increasingly elucidated to allow for experimentation with human salivary gland regeneration. Salivary gland stem cells are being characterized and will play a large role in therapeutic salivary gland regeneration.

Salivary gland regeneration for the purposes of restoring function in xerostomia and irradiated salivary glands is a major focus of research. Mouse models have allowed researchers to propose several methods of salivary gland regeneration [7]. Approaches to regeneration are gene therapy with viral vectors, stem cell therapy, and replacement of native gland tissue with bioengineered salivary glands [7, 123]. Viral vectors have been used to express water channels (aquaporins) into the ductal epithelium via intraductal injection of the vector [124]. Bone marrow stem cells have been transplanted into irradiated mouse salivary glands; the cells secreted a factor which, acted in a paracrine fashion to regenerate epithelia and increased salivary secretions, provided cell protection, increased vascularity, and induced the upregulation of biomarkers responsible for cell regeneration [125–127]. The Coppes lab has demonstrated that the transplantation of cells expressing Kit (a tyrosine kinase growth factor receptor) into mouse salivary glands induced the functional regeneration of gland epithelium. Autologous gland transplantation in humans with Kit+ salivary gland cells biopsied prior to irradiation, and then reimplantation postradiation may be a therapeutic implication of these findings [7, 128, 129].

Stem cells (including embryonic and other types) have complex interaction patterns with their microenvironment, also termed "stem cell niche"; stem cells affect the microenvironment and differentiate under the influence of extrinsic factors. Stem cells have been proposed to reside outside the ducts of salivary glands. The respective salivary gland niche likely impacts the differentiation of salivary gland stem cells [130]. Ono et al. attempted to regenerate salivary gland cells by coculturing embryonic salivary glands and induced pluripotent stem cells [131]. They demonstrated that coculture of embryonic mouse submandibular gland cells resulted in better-developed epithelial structures (acinar-like aggregations) than monoculture of embryonic mouse salivary gland cells. This highlights the significance of the stem cell niche and suggests that induced pluripotent stem cells may be able to accelerate to regeneration and development of salivary glands. These studies provide hope that the functional regeneration of salivary glands will soon be possible in patients. It is unclear at this time how the stem cell microenvironment in human salivary glands is affected after radiation, surgery, or in the setting of both mild and severe autoimmune disease. Future studies may further investigate the impact of these variables on stem cell niche and the potential for human application.

There is also active research interest in the use of bioengineered cells and tissue for functional organ restoration. Ogawa et al. performed orthotopic transplantation of bioengineered salivary gland germ cells into gland-deficient mice and demonstrated functional regeneration of mature salivary glands that produced saliva in response to pilocarpine and gustatory stimulation, protected against bacterial infection, and improved swallowing [132]. Synthetic extracellular matrix has been proposed to serve as a scaffold for implanted cells to form epithelium and other glandular components. Molecular components of the extracellular matrix regulate cell proliferation and development; a variety of synthetic extracellular matrix scaffolds may one day be designed to customize cellular polarity and function. In vitro regeneration of human salivary gland tissue by culturing human salivary gland cells in three dimensions in a collagen and matrigel construct has been described. Single human gland cells are proliferated and assembled into both acinar and ductal structures [133]. Cells may be cultured in this fashion for ultimate implantation into in vivo salivary gland tissue [134, 135].

Three-dimensional printing in resin has been used in murid models to replicate the anatomic shape of soft tissue organs such as the salivary gland based on three-dimensional MRI reconstructions [136]. 3D printing of functional salivary glands has yet to be developed; however, this technology could help shape synthetic matrices for cellular growth and biogenesis of functional salivary glands. These methods will be further explored and validated in nonhuman models; human clinical trials may one day incorporate several of these techniques for salivary gland regeneration.

13.4.1 Implications for the Head and Neck Surgeon

The study of functional restoration of human salivary gland tissue is in its infancy; it has yet to directly translate into the routine care for patients with xerostomia. Discoveries in regenerative medicine may one day allow partial or complete regeneration of atrophic glands by implantation of salivary gland cells or implantation of entire salivary glands generated in vitro. The head and neck surgeon will need to be aware of the implications of these potential treatments. Studies that successfully describe the regeneration of salivary gland tissue are carried out in a controlled and favorable environment. Head and neck surgeons often perform salivary gland surgery on patients with altered anatomy, infected salivary glands, glands containing neoplasms, atrophic glands, and fibrotic salivary glands. Each of these conditions affects the procedural technique and level of difficulty for the surgeon. Regeneration after stem cell or bioengineered tissue implantation may not be successful in human salivary glands that were irradiated in vivo or atrophied after an inflammatory or obstructive process. The irradiated or atrophic gland's blood supply may be scant due to dense fibrosis. This may prevent the proliferation of regenerative cells in humans. Fibrosis also prevents the expansion of tissue, which may limit the growth of regenerating salivary glands. Patients with poor nutritional status and wound healing capabilities may also benefit to varying degrees compared with healthy patients. The stem cell niche may also vary depending on the location within the histologic architecture of the gland; this is important when deciding the anatomic location for cell or tissue implantation. Duct strictures and kinks will continue to require intervention; parenchymal regeneration may be of no use without a functioning collecting duct and drainage system. Adjunctive procedures such as sialendoscopic dilation of ducts or instillation of anti-inflammatory or growth factors may one day support gland regeneration. Sialoceles and salivary fistulas may develop in cases of aberrant repair or regeneration, as in the setting of trauma. The ability to replicate the three-dimensional anatomy of salivary glands in humans is also unclear. The malignant potential of newly generated or bioengineered tissue is also unknown. Future studies will need to evaluate the feasibility of salivary gland regeneration of human salivary gland tissue in the setting of prior surgery and the abovementioned conditions.

Conclusion

The salivary glands may harbor a variety of conditions including obstructive, inflammatory, infectious, and neoplastic disorders. Surgical management of salivary gland disease requires an in-depth understanding of the anatomy, physiology, and common disease processes involving salivary gland tissue. The history of surgical intervention for salivary gland disorders in otolaryngology and maxillofacial surgery features an evolution of techniques from definitive and invasive procedures such as gland extirpation to minimally invasive procedures that may preserve the glands such as salivary gland duct surgery and sialendoscopy. Salivary duct repositioning has allowed gland preservation in patients with sialorrhea and can provide symptomatic relief to patients with xerostomia. An increasing number of surgical approaches to salivary gland extirpation have also developed with a trend toward smaller skin incisions and the pioneering of endoscopic and robotic techniques. It is difficult to predict how the expanding applications of regenerative medicine will impact the future of salivary gland surgery. Head and neck surgeons must be aware of the technological and procedural advances in salivary gland treatment in order to efficiently incorporate new techniques into practice.

Reference

1. Lydiatt DD, Bucher GS. The historical evolution of the understanding of the submandibular and sublingual salivary glands. Clin Anat. 2012;25:2–11. doi:10.1002/ca.22007.
2. Elluru RG. Physiology of the salivary glands. In: Haughey BH, Robbins KT, editors. Cummings otolaryngol, Head Neck Surg. 5th ed. Philadelphia: Mosby-Elsevier; 2010. p. 1134–42.
3. Nadershah M, Salama A. Removal of parotid, submandibular, and sublingual glands. Oral Maxillofac Surg Clin North Am. 2012;24:295–305. doi:10.1016/j.coms.2012.01.005.
4. Walvekar RR, Loehn BC, Wilson MN. Anatomy and physiology of the salivary glands. In: Eibling DE, Newlands SD, editors. Bailey's head neck surgery – otolaryngol. 5th ed. Baltimore: Lippincott Williams & Wilkins; 2014. p. 691–700.
5. Carlson GW. The salivary glands. Embryology, anatomy, and surgical applications. Surg Clin North Am. 2000;80:261–73. xii
6. Du Toit DF, Nortje C. Salivary glands: applied anatomy and clinical correlates. SADJ. 2004;59:65–74. doi:10.1017/S000748530002229X.
7. Holmberg KV, Hoffman MP. Anatomy, biogenesis and regeneration of salivary glands. Monogr Oral Sci. 2014;24:1–13. doi:10.1159/000358776.
8. Calzada GG, Hanna EY. Benign neoplasms of the salivary glands. In: Haughey BH, Robbins KT, editors. Cummings otolaryngol, Head Neck Surg. 5th ed. Philadelphia: Mosby-Elsevier; 2010. p. 1162–77.
9. Frommer J. The human accessory parotid gland: its incidence, nature, and significance. Oral Surg Oral Med Oral Pathol. 1977;43:671–6.
10. Katz P. Endoscopy of the salivary glands. Ann Radiol (Paris). 1991;34:110–3.
11. Kwak HH, Park HD, Youn KH, Hu KS, Koh KS, Han SH, et al. Branching patterns of the facial nerve and its communication with the auriculotemporal nerve. Surg Radiol Anat. 2004;26:494–500. doi:10.1007/s00276-004-0259-6.
12. Cannon CR, Replogle WH, Schenk MP. Facial nerve in parotidectomy: a topographical analysis. Laryngoscope. 2004;114:2034–7. doi:10.1097/01.mlg.0000147943.13052.62.
13. Oh YS, Russel MS, Eisele DW. Salivary gland neoplasms. In: Gourin CG, Pou AM, editors. Bailey's head neck surgery – otolaryngol. 5th ed. Baltimore: Lippincott Williams & Wilkins; 2014. p. 1760–87.
14. Zenk J, Hosemann WG, Iro H. Diameters of the main excretory ducts of the adult human submandibular and parotid gland: a histologic study. Oral Surg Oral Med Oral Pathol Oral Radiol Endod. 1998;85:576–80.
15. Humphrey SP, Williamson RT. A review of saliva: normal composition, flow, and function. J Prosthet Dent. 2001;85:162–9. doi:10.1067/mpr.2001.113778.
16. Edgar WM. Saliva and dental health. Clinical implications of saliva: report of a consensus meeting. Br Dent J. 1990;169:96–8. doi:10.1038/sj.bdj.4807284.
17. Mandel ID. The role of saliva in maintaining oral homeostasis. J Am Dent Assoc. 1989;119:298–304.
18. Walvekar RR, Bowen MA. Nonneoplastic diseases of the salivary glands. In: Eibling DE, Newlands SD, editors. Bailey's head neck surgery – otolaryngol. 5th ed. Baltimore: Lippincott Williams & Wilkins; 2014. p. 702–16.
19. Rogers J, McCaffrey TV. Inflammatory disorders of the salivary glands. In: Haughey BH, Robbins KT, editors. Cummings otolaryngol, Head Neck Surg. 5th ed. Philadelphia: Mosby-Elsevier; 2010. p. 1151–61.
20. Brook I. Diagnosis and management of parotitis. Arch Otolaryngol Head Neck Surg. 1992;118:469–71. doi:file://Z:\References\Text Files\00000003991.txt
21. Brook I. Aerobic and anaerobic microbiology of suppurative sialadenitis. J Med Microbiol. 2002;51: 526–9.
22. Lary BG. Postoperative suppurative parotitis. Arch Surg. 1964;89:653–5.
23. Perry RS. Recognition and management of acute suppurative parotitis. Clin Pharm. 1985;4:566–71.
24. Iro H. Salivary gland diseases in children. GMS Curr Top Otorhinolarynol. 2014;13:1–30. doi:10.3205/cto000109.
25. Capaccio P, Sigismund PE, Luca N, Marchisio P, Pignataro L. Modern management of juvenile recurrent parotitis. J Laryngol Otol. 2012;126:1254–60. doi:10.1017/S0022215112002319.
26. Schneider H, Koch M, Künzel J, Gillespie MB, Grundtner P, Iro H, et al. Juvenile recurrent parotitis: a retrospective comparison of sialendoscopy versus conservative therapy. Laryngoscope. 2014;124:451–5. doi:10.1002/lary.24291.
27. Capaccio P, Torretta S, Ottavian F, Sambataro G, Pignataro L. Modern management of obstructive salivary diseases. Acta Otorhinolaryngol Ital Organo Uff Della Soc Ital Di Otorinolaringol E Chir Cerv-Facc. 2007;27:161–72.
28. Drage NA, Brown JE, Escudier MP, McGurk M. Interventional radiology in the removal of salivary calculi. Radiology. 2000;214:139–42. doi:10.1148/radiology.214.1.r00ja02139.
29. Kelly IM, Dick R. Technical report: Interventional sialography: dormia basket removal of Wharton's duct calculus. Clin Radiol. 1991;43:205–6.
30. Ma Q, Song H. Diagnosis and management of lymphoepithelial lesion of the parotid gland. Rheumatol Int. 2011;31:959–62. doi:10.1007/s00296-010-1617-9.
31. Cheuk W, Chan JKC. Kuttner tumor of the submandibular gland fine-needle aspiration cytologic findings of seven cases. Am J Clin Pathol. 2002;117:103–8. doi:10.1309/G9T3-22MH-Q7KL-G2DL.
32. Kenefick JS. Some aspects of salivary gland disorders. Proc R Soc Med. 1975;68:283–5.
33. Sigismund PE, Zenk J, Koch M, Schapher M, Rudes M, Iro H. Nearly 3,000 salivary stones: some clinical

and epidemiologic aspects. Laryngoscope. 2015;125:1879–82. doi:10.1002/lary.25377.
34. Marchal F, Dulguerov P. Sialolithiasis management: the state of the art. Arch Otolaryngol Head Neck Surg. 2003;129:951–6. doi:10.1001/archotol.129.9.951.
35. Singh PP, Gupta V. Sialendoscopy: introduction, indications and technique. Indian J Otolaryngol Head Neck Surg. 2014;66:74–8. doi:10.1007/s12070-013-0675-1.
36. Kagami H, Wang S, Hai B. Restoring the function of salivary glands. Oral Dis. 2008;14:15–24. doi:10.1111/j.1601-0825.2006.01339.x.
37. Denny MC, Fotino AD. The Heerfordt-Waldenström syndrome as an initial presentation of sarcoidosis. Proc (Bayl Univ Med Cent). 2013;26:390–2.
38. Saarilahti K, Kouri M, Collan J, Hämäläinen T, Atula T, Joensuu H, et al. Intensity modulated radiotherapy for head and neck cancer: evidence for preserved salivary gland function. Radiother Oncol. 2005;74:251–8. doi:10.1016/j.radonc.2004.11.004.
39. Kim JW, Han GS, Lee SH, Lee DY, Kim Y-M. Sialoendoscopic treatment for radioiodine induced sialadenitis. Laryngoscope. 2007;117:133–6. doi:10.1097/01.mlg.0000247776.72484.62.
40. Caglar M, Tuncel M, Alpar R. Scintigraphic evaluation of salivary gland dysfunction in patients with thyroid cancer after radioiodine treatment. Clin Nucl Med. 2002;27:767–71. doi:10.1097/01.RLU.0000031607.42515.B3.
41. Hemenway WG, Bergstrom L. Parotid duct fistula: a review. South Med J. 1971;64:912–8.
42. Lewkowicz AA, Hasson O, Nahlieli O. Traumatic injuries to the parotid gland and duct. J Oral Maxillofac Surg. 2002;60:676–80. doi:10.1053/joms.2002.33118.
43. Work WP. Cysts and congenital lesions of the parotid gland. Otolaryngol Clin North Am. 1977;10:339–43.
44. Aronsohn RS, Batsakis JG, Rice DH, Work WP. Anomalies of the first branchial cleft. Arch Otolaryngol. 1976;102:737–40.
45. Huang S-F, Liao C-T, Chin S-C, Chen I-H. Transoral approach for plunging ranula – 10-year experience. Laryngoscope. 2010;120:53–7. doi:10.1002/lary.20674.
46. Harrison JD. Modern management and pathophysiology of ranula: literature review. Head Neck. 2010;32:1310–20. doi:10.1002/hed.
47. Zhao YF, Jia Y, Chen XM, Zhang WF. Clinical review of 580 ranulas. Oral Surg Oral Med Oral Pathol Oral Radiol Endod. 2004;98:281–7. doi:10.1016/j.tripleo.2004.01.013.
48. Lai JB, Poon CY. Treatment of ranula using carbon dioxide laser – case series report. Int J Oral Maxillofac Surg. 2009;38:1107–11. doi:10.1016/j.ijom.2009.04.024.
49. Delbem AC, Cunha RF, Vieira AE, Ribeiro LL. Treatment of mucus retention phenomena in children by the micro-marsupialization technique: case reports. Pediatr Dent. n.d.;22:155–8.
50. Woo SH, Chi JH, Kim BH, Kwon SK. Treatment of intraoral ranulas with micromarsupialization: clinical outcomes and safety from a phase II clinical trial. Head Neck. 2015;37:197–201. doi:10.1002/hed.23579.
51. Rosow DE, Ward RF, April MM. Sialodochostomy as treatment for imperforate submandibular duct: a systematic literature review and report of two cases. Int J Pediatr Otorhinolaryngol. 2009;73:1613–5. doi:10.1016/j.ijporl.2009.03.007.
52. Dardick I, Burford-Mason AP. Current status of histogenetic and morphogenetic concepts of salivary gland tumorigenesis. Crit Rev Oral Biol Med. 1993;4:639–77. doi:10.1177/10454411930040050201.
53. Adelstein DJ, Rodriguez CP. What is new in the management of salivary gland cancers?. 2011;23. doi:10.1097/CCO.0b013e328344f59c.
54. Sunwoo JB, Lewis Jr JS, McJunkin J, Sequeira S. Malignant neoplasms of the salivary glands. In: Haughey BH, Robbins KT, editors. Cummings otolaryngol, Head Neck Surg. 5th ed. Philadelphia: Mosby-Elsevier; 2010. p. 1178–99.
55. Guzzo M, Locati LD, Prott FJ, Gatta G, McGurk M, Licitra L. Major and minor salivary gland tumors. Crit Rev Oncol Hematol. 2010;74:134–48. doi:10.1016/j.critrevonc.2009.10.004.
56. Mukunyadzi P, Bardales RH, Palmer HE, Stanley MW. Tissue effects of salivary gland fine-needle aspiration. Does this procedure preclude accurate histologic diagnosis? Am J Clin Pathol. 2000;114:741–5. doi:10.1309/NR4U-70L7-MBDG-DXTL.
57. Ethunandan M, Pratt CA, Higgins B, Morrison A, Umar T, Macpherson DW, et al. Factors influencing the occurrence of multicentric and “recurrent” Warthin’s tumour: a cross sectional study. Int J Oral Maxillofac Surg. 2008;37:831–4. doi:10.1016/j.ijom.2008.05.001.
58. Ghosh-Laskar S, Murthy V, Wadasadawala T, Agarwal J, Budrukkar A, Patil N, et al. Mucoepidermoid carcinoma of the parotid gland: factors affecting outcome. Head Neck. 2011;33:497–503. doi:10.1002/hed.21477.
59. Heeneman H. Identification of the facial nerve in parotid surgery. Can J Otolaryngol. 1975;4:145–51.
60. Casani AP, Cerchiai N, Dallan I, Seccia V, Franceschini SS, Unit O. Benign tumours affecting the deep lobe of the parotid gland: how to select the optimal surgical approach. Acta Otorhinolaryngol Ital. 2015;35:80–7.
61. Koch M, Zenk J, Iro H. Long-term results of morbidity after parotid gland surgery in benign disease. Laryngoscope. 2010;120:724–30. doi:10.1002/lary.20822.
62. Guntinas-Lichius O, Gabriel B, Klussmann JP. Risk of facial palsy and severe Frey’s syndrome after conservative parotidectomy for benign disease: analysis of 610 operations. Acta Otolaryngol. 2006;126:1104–9. doi:10.1080/00016480600672618.
63. Li C, Zhang Q, Li L, Shi Z. Interventions for the treatment of Frey’s syndrome. Cochrane Database of Systematic Reviews. 2015;7:CD009959.
64. Baillarger M. Mémoire sur l’oblitération du canal Sténon. Gaz Médicale Paris. 1853;23:194–7.

65. Frey L. Le syndrome du nerf auriculo-temporal. Rev Neurol. 1923;2:97–104.
66. Marchal F, Becker M, Dulguerov P, Lehmann W. Interventional sialendoscopy. Laryngoscope. 2000; 110:318–20. doi:10.1097/00005537-200002010-00026.
67. Atienza G, López-Cedrún JL. Management of obstructive salivary disorders by sialendoscopy: a systematic review. Br J Oral Maxillofac Surg. 2015;53:507–19. doi:10.1016/j.bjoms.2015.02.024.
68. Karl Storz GmbH and Co. KG G. Telescopes and instruments for sialendoscopy. KARL STORZ GmbH & Co. KG, Germany. Catalog of Oral and Maxillofacial Surgery. 3rd ed. 2012.
69. Bruch JM, Setlur J. Pediatric sialendoscopy. Adv Otorhinolaryngol. 2012;73:149–52. doi:10.1159/000334474.
70. Gourin CG, Terris DJ. Surgical robotics in otolaryngology: expanding the technology envelope. Curr Opin Otolaryngol Head Neck Surg. 2004;12:204–8. doi:00020840-200406000-00010 [pii]
71. Baek C-H, Jeong H-S. Endoscope-assisted submandibular sialadenectomy: a new minimally invasive approach to the submandibular gland. Am J Otolaryngol. 2006;27:306–9. doi:10.1016/j.amjoto.2005.11.018.
72. Chen M-K. Minimally invasive endoscopic resection of the submandibular gland. Oper Tech Otolaryngol Neck Surg. 2008;19:33–5. doi:10.1016/j.otot.2008.03.001.
73. Song CM, Jung YH, Sung M-W, Kim KH. Endoscopic resection of the submandibular gland via a hairline incision: a new surgical approach. Laryngoscope. 2010;120:970–4. doi:10.1002/lary.20865.
74. Parente Arias PL, Fernández Fernández MM, Varela Vázquez P, de Diego Muñoz B. Minimally invasive video-assisted submandibular sialadenectomy: surgical technique and results from two institutions. Surg Endosc. 2015; doi:10.1007/s00464-015-4604-0.
75. Beahm DD, Peleaz L, Nuss DW, Schaitkin B, Sedlmayr JC, Rivera-Serrano CM, et al. Surgical approaches to the submandibular gland: a review of literature. Int J Surg. 2009;7:503–9. doi:10.1016/j.ijsu.2009.09.006.
76. Ruscito P, Pichi B, Marchesi P, Spriano G. Minimally invasive video-assisted submandibular sialoadenectomy: a preliminary report. J Craniofac Surg. 2007;18:1142–7.
77. Kim HS, Chung SM, Pae SY, Park HS. Endoscope assisted submandibular sialadenectomy: the face-lift approach. Eur Arch Otorhinolaryngol. 2011;268:619–22. doi:10.1007/s00405-010-1392-y.
78. De Virgilio A, Park YM, Kim WS, Lee SY, Seol JH, Kim S-H. Robotic sialoadenectomy of the submandibular gland via a modified face-lift approach. Int J Oral Maxillofac Surg. 2012;41:1325–9. doi:10.1016/j.ijom.2012.04.008.
79. Terris DJ. Surgical robotics in otolaryngology. In: Flint PW, editor. Cummings Otolaryngol, Head Neck Surg. 5th ed. Philadelphia: Mosby-Elsevier; 2010. p. 38–44.
80. Maan ZN, Gibbins N, Al-Jabri T, D'Souza AR. The use of robotics in otolaryngology-head and neck surgery: a systematic review. Am J Otolaryngol Head Neck Med Surg. 2012;33:137–46. doi:10.1016/j.amjoto.2011.04.003.
81. Haus BM, Kambham N, Le D, Moll FM, Gourin C, Terris DJ. Surgical robotic applications in otolaryngology. Laryngoscope. 2003;113:1139–44. doi:10.1097/00005537-200307000-00008.
82. Terris DJ, Haus BM, Gourin CG, Lilagan PE. Endo-robotic resection of the submandibular gland in a cadaver model. Head Neck. 2005;27:946–51. doi:10.1002/hed.20273.
83. Lee HS, Park DY, Hwang CS, Bae SH, Suh MJ, Koh YW, et al. Feasibility of robot-assisted submandibular gland resection via retroauricular approach: preliminary results. Laryngoscope. 2013;123:369–73. doi:10.1002/lary.23321.
84. Walvekar RR, Tyler PD, Tammareddi N, Peters G. Robotic-assisted transoral removal of a submandibular megalith. Laryngoscope. 2011;121:534–7. doi:10.1002/lary.21356.
85. Razavi C, Pascheles C, Samara G, Marzouk M. Robot-assisted sialolithotomy with sialendoscopy for the management of large submandibular gland stones. Laryngoscope. 2016;126:345–51. doi:10.1002/lary.25396.
86. Walvekar RR, Peters G, Hardy E, Alsfeld L, Stromeyer FW, Anderson D, et al. Robotic-assisted transoral removal of a bilateral floor of mouth ranulas. World J Surg Oncol. 2011;9:78. doi:10.1186/1477-7819-9-78.
87. Emami B, Lyman J, Brown A, Coia L, Goitein M, Munzenrider JE, et al. Tolerance of normal tissue to therapeutic irradiation. Int J Radiat Oncol Biol Phys. 1991;21:109–22. doi:10.1016/0360-3016(91)90171-Y.
88. Zhuang L, Yang Z, Zeng X, Zhua X, Chen Z, Liu L, et al. The preventive and therapeutic effect of acupuncture for radiation-induced xerostomia in patients with head and neck cancer: a systematic review. Integr Cancer Ther. 2013;12:197–205. doi:10.1177/1534735412451321.
89. Kałuzny J, Wierzbicka M, Nogala H, Milecki P, Kopeć T. Radiotherapy induced xerostomia: mechanisms, diagnostics, prevention and treatment – evidence based up to 2013. Otolaryngol Pol. 2014;68:1–14. doi:10.1016/j.otpol.2013.09.002.
90. Hackett KL, Deane KHO, Strassheim V, Deary V, Rapley T, Newton JL, et al. A systematic review of non-pharmacological interventions for primary Sjögren's syndrome. Rheumatology (Oxford). 2015;54:2025–32. doi:10.1093/rheumatology/kev227.
91. Li L, Tian G, He J. The standardization of acupuncture treatment for radiation-induced xerostomia: a literature review. Chin J Integr Med. 2015; doi:10.1007/s11655-015-2145-y.
92. Furness S, Worthington H V, Bryan G, Birchenough S, McMillan R. Interventions for the management of dry mouth: topical therapies. Cochrane Database Syst

Rev. 2011;(12):CD008934. doi:10.1002/14651858. CD008934.pub2.
93. Jha N, Seikaly H, McGaw T, Coulter L. Submandibular Salivary Glanf Transfer Prevents Radiation-Induced Xerostomia. International Journal of Radiation Oncology Biology Physics. 2000;46:7-11.
94. MacLeod AM, Robbins SP. Submandibular gland transfer in the correction of dry eye. Aust N Z J Ophthalmol. 1992;20:99–103.
95. Geerling G, Sieg P, Bastian GO, Laqua H. Transplantation of the autologous submandibular gland for most severe cases of keratoconjunctivitis sicca. Ophthalmology. 1998;105:327–35.
96. Geerling G, Sieg P. Transplantation of the major salivary glands. Dev Ophthalmol. 2008;41:255–68. doi:10.1159/000131094.
97. Geerling G, Raus P, Murube J. Minor salivary gland transplantation. Dev Ophthalmol. 2008;41:243–54. doi:10.1159/000131093\r131093. [pii]
98. Jha N, Seikaly H, McGaw T, Coulter L. Submandibular salivary gland transfer prevents radiation-induced xerostomia. Int J Radiat Oncol Biol Phys. 2000;46:7–11. doi:10.1016/S0360-3016(00)80143-1.
99. Wu F, Weng S, Li C, Sun J, Li L, Gao Q. Submandibular gland transfer for the prevention of postradiation xerostomia in patients with head and neck cancer: a systematic review and meta-analysis. Oral Oncal. 2015;610041:70–86. doi:10.1159/000371854.
100. Su J-Z, Liu X-J, Liu D-G, Ren W-G, Yu G-Y. Sialography of the transplanted submandibular gland. Ocul Surf. 2014;12:215–20. doi:10.1016/j.jtos.2014.02.005.
101. Geerling G, Sieg P, Raus P. Salivary gland transplantation. In: Essentials Ophthalmol. Berlin: Springer Berlin Heidelberg; 2007. p. 208 .Figure 14.1
102. Murube-del-Castillo J. Transplantation of salivary gland to the lacrimal basin. Scand J Rheumatol Suppl. 1986;61:264–7.
103. Gelatt K. Canine lacrimal and nasolacrimal diseases. 2nd ed. London: Lea & Febiger; 1991.
104. Sieg P, Geerling G, Kosmehl H, Lauer I, Warnecke K, von Domarus H. Microvascular submandibular gland transfer for severe cases of keratoconjunctivitis sicca. Plast Reconstr Surg. 2000;106:554–60.
105. Yu GY, Zhu ZH, Mao C, Cai ZG, Zou LH, Lu L, et al. Microvascular autologous submandibular gland transfer in severe cases of keratoconjunctivitis sicca. Int J Oral Maxillofac Surg. 2004;33:235–9. doi:S0901-5027(02)90438-8 [pii]\r10.1006/ijom.2002.0438
106. Kumar PA, Hickey MJ, Gurusinghe CJ, O'Brien BM. Long term results of submandibular gland transfer for the management of xerophthalmia. Br J Plast Surg. 1991;44:506–8.
107. Paniello RC. Submandibular gland transfer for severe xerophthalmia. Laryngoscope. 2007;117:40–4. doi:10.1097/01.mlg.0000246953.44163.81.
108. Jia G, Wang Y, Lu L, Wang X, Li Z. Reconstructive lacrimal gland with free submandibular gland transfer for management of xerophthalmia. Zhonghua Yan Ke Za Zhi. 1998;34:388–90.
109. Hockstein NG, Samadi DS, Gendron K, Handler SD. Sialorrhea: a management challenge. Am Fam Physician. 2004;69:2628–34.
110. Reed J. Surgical management of drooling. Arch Otolaryngol Head Neck Surg. 2009;135:924–31. doi:10.1542/gr.15-2-19.
111. Puraviappan P, Dass DB, Narayanan P. Efficacy of relocation of submandibular duct in cerebral palsy patients with drooling. Asian J Surg. 2007;30:209–15. doi:10.1016/S1015-9584(08)60024-X.
112. De M, Adair R, Golchin K, Cinnamond MJ. Outcomes of submandibular duct relocation: a 15-year experience. J Laryngol Otol. 2003;117:821–3. doi:10.1258/002221503770716287.
113. Panarese A, Ghosh S, Hodgson D, McEwan J, Bull PD. Outcomes of submandibular duct re-implantation for sialorrhoea. Clin Otolaryngol Allied Sci. 2001;26:143–6. doi:10.1046/j.1365-2273.2001.00439.x.
114. Uppal HS, De R, D'Souza AR, Pearman K, Proops DW. Bilateral submandibular duct relocation for drooling: an evaluation of results for the Birmingham Children's Hospital. Eur Arch Otorhinolaryngol. 2003;260:48–51. doi:10.1007/s00405-002-0516-4.
115. Katona G, Csákányi Z, Lorincz A, Gerlinger I. Bilateral submandibular duct relocation by high-frequency radiosurgery. Eur Arch Otorhinolaryngol. 2008;265:1103–8. doi:10.1007/s00405-008-0604-1.
116. Crawford B. Parotid duct transplantation for ocular xerosis. Trans Aust Coll Ophthalmol. 1970;2:92–5.
117. Filatov V, Chevaljev V. Surgical treatment of parenchymatous ophthalmoxerosis. J Opthal. 1951;3:131–7.
118. Pierce M, Goldberg J, Brooks C. A direct approach for transposition of the parotid duct. Arch Ophthalmol. 1960;64:566–70.
119. Zhang H, Zhou Z, Chen Z, Zhao C. Management of the dry eye with parotid duct transplantation: a summary on 40 cases. Yan Ke Xue Bao. 1995;11:67–9.
120. Longo B, Germano S, Laporta R, Belli E, Santanelli F. Stensen duct relocation after cheek mucosa tumor resection. J Craniofac Surg. 2012;23:e250–1. doi:10.1097/SCS.0b013e31824ef809.
121. Mehta S, Agrawal J, Dewan AK, Pradhan T. Parotid duct relocation in buccal mucosa cancer resection. J Craniofac Surg. 2014;25:1746–7. doi:10.1097/SCS.0000000000000998.
122. Sakakibara A, Minamikawa T, Hashikawa K, Sakakibara S, Hasegawa T, Akashi M, et al. Does salivary duct repositioning prevent complications after tumor resection or salivary gland surgery? J Oral Maxillofac Surg. 2015;73:1003–7. doi:10.1016/j.joms.2014.12.006.
123. Redman RS. On approaches to the functional restoration of salivary glands damaged by radiation therapy for head and neck cancer, with a review of related aspects of salivary gland morphology and development. Biotech Histochem. 2008;83:103–30. doi:903247450 [pii]\r10.1080/10520290802374683

124. Baum BJ, Alevizos I, Zheng C, Cotrim AP, Liu S, McCullagh L, et al. Early responses to adenoviral-mediated transfer of the aquaporin-1 cDNA for radiation-induced salivary hypofunction. Proc Natl Acad Sci U S A. 2012;109:19403–7. doi:10.1073/pnas.1210662109.
125. Lombaert IMA, Wierenga PK, Kok T, Kampinga HH, Coppes RP. Cancer therapy: preclinical mobilization of bone marrow stem cells by granulocyte colony-stimulating factor ameliorates radiation-induced damage to salivary glands. Clin Cancer Res. 2006;12:12–5. doi:10.1158/1078-0432.CCR-05-2381.
126. Tran SD, Liu Y, Xia D, Maria OM, Khalili S, Wang RW-J, et al. Paracrine effects of bone marrow soup restore organ function, regeneration, and repair in salivary glands damaged by irradiation. PLoS One. 2013;8:e61632. doi:10.1371/journal.pone.0061632.
127. Pringle S, Van Os R, Coppes RP. Concise review: adult salivary gland stem cells and a potential therapy for xerostomia. Stem Cells. 2013;31:613–9. doi:10.1002/stem.1327.
128. Feng J, van der Zwaag M, Stokman MA, van Os R, Coppes RP. Isolation and characterization of human salivary gland cells for stem cell transplantation to reduce radiation-induced hyposalivation. Radiother Oncol. 2009;92:466–71. doi:10.1016/j.radonc.2009.06.023.
129. Nanduri LSY, Maimets M, Pringle SA, van der Zwaag M, van Os RP, Coppes RP. Regeneration of irradiated salivary glands with stem cell marker expressing cells. Radiother Oncol. 2011;99:367–72. doi:10.1016/j.radonc.2011.05.085.
130. Lombaert IMA, Hoffman MP. Epithelial stem/progenitor cells in the embryonic mouse submandibular gland. Front Oral Biol. 2010;14:90–106. doi:10.1159/000313709.Epithelial.
131. Ono H, Obana A, Usami Y, Sakai M, Nohara K, Egusa H, et al. Regenerating salivary glands in the microenvironment of induced pluripotent stem cells. Biomed Res Int. 2015;2015:293570. doi:10.1155/2015/293570.
132. Ogawa M, Oshima M, Imamura A, Sekine Y, Ishida K, Yamashita K, et al. Functional salivary gland regeneration by transplantation of a bioengineered organ germ. Nat Commun. 2013;4:2498. doi:10.1038/ncomms3498.
133. Joraku A, Sullivan CA, Yoo J, Atala A. In-vitro reconstitution of three-dimensional human salivary gland tissue structures. Differentiation. 2007;75:318–24. doi:10.1111/j.1432-0436.2006.00138.x.
134. Sun T, Zhu J, Yang X, Wang S. Growth of miniature pig parotid cells on biomaterials in vitro. Arch Oral Biol. 2006;51:351–8. doi:10.1016/j.archoralbio.2005.10.001.
135. Cantara SI, Soscia DA, Sequeira SJ, Jean-Gilles RP, Castracane J, Larsen M. Selective functionalization of nanofiber scaffolds to regulate salivary gland epithelial cell proliferation and polarity. Biomaterials. 2012;33:8372–82. doi:10.1016/j.biomaterials.2012.08.021.
136. Cecchini MP, Parnigotto M, Merigo F, Marzola P, Daducci A, Tambalo S, et al. 3D printing of rat salivary glands: the submandibular-sublingual complex. Anat Histol Embryol. 2014;43:239–44. doi:10.1111/ahe.12074.

FSC
www.fsc.org
MIX
Papier | Fördert gute Waldnutzung
FSC® C083411

Kolibri 360 GmbH
Ferdinand-Jühlke-Straße 7
99095 Erfurt, Deutschland
produktsicherheit@kolibri360.de